Opportunities & Challenges
In the 21st Century

Simon S. M. Chin

Published in 2025 by IngramSpark

Design by Fred Wallington

Manufactured in the USA

2025916141

Library of Congress Cataloging-in-Publication Data

Chin, Simon S. M.

Opportunities and challenges in the 21st century

eBook ISBN: 979-8-9924674-9-9
Paperback ISBN: 979-8-9924674-3-7

Contents

Section 7: Game-Changer: Artificial Intelligence

Section 8: Lifelong Learning

Dedication

To my beloved mother

A mother I love wholeheartedly and respect deeply

Preface

This book is for a global audience of mostly young adults and adults. If you want to learn what you can do to fight against the scourge of pandemics, wars, global financial crisis, income inequality, climate change, misuse of artificial intelligence (AI), diseases and unequal opportunities, overwhelming stress and losing your soul, then this book is for you.

After my beloved mother passed away on July 7, 2023, I decided to write this book. She was an inspiration to me and I want to share the insights I gained throughout my life. I believe it is important to learn from each other's experiences. When I saw my mother's soul leaving her body five minutes after she stopped breathing, it was clear to me that our bodies are just vessels for our souls.

My life has been filled with many blessings. I was lucky to be in the right place at the right time. God gave me a photographic memory and I am grateful. In Kindergarten, I learned fast and skipped first grade. When I left Burma to come to the United States, I was ranked first in my class. Having attended a large high school in San Francisco, I became the head of the student body after I ran for student government for the first time against more experienced and popular students.

God has led the right people into my life at the right time to help me achieve many things. University of California, Berkeley was ranked #1 for Chemistry in the country in 1977, and it was the only university to which I applied. I got in and was given the Alumni Scholarship based on merit. My faculty advisor, Dr. Glenn T. Seaborg, Nobel Laureate in Chemistry and Second Chancellor of UC Berkeley, gave me an opportunity to do research with him at the Lawrence

Berkeley National Laboratory starting in my first quarter as a freshman. I was very lucky and grateful.

After graduating from UC Berkeley with a degree in Chemical Engineering, I was hired by the first company that I interviewed with in Silicon Valley. My trajectory there was swift – starting as an engineer, I was promoted to a managerial role overseeing manufacturing within just 9 months, doubling my salary by the end of my first year of employment. Assuming responsibility for the company's largest operation 6 months after the first promotion, we experienced a period of great success and rapid growth. I was blessed to be able to buy my first house less than two and a half years after graduation from UC Berkeley.

I simultaneously pursued an MBA at Santa Clara University while working full-time - juggling employment's demands with graduate studies' rigors. I actually enjoyed the synergy of applying what I learned in school to my work and what I learned at work to my school projects almost on a daily basis.

On my first trip oversea to transfer technologies to Bosch in Germany, Lufthansa Airlines oversold business class and upgraded me to the first class cabin. The technology transfer projects I managed for Du Pont, the eighth-largest company in the US at the time, contributed to uplifting 800 million people out of poverty in China. The first professional company I founded, Iris Biotechnologies Inc., started trading as a public company in 2008.

Iris applied for patents all over the world, and all our patent applications were granted as patents. I have three patents on Artificial Intelligence Systems for Genetic Analysis. I completed a marathon the first time I ran one, raising money for leukemia and lymphoma cancer research. I then founded Iris Wellness Labs to help patients and physicians with complex medical cases by analyzing the whole human genome (3 billion base-pairs), microbiome, and about 900 metabolites.

My goal is to share my more than forty years of experience in high technology, biotechnology, medicine, and artificial intelligence and fifty years studying Christianity, Buddhism, and other religions. I hope to offer clear ideas to empower people to help themselves and make a positive impact.

Most scholars believe that Darwin withheld from publishing his book, On the Origin of Species, for twenty years because he was unsure about his theory. He was forced to publish because he was afraid that Alfred Russel Wallace was going to publish a competing book first. If Darwin knew what we know now, would he have published his book? Some leading scientists believe that mutations reflect adaptation, not the creation of new animals or plants.

Many people today are dealing with multiple, unprecedented crises at the same time. Some people are trying their best, while others are using unhealthy ways to cope, like drugs and alcohol. We live in challenging times. I want to give back, inspire hope and encourage people to work towards a better future. My intent is to provide evidence of what the truth tells about what has been going on in the world.

I am very grateful for the amazing opportunities I've had in life. Many people have shared their wisdom and showed me love and kindness. I am grateful to all my mentors. My life's journey also had many hardships and injustices that made me stronger and prepared me for the future. As a Christian who was raised in a Buddhist family and lived in the United States, Europe and Asia, my unique experience helped me to understand and appreciate many different cultures and viewpoints. The United States became the most innovative and successful nation in human history by attracting some of the best minds from around the world.

God is real and I want to share my personal experience, beyond reasonable doubt, about him. God gave you a soul and it is the very essence of who you are. This is your eternal soul's journey in this world. I hope this truth will set you free. God gave us the freedom to choose, and we must choose wisely.

I am grateful to Dr. Wing Hsieh, Mr. James Moshofsky, my sisters Grace Osborne and Catherine Cheah, my nephews Joshua and Daniel Osborne, Amazon Publishing Portal's professional book editing team, and others for helping me to review different parts of the book and publish this work.

Introduction

This book aims to share the truth. It draws from my experiences over 16 years as an engineer and manager in semiconductor technology and 25 years as an entrepreneur in biotechnology, artificial intelligence, and medicine to make life better for people. After my beloved mother passed away in 2023, I felt a strong need to share my story and insights.

In 1999, I founded Iris Biotechnologies to improve breast cancer diagnosis and treatment. Iris started trading as a public company in 2008. In 2014, I founded a subsidiary, Iris Wellness Labs, to provide in-depth scientific analysis to give insights to patients and doctors facing complex medical conditions. Sharing my experiences and perspectives in this book can uplift and inspire others.

Caring for my mom during her last three years was an honor for me. Witnessing her soul depart minutes after her last breath reaffirmed my belief that the soul is eternal in nature. Her passing left an immense void in my life, but it also made me reflect deeply on my unusual journey from childhood in Burma to becoming an entrepreneur in Silicon Valley in the United States. I also reflected on studying different religions and history for 50 years. God has always been with me, even when I did not realize it.

We live in a Dopamine Culture of clickbaits on reels of short videos on TikTok, YouTube, and other platforms, scrolling and sending short texts on iPhone, Android, and other smart phones and tablets. We also swipe on various apps and gamble with ease on our choice of communication devices. This culture is a far cry from reading books and newspapers, writing letters, and having real and lasting interpersonal relationships such as long-term friendships and marriage.

Dopamine is pleasurable and addictive. Having dopamine easily all the time is not good for us physically, mentally, or spiritually.

Charles Darwin's book "On the Origin of Species" has changed the world for 175 years. I have diligently studied his theory and had discussions with people, including leading scientists in biology and chemistry, for decades. I patiently searched and waited for proof of Darwin's Theory, and I have come to the clear conclusion that life evolving from non-life by chance is impossible. We must teach our assumptions truthfully and clearly. Did we evolve from rodents, or did God create humans?

I am struck by the many perilous challenges facing humanity today – challenges like the pandemic, climate change, tensions between countries, global debt crisis and disruptions caused by new technologies such as artificial intelligence. We are all in the same ship and each of these gaping holes must be fixed concurrently. In this pivotal moment, I feel called to offer insights into some of life's most profound questions about our origins, our purpose, and our ultimate destiny.

The ability to discern is one of the most important skills you'll ever learn. It involves seeing, recognizing, or apprehending something by sight, another sense, or the intellect. Having the skill to discern will help you to succeed in life. We can learn to discern and then teach our children how to discern, who to trust, who not to trust, and why truth matters. Children are our future and we must help them to succeed.

My study of prophecies from the Bible and my analysis of trends in technology, politics, and society have led me to believe we are heading toward the end times foretold in the

Book of Revelation. I don't know the precise timing, but I believe humanity stands at a critical spiritual crossroads. The choices we make collectively will shape the fate of our species. This is the last call.

At age 66, my primary allegiance is to truth – the truth of God's love, the truth of science, the truth of history, and the truth about our present circumstances. My aim is not to prove anything or please everyone but simply to bear witness to what God has revealed in my life. I hope it enriches others' spiritual understanding and personal journeys. I consider this book to be the most important work of my life. In this book, I will discuss topics like:

- The entity that enabled wars resulting in 100 million deaths
- The story of my blessed and grateful life from a humble beginning
- The limitations of evolutionary theory in explaining the origin of DNA
- The impact of climate change on past and present civilizations
- The importance of entrepreneurship and lifelong learning
- The potential of medicine and technology to improve lives
- The game-changing nature of artificial intelligence (AI)
- The nature and importance of our souls' journeys
- The fulfillment of end-time prophecies from religious texts

My perspective comes from my own lived experiences, and your perspective may differ from mine. I do not intend to judge but simply to share what I have learned in the hopes that it inspires you to re-examine your priorities and life's purpose in light of the precious gift of life itself.

All humans begin their lives at conception. From a biological standpoint, conception at fertilization marks the beginning of a new genetic individual. At this point, a unique genome is formed, setting the stage for potential human development. There is no debate or controversy about that.

Between conception and birth, there are disagreements as to when the growing cells can be called a human. Abortion ends the development of cells that could result in the birth of a baby. There is also no debate or controversy about that.

Whether people argue life began at birth, at 120 days, or at any other time between conception and birth, DNA already defines a person's genetic identity at conception. That is a fact. People can believe whatever they choose to believe, but that won't change the truth. We live in a time of deceit, and we must be extra vigilant, especially with AI used to create "Untruthful clickbaits" online.

I am a scientist and an engineer who values evidence in our observations. I am a first-hand witness to my mother's living soul continuing its journey by departing through the right side of her mouth five minutes after she stopped breathing. Your body is just a temporary vessel for your soul, which lives on after your body dies.

Your spirit is your conscience. You have "free will" because of the human spirit, which asks whether what you intend to do is right or wrong. It is the human spirit that gives us a consciousness of self, intellect, emotions, fears, courage, resilience, passions, creativity, and the unique ability to comprehend and understand. We can think, feel, love, design, create, and enjoy music, humor, and art because of the spirit. It gives us the ability to connect and have an intimate relationship with God.

Jesus said, "God is spirit, and those who worship him must worship in spirit and truth" (John 4:24). Our spirits have an innate knowledge of right and wrong (Romans 2:14–15). The soul of a person is the courtroom where life decisions are made, and I discuss the soul in depth. Identical twins have exactly the same DNA but different souls and spirits with unique personalities, talents, likes, and dislikes.

Do you think God is real? I had a profound experience, beyond a reasonable doubt, that God is real. It happened when I least expected it: what you believe regarding God and what you think have profound consequences for your life. If God is real and everywhere at the same time for all times, then we are all in God, and he knows all our thoughts and memories. We, on the other hand, don't know all that God knows.

As humans, we have limitations. According to our understanding of biology, human life, in our times, is limited to approximately 120 years. Research has shown that when all of the cells ever created in the human body are multiplied by the average time it takes for cells to reach the end of their lives, you get roughly 120 years. This is called the "Hayflick limit, the maximum number of years a human can expect to live." Leonard Hayflick is a professor of anatomy at the University of California, San Francisco, and formerly a professor of medical microbiology at Stanford University.

Our universe consists of 5% visible matter, 27% dark matter, and 68% dark energy. Hubble Telescope and James Watt Space Telescope (JWST) show that our sun is one of approximately 100-400 billion stars in the Milky Way galaxy. Our Milky Way is one of 200 billion to 2 trillion galaxies in the universe.

It is improbable to think that God created life only on Earth. The human life span, less than 120 years, is just a blink of an eye in cosmic time. Humans cannot experience the fullness of life without Earth or something like Earth. That's why we must do what we can to preserve our world. A human soul is an eternal luminous essence and can go anywhere in the universe.

Section 1: Take Actions While You Can

Chapter 1
The World at a Crossroads

Our world stands at a critical juncture, facing unprecedented challenges that threaten the very foundations of our global society. From the ongoing pandemic to the looming specter of climate change, from economic instability to rising social unrest, we are witnessing a convergence of crises that demand our immediate attention and action.

But these are not isolated developments they are deeply interconnected symptoms of a deeper fracture: a spiritual and structural unraveling of trust, truth, and equity on a global scale.

Perhaps most alarming is the growing chasm between the ultra-wealthy and the rest of society. As of April 2024, there were 2,781 billionaires globally with a combined worth equal to $14.2 trillion. The U.S. has 813 billionaires, followed by China, which has 473, and India, which has 200. This rapid accumulation of wealth at the top has far-reaching implications for social mobility, political influence, and economic stability.

The numbers are stark but more importantly, they reflect a global architecture where wealth accumulation has become disconnected from value creation, ethics, or human progress. Instead, the system rewards speculation, monopolization, and financial manipulation.

As of June 2024, the three richest individuals in the world Jeff Bezos, Elon Musk, and Bernard Arnault held a staggering $607 billion in wealth. This concentration of resources in the hands of so few stands in stark contrast to the struggles of billions who face economic hardship and uncertainty.

A handful of individuals now wield more economic power than many nations, influencing everything from markets and media to space exploration and climate strategies.

As of December 2024, Elon Musk is the wealthiest person in the world, with an estimated net worth of US$486 billion, according to the Bloomberg Billionaires Index, and $464 billion according to Forbes, primarily from his ownership stakes in Tesla, Inc. and SpaceX.

This era has produced techno-oligarchs entrepreneurs turned de facto sovereigns who often operate with little oversight and immense influence over public policy and collective imagination.

The power wielded by massive financial institutions further complicates this picture. BlackRock, the world's largest asset manager, oversees an astounding $11.5 trillion in assets as of November 2024. While positioning itself as a leader in environmental, social, and corporate governance (ESG), the company has faced criticism for investments in controversial industries and its close ties to the Federal Reserve during the COVID-19 pandemic. The sheer scale of BlackRock's influence raises questions about the concentration of economic power and its impact on global financial markets.

The very institutions that claim to champion sustainability and social responsibility are often deeply entangled in the engines of extraction, inequality, and political influence.

The three largest asset managers BlackRock, Vanguard, and State Street combined are the largest owners of 438 out of the 500 largest corporations in the US. These managers are also the largest shareholders in each other's companies. Since March 13, 2009, the S&P 500 has grown from 757 to 6,389 on July 25, 2025. Since March 9, 1990, the NASDAQ has grown from 437 to 21,108 on July 25, 2025.

Yet, this spectacular market growth primarily benefits the financial elite. The gains are concentrated, while the risks job losses, inflation, housing instability are distributed across the working and middle class.

Oligarchs and the Israel lobby control US politicians, and no matter who wins the US presidential election, Wall Street always wins ultimately. The wealthiest 10% of Americans own 93% of US equities, while the bottom 50% hold just 1%. From 1979 to 2019, the wages of the top 1% rose by 160% after inflation, while wages rose 345% for the highest 0.1% of earners. Wage growth in the middle has been sluggish, with median pay rising just 13.7%.

What we see is not capitalism, but a plutocratic shell where politics is theater and policy is engineered to maintain the wealth concentration. Democracy without economic justice is a hollow ideal.

Blackstone, with over $1 trillion in assets under management, has become the largest alternative investment firm globally and the largest commercial landlord in history. In the US, it is by far the largest landlord, owning almost 350,000 units of rental housing.

The roots of financial giants BlackRock and Blackstone can be traced back to the same institution that played a pivotal role in the 2008 global financial crisis, raising questions about the lessons learned from that near-catastrophic event.

Larry Fink, the CEO of BlackRock, and Ralph Schlosstein, two of BlackRock's founders, previously ran the mortgage-backed securities divisions at First Boston and Lehman Brothers, respectively. They initially joined Blackstone to manage an investment fund and provide advice to financial institutions.

This tight-knit circle of financial elites has not only recovered from past crises they've been rewarded for them. They now own more real estate, control more capital, and influence more government policy than ever before.

Blackstone was founded in 1985 as a mergers and acquisitions firm by Peter G. Peterson and Stephen A. Schwarzman, who had previously worked together at Lehman Brothers. Peterson was the former chairman and CEO of Lehman Brothers, and Schwarzman served as head of global mergers and acquisitions business. Schwarzman, the CEO of Blackstone, was briefly chairman of former president Donald Trump's Strategic and Policy Forum. Shortly after the 2020 election, President Biden hired two BlackRock executives for senior roles on his economics team. Brian Deese runs the National Economic Council, and Adewale Adeyemo is the deputy Treasury secretary. This revolving door between finance and government continues unchallenged, and bipartisan participation reveals that this is not a partisan issue it's a structural one.

The 2008 crisis, centered around the collapse of Lehman Brothers, brought the global financial system to the brink of total meltdown, requiring worldwide massive government bailouts, which some call the most significant generational theft in the history.

In 2009 and 2010, massive unemployment and sovereign debts threatened to destabilize economies, particularly in

Europe, resulting in civil unrest in various countries. Millennials, born from 1981 to 1996, were severely impacted by this generational theft. Their dreams of homeownership, job security, and upward mobility were shattered. Many became the first generation to expect a lower quality of life than their parents.

Wall Street CEOs and Big Bank CEOs were the most responsible for causing the Great Recession. President Barack Obama did not hold any of them accountable. None of them went to jail, including the CEO of Lehman Brothers, the fourth-largest investment bank in the United States that filed for Chapter 11 bankruptcy protection on September 15, 2008, with $613 billion in debt. Lehman was operational for 158 years, from its founding in 1850 until 2008. This impunity set a dangerous precedent: that systemic fraud, when done at scale, is immune from justice.

Lehman's bankruptcy filing remains the largest in US history and played a major role in the unfolding of the financial crisis of 2007–2008. Dick Fuld, the CEO of Lehman Brothers, walked away with half a billion dollars and three homes when Lehman failed. The US legal system did not hold him accountable for all the pain, suffering, and financial losses he caused. The collapse was largely due to Lehman's involvement in the subprime mortgage crisis and its exposure to less liquid assets. The ripple effects of this crisis continue to shape our economic landscape today. Many argue that the subsequent bailouts enabled by the Federal Reserve and policy decisions have only exacerbated inequality and set the stage for future instability.

In essence, the crisis wasn't resolved it was deferred, deepened, and reloaded into a new system of dependency.

The Federal Reserve has also enabled the financing of wars since WWI. US military spending rose from hundreds of millions pre-WWII to $85 billion in 1943 and $91 billion in 1944 with the Fed's help. The Treasury and the Federal Reserve devised plans for financing the war, meeting frequently to determine how to finance it through taxation and issuance of government bonds.

The 16th Amendment, ratified on February 3, 1913, established Congress's right to impose federal income taxes. Currently, the worst part of the unfair tax law is that it enables major corporations and very rich people to avoid paying taxes.

War and finance have long been partners funded by the labor of ordinary citizens, benefiting those with power, and framed as necessary for national security. But who really profits from war?

As we grapple with these challenges, it becomes increasingly clear that understanding our history is crucial to charting a path forward. The patterns of the past, the complex interplay of human nature and power dynamics, and the lessons of both triumphs and failures all offer vital insights for addressing our current predicaments. Learning about history is very important if we want to understand the present and create a better future.

History helps us avoid making the same mistakes again. In our fast-paced world, knowing history is more important than ever because many of today's challenges have deep roots in the past. But history is not merely academic it is personal. The choices of prior generations have built the systems we live under today, and we must choose whether to continue in their image or forge a new, more ethical path.

Chapter 1
The World at a Crossroads

When we study history, we have to face hard truths so we can build a fairer world. History shows what drives social change. It also shows how resilient the human spirit is when facing impossible challenges. As we navigate the 21st century, history will be an essential guide. It gives us the context and perspective we need to understand our challenges and find effective solutions.

It reminds us of the power of people working together to make positive changes to our shared humanity. And it teaches us this: nothing changes without courage. No empire reformed itself from within without pressure. No elite ever surrendered privilege without resistance. Therefore, the people must rise not in rebellion, but in reformation.

Moreover, the psychological toll of these ongoing crises cannot be understated. The COVID-19 pandemic, in particular, has exposed and exacerbated existing mental health challenges, leading to widespread increases in anxiety, depression, and post-traumatic stress disorder.

The American Psychological Association's Stress in America surveys paint a picture of a nation grappling with prolonged stress and uncertainty. Collective traumas disrupt daily life, create feelings of uncertainty and fear, and challenge our sense of safety. The pandemic has increased rates of anxiety, depression, and PTSD across all age groups while making existing mental health inequalities worse.

Social isolation and disruption have added to this toll, showing how important social connection is for well-being. Mental health is not a secondary issue it is the foundation of a functional society. When the mind suffers, everything suffers: our creativity, our productivity, our empathy, and our moral clarity.

Economic hardship further compounds these mental health challenges. The shrinking middle class and growing income inequality create a pervasive sense of insecurity and despair. Families struggle to make ends meet, facing the constant threat of job loss or financial ruin. This stress ripples through relationships, impacting family dynamics and potentially setting the stage for long-term adverse childhood experiences. In the U.S., the share of adults in the middle class fell from 61% in 1971 to 50% in 2021.

At the same time, the share in the upper-income tier rose from 14% to 21%, and the share in the lower-income tier increased from 25% to 29%. In the past, some societies with these trends have collapsed. When prosperity is reserved for the few and despair becomes the default for many, social fabric unravels. This is how civilizations crumble not through bombs, but through broken homes and hopeless hearts.

The impact of economic hardship on relationships and family dynamics can be significant. Financial stress can strain even the strongest relationships, leading to conflict, resentment, and even divorce.

Children who grow up in families struggling with economic hardship may be more likely to face adverse childhood experiences, which can have lasting effects on their mental and physical health. This generational trauma is not theoretical it is measurable. It manifests in declining educational outcomes, poor health, increased crime, and reduced life expectancy. If we do not intervene, we risk raising a generation that inherits not only our debt but also our dysfunction.

As we stand at this crossroads, it's clear that business as usual is no longer an option. We must confront these challenges head-on, armed with knowledge of our past and a commitment to building a more equitable and sustainable future.

The path forward will require bold action, innovative thinking, and a renewed sense of global cooperation. To address the psychological impact of collective traumas and economic hardship, we must prioritize mental health in how we respond and recover.

This means making mental health services more accessible, providing support to vulnerable groups, and building communities that are more resilient. It also means dealing with the root causes of crises and working to create a society that values sustainability, fairness, and well-being. This is not merely a social agenda it is a spiritual one. Because to care for the soul of society, we must care for the souls within it.

In early April 2023, the National Opinion Research Center at the University of Chicago and the Wall Street Journal provided a side-by-side snapshot of the decline in the latest in a series of surveys tracking Americans' attitudes towards civic virtues, and the comparison is not flattering. "The NORC-WSJ survey reports, a lot has happened in America, but none of it is good."

In the last 25 years, the number of respondents who say that patriotism is "very important" to them has declined from 70 percent to 38 percent." For "religion," the number declined from 62 percent to 39 percent; "having a child" halved from 59 percent to 30 percent; "hard work" is down from 84 percent to 67 percent; and, after a brief spike in 2019, "community involvement" fell from 47 percent to 27 percent.

The only thing that has increased in importance for Americans, from 31 percent to 43 percent, is "money." We are witnessing a value inversion. What was once central to meaning faith, community, and purpose has been replaced by individualism, wealth accumulation, and digital distraction.

Most humans are slaves to money. It's okay to have money, but the love of money is the root of all evil. The Global Financial Crisis highlighted by the stock market crash of 2008 inflicted suffering on billions of people. The unbridled greed of Wall Street and The Great Recession forever negatively changed the global confidence in the US financial system.

Countries have started dumping the US dollar, which could ultimately lead to a new world order. Central banks have been buying gold in the past few years because they worry their fiat currencies will lose value. These movements signal not just market anxiety but also a shift in global trust and power. The world is hedging against American exceptionalism.

James Madison, America's fourth President (1809–1817), was known as the "Father of the Constitution." He said, "History records that the money changers have used every form of abuse, intrigue, deceit, and violent means possible to maintain their control over governments by controlling money and its issuance." This quote rings out across centuries as both warning and prophecy. We ignored it once can we afford to ignore it again?

This is a wake-up call. People are trapped in complacency and addiction to the rising stock market, video games, drugs, television, the Internet, and other allures. Children are our hope, and we must rescue them. Because if we lose the minds and hearts of the next generation, we don't just lose culture we lose continuity. We lose the very idea of a better tomorrow.

The stakes could not be higher, but within these challenges are opportunities for transformative change. By understanding the interconnected nature of these issues from wealth inequality to mental health, from historical patterns to current crises we can begin to craft comprehensive solutions that address the root causes of our societal challenges.

The path ahead is difficult, but with collective effort, informed decision-making, and a commitment to equity and sustainability, we can work towards a future that offers prosperity and well-being for all. This is the moment. Not to panic but to prepare. Not to retreat but to rise.

Chapter 2
Federal Reserve and America's Economic Quandary

In the 1912 United States presidential election, Woodrow Wilson, a Democratic Governor from New Jersey with only two years of political experience, managed to defeat the incumbent Republican President William Howard Taft and the popular former Republican President Theodore Roosevelt, who served from 1901 to 1909. Wilson won with just 41.8% of the popular vote, the third-lowest winning margin in history, and he was the first Democrat to win a presidential election since 1892.

This victory was less about Wilson's popularity and more about a fractured political landscape. The Republican vote split between Taft and Roosevelt, handing Wilson the presidency despite a minority of support.

Taft and Roosevelt, who ran under the newly formed Progressive/"Bull Moose" party, split 50.5% of the popular votes. Roosevelt was shot on October 14, three weeks before the election, with a bullet that remained in his chest for the rest of his life. On October 30, 1912, Vice President James S. Sherman died, leaving Taft without a running mate less than a week before the election. Eugene V. Debs, the fourth-place finisher, won 6% of the popular vote, which remains the highest percentage ever won by a socialist candidate in U.S. presidential elections.

This chaotic election cycle unfolded during a time of deep political and economic unrest conditions that would soon reshape America's financial future.

Chapter 2
Federal Reserve and America's
Economic Quandary

In November 1910, six men, using only their first names, met secretly at the Jekyll Island Club to write the foundational plan for the Federal Reserve Act. For twenty years, they denied their meetings to draft the Aldrich plan, which formed the basis of the Federal Reserve, well after the key provisions of their plan were approved as the Federal Reserve Act.

The secrecy of this gathering, coupled with its immense impact on the future of American finance, has fueled debate and suspicion ever since.

Why did these six men, who were not members of the most expensive and exclusive club, have access to it for several days of meetings? Club members included prominent figures such as J.P. Morgan, Joseph Pulitzer, William K. Vanderbilt, Marshall Field, and William Rockefeller.

This begs a deeper question: how did a small group of mostly unelected elites come to wield such significant influence over America's monetary future, in a setting historically reserved for the financial aristocracy?

The six men included Republican senator Nelson Aldrich, chair of the Senate Finance Committee; Henry Davison, a partner at J.P. Morgan; Abraham Piatt Andrew, Assistant Treasury Secretary; Aldrich's private secretary Arthur Shelton; Frank A. Vanderlip, president of National City Bank, the largest bank in the U.S. and a former Treasury official; and Paul M. Warburg, regarded by many as the chief driving force behind the establishment of the Federal Reserve.

Together, these men represented a cross-section of government, finance, and industry entities now seen as entwined in shaping policy behind closed doors.

14

President Wilson appointed Warburg to the Federal Reserve Board from 1914 to 1918. Wilson opposed the Aldrich Plan because it gave the most authority to bankers. Wilson believed the plan needed oversight and that neither Congress nor the public would support a proposal that gave the government hardly any control.

Aldrich's bill faced strong opposition from politicians, who accused him of bias due to his close ties to wealthy bankers such as J. P. Morgan and John D. Rockefeller Jr., his son-in-law. Wilson's initial resistance reflected democratic concerns but even his version of reform would later be altered.

Although Congress rejected the "Aldrich plan," it laid the foundation for the Federal Reserve Act of 1913, which created the Federal Reserve System. The bill passed on December 22, 1913, and President Wilson signed it into law the next day.

The Federal Reserve Act that Wilson signed was altered after his signature, making it easier for the Federal Reserve to create money. This post-signature alteration weakened democratic checks and increased the Fed's autonomy, allowing it to operate more like a private entity than a government-controlled institution.

Wilson, lacking financial sophistication, likely didn't foresee how the Federal Reserve could evolve beyond his recognition. President Woodrow Wilson's deep Christian faith influenced him, and he ran for re-election in 1916 as the candidate to "Keep the US out of WWI."

Ironically, just a year later, America would be drawn into the very conflict he vowed to avoid triggered, in part, by the financial entanglements made possible through the system he helped create.

On October 15, 1915, American bankers organized under J.P. Morgan & Company authorized a $500 million loan to the British and French governments to finance their WWI expenses. On April 4, 1917, the U.S. Senate voted to support the measure to declare war on Germany, with the House concurring two days later. These were the initial steps for the U.S. to enter WWI. Financial interests, particularly through private lending to foreign governments, created a conflict of interest that made neutrality politically and economically unsustainable.

World War I resulted in approximately 40 million casualties, including 20 million deaths. I believe Wilson was a tortured soul burdened by guilt for his role in creating the Federal Reserve, which facilitated U.S. entry into WWI. He attempted to establish the League of Nations to prevent future wars but failed. In 1919, Wilson suffered a severe stroke that left him incapacitated until the end of his presidency in 1921.

Wilson's health decline seemed to mirror the crumbling peace he had tried too late to secure. His legacy remains deeply conflicted: architect of idealism, yet catalyst for mechanized warfare.

WWI and its aftermath led to WWII, in which an estimated 70–85 million people perished, accounting for about 3% of the global population in 1940. WWI and WWII also caused many people to become atheists. Prior to the creation of the Federal Reserve that facilitated wars, the United States stayed out of foreign war since the country's founding in 1776.

The 20th century introduced a pattern: centralized banking enabled militarized foreign policy, which in turn fractured spiritual and moral confidence worldwide.

If the Federal Reserve is allowed to facilitate WWIII, as it had done with WWI, WWII, and all the other wars that the US was involved in since WWI, billions of people will die. We must take action now to stop the Federal Reserve before it is too late. This is not hyperbole it is a warning grounded in historical precedent. Monetary policy and militarism have become twin engines of destruction, and history shows us the devastating cost of silence.

After the Stock Market Crash of 1929 and the economic problems that followed, millions of people suffered during the "Great Depression." According to Federal Reserve Chairman Ben Bernanke, the Federal Reserve's monetary policy played a significant role in causing the "Great Depression."

This acknowledgment from within the Fed itself reveals a haunting truth: that the institution's missteps can devastate entire generations.

The Federal Reserve also enables the rich to get richer and creates a wider and wider wealth gap between the very rich and the rest of the people. The Federal Reserve, a private bank and the central banking system of the USA, has enormous power and influence over the American and global economy. Yet, it operates with limited transparency and accountability.

The Fed's dual identity private in character, public in function makes it uniquely insulated from democratic scrutiny, even as it shapes the financial lives of every citizen.

Chapter 2
Federal Reserve and America's
Economic Quandary

Through its control of interest rates and ability to create money through quantitative easing (creating money out of nothing), the Fed's monetary policies have far-reaching impacts on economic growth, inflation, asset prices, and wealth distribution. Critics argue that the Fed's actions, especially after the 2008 crisis, have mainly benefited the wealthy by inflating asset bubbles while doing little for regular Americans. It rewards capital over labor, speculation over savings, and short-term profit over long-term stability.

The Fed has also played a critical role in financing U.S. government debt and deficit spending, effectively enabling the expansion of American militarism. The Fed helped finance World War I, World War II, the Korean War, the Vietnam War, the Middle East Wars, the Afghanistan War, the Ukraine War, the Israel-Gaza War, and other wars that took the lives of more than 100 million people and injured many more.

This unbroken chain of conflicts spanning over a century underscores how central banking has become a financial pillar of perpetual war.

Based on the Federal Reserve Economic Data, from 1901 to 1914, the average annual federal surplus was 5.79 million. After the creation of the Federal Reserve, from 1915 to 1963, the average annual federal deficit was 5.13 billion. Since the surge of the Vietnam War in 1964 to 2023, the average annual federal deficit was 409.38 billion.

During the Global Financial Crisis recovery years from 2009 to 2012, the average annual federal deficit was $1.27 trillion. During the COVID-19 recovery years from 2020 to 2023, the average annual federal deficit was $2.24 trillion. This exponential rise in deficit spending reveals a system addicted to debt that enables war, distorts markets, and burdens future generations.

As of July 25, 2025, our national debt was over 37 trillion. The interest payment alone on this debt exceeds $1 trillion annually, limiting the government's ability to invest in critical areas such as infrastructure, education, and healthcare. This debt crisis is not a distant threat it is a present reality already constraining our national priorities and eroding trust in public institutions.

The intertwining of the Fed with the interests of Wall Street and the military-industrial complex raises serious questions about whose interests it really serves. Critics argue that the Fed's low interest rate policies and quantitative easing after the 2008 crisis have contributed to asset bubbles, rising inequality, and increased risk-taking by financial institutions.

Lack of public understanding and democratic oversight of the Fed's operations is a major concern. Decisions that profoundly impact millions of lives are made by a small group of unelected officials, often with close ties to financial elites. True accountability requires sunlight. Yet the Federal Reserve remains cloaked in complexity, inaccessible to the average citizen, and virtually immune to meaningful reform.

The Federal Reserve controls the amount of money in circulation. Part of the money is grouped under M1 Money, which includes coins and currency in circulation + checkable (demand) deposits + traveler's checks + saving deposits.

The other money group is called M2, which includes M1 + money market funds + certificates of deposit + other time deposits. These two categories form the backbone of monetary liquidity in the U.S. economy, directly affecting interest rates, credit availability, and inflationary pressures.

In 2023, the money supply in the U.S. shrank for the first time in 74 years. M2 money supply has dipped from a peak of $21.7 trillion in July 2022 to $20.87 trillion in December 2023. On Jan. 23, 2024, M2 is down a little over 2% on a year-over-year basis and 4.31% from its all-time high set in mid-2022.

When M2 shrinks, there are several implications. Interest rates rise. Economic growth slows. Unemployment increases. A shrinking money supply can trigger a chain reaction of economic pain particularly for small businesses, borrowers, and the working class while tightening access to capital across sectors.

Calls for Fed reform have grown louder in recent years. Ideas include increasing transparency and accountability, diversifying the Fed's leadership, and limiting its powers and role. Some argue for getting rid of the Fed altogether and returning to a money system backed by real assets. At the very least, we need a robust public debate about the proper role and governance of the Fed.

As we deal with the economic challenges of the 21st century, we must ensure our financial system serves the needs and interests of all Americans, not just a privileged few. A growing number of Americans now question whether a system designed over a century ago under vastly different economic and political realities can still meet the needs of modern society.

At the heart of America's economic challenges lies the Federal Reserve, a private institution with the extraordinary power to create unlimited currency for government borrowing and spending. This system, which places the burden of repayment on taxpayers, has undergone significant changes since its inception in 1913.

Understanding the role and evolution of the Federal Reserve is crucial to grasping the complexities of America's current economic situation. The Fed operates with unique and far-reaching authority blurring the line between fiscal and monetary policy in ways that directly affect national sovereignty and economic equity.

A pivotal moment came in 1971 when President Nixon severed the link between the US dollar and gold, transforming our currency into fiat money. This shift has had profound implications, not least of which is the skyrocketing national debt. The Federal Reserve's policies have enabled nearly perpetual wars and other costly endeavors, contributing significantly to this mounting debt.

The ability to create money without the constraint of gold backing has allowed for unprecedented levels of government spending, often with little regard for long-term fiscal sustainability. By removing the anchor of gold, the government removed its own restraints opening the door to debt-fueled policy decisions that benefit the few at the expense of future generations.

The erosion of the dollar's purchasing power is staggering. In 2024, a dollar is worth less than 4 cents compared to its value in 1913 when the Federal Reserve took control of the US banking system. To put this in perspective, prior to the Federal Reserve Act of 1913, an ounce of gold cost around $18.92.

As of April 22, 2025, that same ounce costs over $3,400 a stark illustration of the dollar's decline. This devaluation has significant implications for savings, investments, and the overall economic well-being of American citizens. Long-term savers are punished, wages lose real value, and financial security becomes more elusive particularly for the middle and working class.

This devaluation of currency is not without historical precedent. The hyperinflation in 1920s Germany serves as a cautionary tale. In early 1922, one US dollar was worth 160 German marks. By November 1923, this had plummeted to 4.2 trillion marks to the dollar.

While the US has not experienced hyperinflation of this magnitude, the 88% loss in the dollar's purchasing power since 1971 is cause for serious concern. It raises questions about the long-term stability of our monetary system and the potential for future economic crises. Unchecked money creation, whether driven by war, crisis, or politics, leads nations down a dangerous road. History has shown us where that road can end.

Had President John Kennedy not been assassinated in 1963, we may have avoided the high costs of the Vietnam War that later compelled Nixon to end the gold standard. The Vietnam War cost more than $120 billion and 58,220 American casualties.

Kennedy's economic vision focused on reforming the Federal Reserve and reducing military entanglements posed a direct threat to entrenched interests. His absence reshaped America's fiscal trajectory.

During the Great Depression, Congress enacted the Glass-Steagall Act to protect investors in 1933. The Banking Act of 1933 prohibited commercial banks from underwriting securities in order to prevent conflicts of interest and protect investors.

This law served the U.S. well until November 12, 1999, when the Gramm-Leach-Bliley Act repealed its provisions prohibiting bank holding companies from owning other financial companies. This repeal marked a turning point dismantling the firewall that had protected the economy from reckless speculation for over six decades.

The repeal of the Glass-Steagall Act dismantled this Depression-era law. The removal of this safeguard allowed for the creation of financial behemoths that were deemed "too big to fail," setting the stage for the moral hazard that would come to define the global financial crisis.

Also, while the Federal Reserve's monetary policy provided cheap money that fueled the real estate bubble, the government mandating lenders to issue more sub-prime mortgage loans accelerated the economic meltdown. This repeal significantly contributed to the near-collapse of the global financial system in 2008–2009. The cost of deregulation was paid not by those who engineered it but by millions who lost homes, jobs, and retirement savings.

If you research the causes of our key sufferings in the 21st century, it is quite evident that President Clinton is a key contributor to causing many of our miseries. First, he enabled the repeal of the Glass-Steagall Act, leading to the 2008–2009 Global Financial Crisis (Great Recession), resulting in the suffering of billions of people globally. His permissive handling of the Internet allowed the proliferation of pornography on the Internet. His push to expand NATO and NATO's ambition could eventually lead to WWIII in this decade. Clinton's economic and foreign policies widely praised at the time unleashed forces that have now become dangerously destabilizing.

In 1994, President Bill Clinton initiated the process to expand NATO. This process eventually resulted in the undeclared NATO-Russian war in Ukraine. Under President Biden, the United States Congress has approved more than $174 billion for Ukraine as of May 2024 to fight against Russia. Can you imagine the US government spending $174 billion to help homeless people within its borders instead? The comparison is sobering revealing where national priorities lie, and whose lives are deemed worthy of investment.

Citigroup and other firms spent $200 million lobbying for the repeal of the Glass-Steagall Act with the blessing of the President of the United States, Bill Clinton, the Chairman of the Federal Reserve System, Alan Greenspan, and the Secretary of the Treasury, Robert Rubin.

Between 1999 and 2009, Rubin received total compensation, including employee stock options, of $126 million from Citigroup. This revolving door between government and Wall Street is not just corruption it's legalized betrayal of the public trust.

In the wake of the Great Recession, the Federal Reserve and the US government's response bailing out big banks and insurance companies while forcing the public to absorb toxic debts has been a subject of intense debate.

The surviving banks emerged even larger and more powerful, raising concerns about the potential for an even more devastating crisis in the future. This concentration of financial power in fewer, larger institutions has implications for market competition, systemic risk, and the overall stability of the financial system. Instead of reforming the system, the crisis entrenched it further. Risk was socialized; reward remained privatized.

The scale of government spending and debt accumulation is staggering. The US military budget in 2023 stood at $847 billion, larger than the next ten countries combined. This accounted for about 13% of the federal annual budget of $6.4 trillion, which represented 23.83% of the US GDP. For comparison, in 2007, before the Great Recession, the US budget was $2.568 trillion, or 17.9% of GDP.

This dramatic increase in government spending raises questions about fiscal sustainability and the long-term economic health of the nation. We are now spending more than ever while growing weaker in core areas such as health, infrastructure, and education. This imbalance cannot last.

It's worth noting that in January 1835, for the first and only time in American history, all government interest-bearing debt was paid off. The contrast between this historical moment and our current situation is stark, highlighting the dramatic shift in fiscal policy and economic management over the past two centuries. From debt freedom to debt slavery this arc is both economic and moral.

The challenges facing the US economy are complex and deeply rooted. Addressing them will require a clear-eyed assessment of the Federal Reserve's role, a willingness to learn from historical missteps, and the courage to implement sweeping reforms.

The Federal Reserve's dual mandate of maximizing employment and stabilizing prices has been criticized as conflicting goals that can lead to policy decisions that benefit some sectors of the economy at the expense of others. We must ask whether the Fed's objectives serve the people or merely preserve the status quo.

Moreover, the Fed's role in managing economic crises has expanded significantly since its inception. During the 2008 financial crisis and the COVID-19 pandemic, the Fed took unprecedented actions, including large-scale asset purchases (quantitative easing) and direct lending to businesses.

While these measures helped stabilize the economy in the short term, they have also raised concerns about moral hazard, asset bubbles, and the long-term consequences of such interventions. This trend of borrowing money to boost the US economy is unsustainable. Short-term fixes have become long-term habits and the bill is coming due.

The two largest US debt holders, China and Japan, have lost confidence in the US financial system. China's holding of US debt peaked in 2013 at $1.59 trillion, and Japan's holding of US debt peaked in 2014 at $1.53 trillion.

As of October 2023, Japan held United States Treasury securities totaling about 1.1 trillion U.S. dollars, and China held only $769.6 billion. In less than 10 years, Japan reduced its holding of US Treasury securities by $430 billion, and China sold over $768 billion. This retreat signals a global recalibration one in which the U.S. dollar's dominance is no longer guaranteed.

From Q3 2020 to Q4 2023, China's gold reserve went up by almost 2,000 tons, which is more than the UK, Japan, and India's gold reserves combined. It's like buying a quarter of the US gold reserves in three years. Only five countries have more gold than China. As of Q2, 2024, the US gold reserve was 8,133 tons, Germany's gold reserve was 3,352 tons, followed by Italy, France, Russia and China with 2,452,

2,437, 2,336 and 2,264 tons, respectively. Gold is making a comeback as a hedge against fiat collapse. The world is preparing for a post-dollar order are we?

The story of the Federal Reserve and its impact on the US economy is inextricably linked to broader questions about the nature of money, the role of government in the economy, and the balance between short-term stability and long-term sustainability.

As we grapple with these issues, the need for financial literacy and civic engagement has never been more pressing. The decisions made in the coming years will shape the economic landscape for generations to come. The Fed is not just a financial institution it is a mirror of our national values, our priorities, and our collective future.

We must remember that the 2008 financial crisis brought the world to the brink of economic catastrophe; billions of people suffered, and the Millennials, especially, had to face diminished job prospects, stagnant wages, and mounting student debt. If we are to avoid another generational betrayal, we must act not with fear, but with foresight.

Chapter 3
America's Wars, Interventions, and Their Consequences

Since World War II, the United States has been involved in numerous wars and military actions across the globe. This pattern of intervention, enabled by the Federal Reserve's ability to finance deficits and fund military operations, has had far-reaching consequences both at home and abroad. Rather than being isolated incidents, these interventions form a continuous thread woven into America's post-war foreign policy a policy shaped by power projection, ideological containment, and resource control.

From the proxy battles of the Cold War to more recent conflicts in Afghanistan, Iraq, Syria, Ukraine, and Palestine, US military actions have often resulted in devastating human and economic costs for the countries involved.

While these interventions may have advanced some American strategic and economic interests, they have also raised questions about the true price of global hegemony. Each war leaves behind a trail of trauma, regional instability, and long-term resentment, often sowing the seeds of future conflict.

The ongoing "War on Terror," launched in the wake of the September 11, 2001 attacks, has seen the US become entangled in conflicts in Afghanistan, across the Middle East and Africa.

The 2003 invasion of Iraq, predicated on false claims of weapons of mass destruction, not only destabilized the region but also contributed to the rise of ISIS, creating new security challenges that persist to this day. These actions have left generations of civilians displaced or dead, infrastructures ruined, and extremism fueled not diminished.

After we invaded Iraq, the only thing we protected was oil. Iraq was the cradle of civilization, but museums with priceless artifacts were left unprotected for looters. One of the top profiteers from the Iraq War was the oil field services corporation, Halliburton. Halliburton gained $39.5 billion in "federal contracts related to the Iraq war."

Vice President Dick Cheney was previously CEO of Halliburton. This glaring conflict of interest epitomizes how modern warfare often enriches elites while devastating nations. It raises deep moral questions about whom war really serves and who pays the price.

In the ongoing Israel-Palestine conflict, the US has consistently provided billions in annual military aid to Israel, even as Israel's policies in the Occupied Territories have been widely condemned as violations of international law. This unwavering support has complicated efforts to resolve the conflict and has damaged America's credibility as an honest broker in the region. For decades, this policy has alienated much of the Global South, undermining America's reputation as a defender of democracy and human rights.

The recent escalation of violence following Hamas' attack on Israel on October 7, 2023, has brought these issues into sharp focus. The subsequent Israeli military response in Gaza has resulted in over 53,000 Palestinian death, mostly civilians, over 119,000 injured, and has displaced 1.8 million people as of May 7, 2025. Approximately 1,700 Israelis also died.

The conflict has spread beyond Gaza, involving Lebanon and Yemen, with the US and UK directly engaged in military actions against Yemeni targets. Israel has won almost every battle in Gaza, but it is losing the war in how it is being perceived by the vast majority of people in the world. This crisis has become a test not just of military strength but also of moral standing and the international tide is turning.

Critics argue that this pattern of militarism not only contradicts the principles upon which America was founded but has also eroded its moral standing in the world. The enormous financial cost of these interventions has diverted trillions of dollars away from pressing domestic needs while potentially contributing to the erosion of civil liberties at home. What America spends on projecting power abroad is often money not spent on education, healthcare, infrastructure, or its own veterans ironically, the very people sent to fight these wars.

The US has also been implicated in dozens of coups and regime changes around the world. For example, in 1967, Indonesian President Sukarno was forced to resign under pressure from a CIA-backed takeover led by Suharto.

Within a month, Suharto had signed away Indonesia's rights to the Grasberg mine, the world's largest gold mine, for free to a US company. This pattern reveals a familiar formula: a resource-rich country, a nationalist leader, and then a foreign-sponsored regime change favoring Western corporate interests.

However, in recent years, under President Joko Widodo's leadership, Indonesia successfully took over the Freeport Grasberg gold mine with 51% ownership, signaling a new era of assertiveness and sovereignty over its resources. This marks a rare case where a nation reclaimed what was taken under imperial leverage setting an example for others seeking economic self-determination.

Similarly, on September 11, 1973, the US supported Augusto Pinochet's military coup in Chile, which overthrew the democratically elected government of Salvador Allende. Pinochet's subsequent 17-year rule was marked by widespread human rights abuses, including the execution of thousands and the torture of tens of thousands.

At his death in 2006, about 300 criminal charges were still pending against him in Chile for numerous human rights violations. This intervention, justified in the name of anti-communism, left a lasting wound on Chilean society and set a grim precedent for Cold War-era foreign policy.

More recently, the US has become deeply involved in the conflict in Ukraine. In 2014, the US-backed coup in Ukraine shifted Ukraine's government from Russia to the West. This also resulted in Russia annexing Crimea and part of the Donbas region. This is similar to the US annexing Texas and California from Mexico in the nineteenth century. Geopolitical hypocrisy damages credibility what is condemned in others is often justified when done by the West.

Since the war began in 2022, NATO has supplied escalating lethal weapons to Ukraine and is currently risking a nuclear war. Each shipment of weapons prolongs the conflict and raises the stakes, pushing the world closer to a catastrophic confrontation.

NATO has also applied heavy sanctions against Russia, but they did not work as intended. More than a million soldiers and civilians have died in the war, and millions of Ukrainians are exiled.

Citizens in many EU countries are now paying much higher energy and food prices. The high price of fuel practically killed the German economy. The collateral damage of this war reaches far beyond the battlefield, crippling economies and shaking democratic stability across Europe.

Although Ukraine and Russia reached an agreement for peace in April 2022, the US, through the intervention of UK Prime Minister Boris Johnson, forced Ukraine to decline the deal. This is a tragic war that could have been easily prevented. The roots of many current conflicts can be traced back to past US-led NATO expansions eastward.

In 2025, US Secretary of State Mark Rubio and former UK Prime Minister Boris Johnson admitted on camera that the war in Ukraine was a proxy war against Russia. When leaders admit proxy motives after denying them for years, it confirms what many feared that this war is being fought not for peace, but for power.

The 1953 CIA-backed coup in Iran, which overthrew the democratically elected government of Mohammad Mosaddegh, set the stage for decades of tension between the two nations.

The subsequent support for the Shah's regime, followed by the 1979 Iranian Revolution and hostage crisis, has shaped US-Iran relations to this day. This episode illustrates how American interference often leads to long-term blowback, undermining the very stability it claims to protect.

The United States' conflict with Iran started when Britain's Prime Minister Winston Churchill asked for our help to overthrow the democratically elected government under Prime Minister Mohammad Mosaddegh in Iran. Britain was mad at Mosaddegh for his policy that nationalized the oil companies, and he was overthrown in 1953.

We helped Mohammad Reza Pahlavi, who had been King since 1941 and was widely known as the Shah, who was the last Shah of the Imperial State of Iran, to centralize power until his overthrow in the Iranian Revolution in 1979. In exchange for cheap oil and strategic partnership, the U.S. turned a blind eye to human rights abuses under the Shah until it could no longer ignore the consequences.

In return for providing the US with a steady supply of oil, the Shah received economic and military aid from eight American presidents. The Shah was superseded in 1979 by the theocratic government of Ayatollah Ruhollah Khomeini, a religious cleric who was exiled to France.

On November 4, 1979, Iranian militants stormed the United States Embassy in Tehran and took approximately seventy Americans as hostages. This terrorist act lasted 444 days. This pivotal moment reshaped U.S.-Middle East policy and continues to influence diplomatic tensions to this day.

The Cold War with the Soviet Union almost resulted in a global nuclear war during the 1962 Cuban Missile Crisis. The U.S. embargo against Cuba has been condemned by the UN for 32 consecutive years, with 185 countries demanding its lifting in 2023. Despite global opposition, the embargo persists, serving as a symbol of Cold War-era rigidity that no longer reflects current geopolitical realities.

This longstanding policy has been a source of tension in the Western Hemisphere and has complicated US relations with many Latin American countries. Instead of fostering cooperation, it reinforces distrust and anti-American sentiment in a region that craves mutual respect.

The 2003 U.S. invasion of Iraq, based on false claims of weapons of mass destruction, cost over $2.89 trillion and more than half a million lives. On April 24, 2024, President Biden signed a $95 billion war aid measure that includes aid for Ukraine, Israel, and Taiwan. These ongoing commitments to foreign military aid have raised questions about the prioritization of domestic needs versus international military engagements.

With soaring inflation, homelessness, and healthcare crises at home, many ask: Who benefits from this spending and at what cost to American society?

President Donald J. Trump did not start a new war during his first term. He has said prior to the November 5, 2024, US election that he would quickly end the war in Ukraine if elected president. His claim reflects a broader public fatigue with endless wars and a growing desire for a foreign policy centered on restraint and diplomacy.

As we reflect on this history of intervention and its consequences, it becomes clear that a fundamental reassessment of US foreign policy is needed. Approaches based more on diplomacy, conflict prevention, and respect for international law could better serve America's long-term interests and contribute to building a more peaceful world. The choices made in the coming years will shape the global landscape for generations to come. If the U.S. is to be a true leader in the 21st century, it must lead by example not by domination.

On the subject of conflict resolution, a long time ago, I was chosen to lead eleven newly formed teams comprising about 300 people at DuPont. One of the teams has worked together for many years and there was a great deal of animosity among approximately 25 people in the team. Though on a much smaller scale, their discord reflected the same human factors behind many global conflicts: pride, fear, and mistrust.

So, I asked them to name each of the grievances that they had and I wrote them down on big sheets of paper and put them up on the walls so that everyone could see. There were many pages, and the complaints totaled more than 300. Think about the conflicts you have had with someone or a group. How did you, if you did, resolve the conflict? Behind every war or dispute lies a human need recognition, dignity, clarity, or respect that went unmet.

As for the team above, they voted on each item and narrowed it down to the top 20 complaints, then 10, then 5, and ultimately 1. Some of the top issues involved honesty, trust, respect, communication, effort, balance of power, compromise, decision-making, retribution, limited resources, incompatible goals, unclear responsibilities, perception, pressure, outcome, and independence.

The number one cause of their conflict was respect. Once they started respecting each other more, the team felt better, performance improved greatly, and the bottom line improved significantly. The transformation was simple, yet profound: listen deeply, treat others with dignity, and focus on shared goals. These are not just corporate strategies they are blueprints for peace.

If you improve your "Problem-Solving Skills," you can go a long way. Whether in boardrooms or on battlefields, the true path to peace begins with the willingness to understand.

Chapter 4
Israel-Gaza Conflict – A Powder Keg in the Middle East

The Israel-Gaza conflict, with its deep historical roots and complex geopolitical implications, stands as one of the most intractable and volatile situations in the modern world. The recent escalation of violence has brought this long-standing issue back to the forefront of global attention, raising urgent questions about peace, justice, and the future of the region.

It is not merely a regional crisis it is a global flashpoint that threatens to pull major powers into direct confrontation and destabilize the entire Middle East.

WWIII has not yet come, but danger looms in the air. In 1948, the Jewish State of Israel was established in Palestine, beginning a Jewish-Palestinian war that continues to the present day. In 1967, the Israelis captured Jerusalem in the "Six-Day War." Since then, repeated cycles of violence, occupation, and resistance have prevented any lasting peace.

The horrific attack by Hamas on October 7, 2023, which resulted in the deaths of 1,200 people in Israel, marked a new and tragic chapter in this conflict. Hamas, a Palestinian Sunni Islamist political and military organization that governs parts of the Gaza Strip, has long been at odds with Israel, but the scale and brutality of this attack shocked the world.

According to the U.S. State Department and the European Union, Hamas is designated as a terrorist organization. However, many in the Arab world and parts of the Global South view it as a resistance movement against occupation highlighting the deep division in global narratives.

The Israeli response to this attack has been devastating. As of May 7, 2025, the Israel-Gaza war has claimed over 53,000 Palestinian lives, the majority of whom were civilians, including a large number of children and women. More than 120,000 people have been injured, and an astounding 1.8 million have been displaced. The humanitarian crisis in Gaza has reached catastrophic proportions, with widespread destruction of homes, infrastructure, and essential services.

Reports from international humanitarian organizations such as UNRWA and Médecins Sans Frontières confirm the collapse of healthcare, sanitation, and food access in Gaza, describing it as one of the worst humanitarian disasters of the 21st century.

The conflict has sparked global outrage and led to widespread protests, particularly on college campuses around the world. Demonstrators are demanding divestment from Israel and certain Israeli companies, leading to tensions with university administrations. Some institutions have begun expelling students, further escalating the situation.

In April 2024 alone, there were approximately 2,000 arrests related to these protests. Many student-led movements have invoked parallels to the anti-apartheid movement in South Africa, increasing pressure on institutions to take moral stances amid political controversy.

The international community has struggled to respond effectively to the crisis. On May 20, 2024, following a thorough investigation, the International Court of Justice announced its intention to seek the arrest of Hamas leaders for war crimes committed in Israel, as well as Israeli Prime Minister Benjamin Netanyahu for alleged war crimes in Gaza.

This development underscores the complex and contentious nature of the conflict, with atrocities committed on both sides. This is the first time in modern history that a sitting Israeli prime minister has faced such charges from an international body signaling a possible shift in global accountability standards.

The US role in this conflict has been a subject of intense debate. As a long-standing ally of Israel, the US has consistently provided massive military aid and diplomatic backing. This support has complicated efforts to reach a peaceful resolution.

The US has supplied Israel with over $150 billion in aid since its founding, including $3.8 billion annually under a 10-year memorandum signed during the Obama administration. Critics argue that unconditional support limits America's ability to act as a neutral mediator.

However, there have been moments of progress in the past. The 1978 Camp David Accords, mediated by President Jimmy Carter, led to a peace deal between Egypt and Israel. The 1993 Oslo Accords established the Palestinian Authority and seemed to offer a path towards a two-state solution.

Yet, the assassinations of Egyptian President Anwar Sadat and Israeli Prime Minister Yitzhak Rabin by hardliners opposed to highlight the challenges facing those who seek peace. Both leaders paid the ultimate price for choosing diplomacy over war underscoring how extremism on both sides has repeatedly derailed hope.

The US, given its influence and resources, has the potential to play a crucial role in shaping the future of this conflict. Advocates argue that it must use its leverage, including the billions in annual aid to Israel, to discourage actions that undermine peace, such as the expansion of settlements in the occupied territories. They also call for holding all parties accountable for violence and human rights violations. Many now believe a new framework is needed one rooted not in historical favoritism, but in international law and genuine equality.

The Middle East is increasingly interconnected by overlapping alliances and proxy dynamics, meaning one spark in Gaza can ignite flames across the entire region.

Since November 2023, the Houthis have attacked many commercial boats and U.S. ships. In most cases, the U.S. is launching $2 million defense missiles to stop $2,000 Houthis drones. The Houthis have also taken down 7 $32 million US Reaper drones in addition to diverting US warships, including a carrier, from Asia to Yemen. Moreover, US has lost 3 F/A-18 fighter jets, each costing about $67 million, for various reasons. This asymmetric warfare has exposed the vulnerabilities of traditional military might and further burdened U.S. taxpayers in a war with no clear endgame.

The conflict with Yemen in the Red Sea has increased freight costs, shipping time, and the expenses of insuring commercial trade goods. Lebanon has fired a ballistic missile into Israel for the first time after Israel's massive bombing in Lebanon. This war is escalating fast. The Red Sea disruptions have already impacted global supply chains, pushing inflation higher and affecting food and fuel prices worldwide.

The conflict has also exposed deep divisions within the United States. While the powerful Israel lobby, including organizations like AIPAC, has maintained strong support for Israel in Congress, there has been growing criticism of

Israeli actions, particularly among younger Americans and progressive politicians.

Recent polls by Pew Research show a significant generational shift: while 61% of older Americans sympathize with Israel, only 32% of Millennials and Gen Z do indicating a potential future realignment in U.S. foreign policy.

As the death toll mounts and the humanitarian crisis deepens, the need for a just and lasting solution becomes ever more urgent. Yet the path to peace remains elusive. Any resolution must grapple with fundamental issues of land rights, security, and self-determination for both Israelis and Palestinians.

A two-state solution, while long envisioned, now seems more distant than ever, as settlements expand and extremist voices grow louder on both sides.

The Israel-Gaza conflict serves as a stark reminder of the consequences of unresolved historical grievances and the dangers of entrenched positions. It highlights the need for courageous leadership, innovative diplomacy, and a willingness to challenge long-held assumptions.

As the world watches events unfold in this troubled region, the decisions made in the coming months and years will have profound implications not just for Israelis and Palestinians but also for global peace and stability. With the Middle East at a boiling point, inaction is no longer an option especially for a world that claims to value justice, human dignity, and peace.

As we look back on our own history, we can see that the genocide and ongoing ethnic cleansing in Gaza is a smaller, contemporary version of what the United States military did to the Native Americans. Though the contexts differ, the moral parallels force us to reflect: how can a nation born from overcoming injustice remain silent when others endure it?

However, the US has come a long way. The United States has the most generous and innovative people enjoying freedom that most people on Earth can only dream of. We must do our part to preserve that privilege in a fair way. The time to act is now. If America is to remain a beacon of hope, it must lead not by force, but by example. True leadership is not domination it is compassion, courage, and the pursuit of justice.

Chapter 5
NATO's Expansion and the War in Ukraine

The ongoing conflict in Ukraine represents one of the most significant geopolitical crises of the 21st century, with far-reaching implications for global security and the international order. At its core, this crisis is deeply intertwined with the history of NATO expansion and Russia's perception of threats to its security.

The roots of the current conflict can be traced back to the end of the Cold War. On February 9, 1990, US Secretary of State James Baker made a crucial promise to Soviet leader Mikhail Gorbachev: NATO would not expand eastward if Russia accepted Germany's unification. German Chancellor Helmut Kohl and NATO's secretary general reiterated this assurance in the following months.

For Gorbachev, these assurances were pivotal in clearing the way for a compromise on German reunification. However, this promise was not formalized in the treaty signed on September 12, 1990. Declassified documents later revealed that multiple Western officials gave verbal assurances to Gorbachev, but no legally binding commitment was made.

Following the collapse of the Soviet Union in 1991, many former Warsaw Pact and post-Soviet states sought to join NATO. Despite Russian opposition, Poland, Hungary, and the Czech Republic became NATO members in 1999. This was followed by the accession of seven additional countries in 2004: Bulgaria, Estonia, Latvia, Lithuania, Romania, Slovakia, and Slovenia. Subsequent expansions included Albania and Croatia (2009), Montenegro (2017), North Macedonia (2020), and Finland (2023).

Russian President Vladimir Putin, in his now-famous 2007 Munich speech, expressed Russia's concerns about NATO expansion: "I think it is obvious that NATO expansion does not have any relation with the modernization of the Alliance itself or with ensuring security in Europe. On the contrary, it represents a serious provocation that reduces the level of mutual trust. And we have the right to ask: against whom is this expansion intended?"

The situation reached a critical point in 2008 when NATO welcomed Ukraine and Georgia's aspirations for membership. Russia saw this move as crossing a red line. At the Bucharest Summit, NATO officially stated that Ukraine and Georgia "will become members of NATO," though no specific timeline was offered.

In 2014, a Western-backed coup in Ukraine led to a pro-Western government, further alarming Russia. The Ukraine war started in 2014 when Victoria Nuland, the assistant secretary of state for European and Eurasian Affairs, and neocons instigated the coup that illegally overthrew the Yanukovich government. Russian intelligence intercepted and leaked to the international media a Nuland telephone call with U.S. ambassador to Ukraine Geoffey Pyatt discussing in detail their preferences for specific personnel in a post Yanukovych government while Viktor Yanukovych was still President.

However, many scholars and Western analysts describe the 2014 change in government as the result of a popular uprising (Euromaidan) against widespread corruption and the rejection of an EU trade deal not an orchestrated coup. The tensions finally erupted into full-scale war in 2022 when Russia invaded Ukraine, viewing it as an intolerable step towards NATO membership. The invasion was widely condemned internationally and met with Western sanctions and military aid to Ukraine.

Professor John Mearsheimer of the University of Chicago has consistently warned about the dangers of NATO expansion, arguing that it was the root cause of the war in Ukraine. He was a United States Military Academy West Point graduate, and he served in the US Army and Air Force for ten years. His analysis suggests that Russia would go to great lengths to prevent Ukraine from joining NATO, even if it meant decimating the country. In his 2014 article "Why the Ukraine Crisis Is the West's Fault," Mearsheimer argued that Western policymakers mistakenly assumed Russia would passively accept NATO's eastward expansion.

The conflict has drawn in major global powers. As of May 2024, the United States has been a key supporter of Ukraine, with Congress approving over $174.8 billion in aid since 2022. Of that amount, approximately $83.4 billion had been disbursed by early 2025, including military, humanitarian, and financial support (USA Facts). This involvement goes beyond mere support for Ukrainian sovereignty. According to US Senator Lindsey Graham, Ukraine possesses an estimated 10 to 12 trillion dollars' worth of critical minerals that could be used to repay the US.

While this figure is widely cited, independent estimates from the Ukrainian government and mining analysts suggest Ukraine's recoverable critical mineral reserves are valued in the lower trillions range significant, but difficult to precisely quantify.

Moreover, Ukraine has already sold off $400 billion in assets on Wall Street despite its national GDP being only $200 billion. This claim is difficult to verify and likely conflates long-term investment opportunities, future valuations, and speculative transactions rather than actual asset liquidation of that scale. Ukraine's privatization and foreign investment policies have opened its market to international entities, but exact figures remain contested.

The strategic importance of Ukraine extends beyond its mineral wealth. The US needs the critical minerals that Ukraine has for its military and industrial applications. Similarly, Taiwan, another flashpoint in international relations, is crucial for the US due to its advanced semiconductor industry, which is vital for military, artificial intelligence, and other high-tech applications. I visited most of the semiconductor companies, such as TSMC in Taiwan, for business during the late 90's.

There's a chance the Ukraine war could trigger WWIII, with the US and its allies on one side and Russia and its allies on the other. The West, led by the US and its allies, has given Ukraine hundreds of billions of dollars in cash, weapons, and other forms of assistance while its own citizens have suffered high inflation in food, energy, and other costs. Inflation rates in the US and Europe surged post-2022, partly due to the war's effects on global energy and grain markets. However, inflation has gradually declined in 2023–2024 according to central banks.

There have been hundreds of thousands of deaths in Russia and Ukraine, and millions have fled because of the Ukraine war. As of early 2024, estimates suggest over 500,000 soldiers have been killed or wounded on both sides, and more than 6 million Ukrainians have become refugees (UNHCR). As a result of the disruption of oil and gas flow from Russia, many European countries are burning more coal, causing climate change to worsen. Germany, for example, reactivated coal plants to stabilize its energy supply, despite long-term commitments to renewable energy.

The US-led, NATO-backed coup in 2014 replaced the Ukrainian government loyal to Russia with one on the NATO side. Right after the coup, Russia annexed Crimea to protect its naval base there. Russia shares a 1,200-mile border with Ukraine, and it wanted to prevent NATO from installing weapons at this border. Russia also wanted to protect ethnic Russians in Ukraine, who were persecuted by the new regime.

So, Russia supported Russians in Ukraine's eastern region to claim independence and then become a part of Russia. International human rights organizations have noted tensions between ethnic Russians and the post-2014 Ukrainian government, though claims of systematic persecution have been heavily disputed and used by Russia to justify military intervention.

It didn't make sense as to why NATO and the EU seemingly sacrificed the welfare of their citizens by so easily giving up cheap gas from Russia and providing hundreds of billions of aid to Ukraine until you realize that NATO, EU, and the US may be interested in securing mining contracts from Ukraine.

Ukraine is home to a vast array of critical minerals, including the largest titanium reserves in Europe, accounting for 7 percent of the world's reserves, and almost 500,000 tons of lithium, with an estimated value in excess of US$26 trillion. Ukraine also holds significant untapped rare earth deposits, graphite, cobalt, and other minerals essential for green technologies and defense systems.

Ukraine's reconstruction is projected to cost $486 billion. In October 2024, Ukrainian President Volodymyr Zelenskyy announced a partnership with global asset manager BlackRock to coordinate international investment in rebuilding Ukraine's economy (Reuters). BlackRock's Financial Markets Advisory division is tasked with overseeing investor strategy, signaling that Western corporate and financial interests will likely secure the largest reconstruction contracts.

Germany's new Chancellor Fredrick Merz was formerly the Chairman of BlackRock Germany. He was also a senior counsel for Mayer Brown, the law firm for Lehman Brothers. Lehman was the most responsible for causing the near total meltdown of the global financial system in 2008. On May 14, 2025, Merz pledged that Germany would have the "strongest conventional army in Europe."

Zelenskyy's presidential term expired on May 20, 2024, yet he remains in power under wartime emergency law. This has raised constitutional concerns, especially since no election has been scheduled. Ukraine has at least five million refugees abroad, half of whom are eligible to vote, and an additional six million internally displaced persons. The logistics of organizing a national election under martial law and active war conditions remain unresolved. It is difficult to imagine how the hundreds of thousands of Ukrainian men and women currently serving in the armed forces could meaningfully participate in the election.

The Ukraine conflict has had global consequences. It disrupted energy markets, drove up inflation, worsened food insecurity in developing nations, and strained diplomatic relations between East and West. A clear contrast was seen in April 2024, when the US and its allies intercepted over 180 ballistic missiles and drones launched by Iran toward

Israel, a demonstration of high-tech air defense and unified action. By contrast, Ukraine has endured daily missile and drone attacks for over two years with less coordinated Western military intervention to shield civilians. This disparity has led some to argue that Ukrainians are being treated as expendable geopolitical pawns rather than valued allies.

Critics assert that the US-led NATO alliance is more interested in Ukraine's vast mineral wealth than in the lives of its citizens. Ukraine possesses massive deposits of critical minerals, including the largest known titanium reserves in Europe, substantial lithium, cobalt, rare earth elements, and other materials crucial to defense, energy, and tech industries (BBC). The sooner the Ukrainian people understand this, the argument goes, the sooner they can negotiate peace with Russia and secure a future based on neutrality and non-alignment.

Resolving this conflict and the broader NATO-Russia divide requires a frank assessment of how actions on all sides escalated tensions. NATO expansion, Western support for Ukraine's political transformation, and Russia's military aggression have all contributed to this breaking point. The West's idealistic narrative about "freedom and democracy" masks the realpolitik motivations underlying the conflict, particularly resource acquisition and strategic positioning.

Professor Jeffrey Sachs of Columbia University, a Jewish American economist and longtime advisor to numerous world leaders, delivered a major speech at the European Parliament on February 19, 2025, addressing the wars in Ukraine, Gaza, and elsewhere. He accused the United States, the EU, and NATO of decades-long warmongering and called for a radical shift in foreign policy. Sachs holds a BA, MA, and PhD from Harvard University and formerly taught there before joining Columbia. His views, though controversial, are echoed by a growing number of voices calling for multipolar diplomacy and international accountability.

A sustainable solution may require creative diplomacy to establish a new European security framework that balances NATO and Russian concerns. Simply reverting to the pre-2022 status quo is unlikely to produce lasting peace. One possible approach involves guaranteeing Ukrainian neutrality, similar to Austria or Finland during the Cold War. This would include arms control agreements, conflict-resolution mechanisms, and economic cooperation initiatives, creating a buffer zone that honors Ukrainian sovereignty while reducing NATO-Russia friction.

However, this won't be easy. Years of conflict have deepened mutual mistrust and created hardened positions on all sides. Yet the stakes could not be higher. Continued conflict increases the risk of NATO's direct involvement, which could spiral into a larger European or even global war. While the use of nuclear weapons remains unlikely, it cannot be entirely ruled out, particularly if Russia perceives an existential threat from NATO actions.

The Ukraine crisis is a cautionary tale. It shows what happens when history is ignored, diplomacy is sidelined, and power politics override empathy. Going forward, leaders must embrace a more nuanced, inclusive vision of global security one that prioritizes people over profit, peace over posturing, and cooperation over conquest. The people of Ukraine, Russia, and Europe deserve no less.

Chapter 6
Persecution & Achievements
of Jewish People

Jewish people have endured persecution for thousands of years, from 400 years of slavery in Egypt and long captivity in Babylon in ancient times to various forms of anti-Semitism worldwide, culminating in the Holocaust in the past few thousand years. Anti-Semitism persisted even after the establishment of the State of Israel. According to the Anti-Defamation League, anti-Semitic incidents in the U.S. increased from fewer than 100 in 1979 and 751 in 2013 to 3,697 in 2022.

The ADL also reported that there were "9,354 antisemitic incidents across the United States. This represents a 5% increase from the 8,873 incidents recorded in 2023, a 344% increase over the past five years and a 893% increase over the past 10 years. It is the highest number on record since ADL began tracking antisemitic incidents 46 years ago. "

Despite past persecution, Jewish people have done well. They value education, work hard, and are smart. We should celebrate their achievements and learn from them. I have Jewish friends in the U.S. and Israel. Some of them are relatives of the survivors of the Holocaust. Today, Jewish people are some of the wealthiest and most powerful groups of people on Earth. In 2023 and 2024, six of the 12 richest Americans are Jewish. According to Forbes, there were 267 Jewish billionaires with a combined net worth of 1.7 trillion dollars in 2022.

It is common knowledge that Jews dominate much of the finance industry, communication industry, entertainment industry, social media, Internet search, and many key government officials in powerful countries. Some examples of the Jewish-owned corporations that rule the world include Goldman Sachs, JP Morgan Chase Bank, BlackRock, Blackstone, DreamWorks, ViacomCBS, Facebook (Meta), Google (Alphabet), Oracle, and countless other corporations.

Powerful owners of these corporations include David Solomon, Jamie Dimon, Larry Fink, Stephen Schwarzman, Steven Spielberg, Sumner Redstone (originally Rothstein), Mark Zuckerberg, Larry Page, Sergey Brin, and Larry Ellison. I once asked Sumner Redstone who was the most influential person in his life, and he said it was his mother, who emphasized "Education, education, and education."

Key Figures and Statistics

- At year-end 2018, JPMorgan Chase Bank owned 29.5 percent of the Federal Reserve. In 2024, JPMorgan Chase Bank was the largest bank in the world by market capitalization. J.P. Morgan was instrumental in the founding of the Federal Reserve, which is a private bank. The Federal Reserve could keep the perpetual wars going in the Middle East, Ukraine, and other locations, killing many innocent people and increasing national debts that will be a burden not only to current but future generations. The Federal Reserve enabled the financing of WWI, WWII, the Korean War, the Vietnam War, the Middle East Wars, the Afghanistan War, the Ukraine War, the Israel-Gaza War, and other wars that took the lives of more than 100,000,000 people and injured many more. If and when there is a WWIII, the Federal Reserve will certainly be involved in financing the war that will wipe out human civilization. It is an existential threat to the entire world. The Federal Reserve has also enslaved billions of people to poverty and must be abolished. It is the worst financial weapon ever created against humanity. Moving forward, the world is going through major transitions in many areas. Take action while you can to save yourself and mankind.

- Digital fiat currency will be the new money. We need to make sure we understand what it is and how it will work, especially its use in surveillance where privacy is compromised.

The achievements of the Jewish people are simply remarkable and astonishing. Their accomplishments should be acknowledged.

Since 1995, Jewish individuals who have served as Treasury Secretary include Robert Rubin, Larry Summers, Jacob Lew, Steven Mnuchin, and Janet Yellen. Secretaries of State Henry Kissinger, Madeleine Albright, and Anthony Blinken are Jewish. From 1979 to 2018, Federal Reserve Chairmen Paul Volker, Alan Greenspan, Ben Bernanke, and Janet Yellen were all Jewish.

Supreme Court Justices Louis Brandeis, Ruth Bader Ginsburg, Stephen G. Breyer, and Elena Kagan are Jewish. The last three served simultaneously as one-third of the nine Supreme Court justices for a decade. Justice Elena Kagan is the first person appointed to the Court without prior judicial experience since William Rehnquist and Lewis F. Powell Jr. She is the eighth Jewish justice in the court's history.

Influential senators such as former Senate Majority Leader Chuck Schumer, Bernie Sanders, and Dianne Feinstein are all Jewish. In the current 119th U.S. Congress, 10% of senators are Jewish. Of the recipients of the Nobel Prize and the Nobel Memorial Prize in Economic Sciences between

1901 and 2023, at least 214 out of 965 individuals were Jews or people with at least one Jewish parent, representing 22% of all recipients. Most Ivy League and other top universities have Jewish presidents, professors, and significant Jewish student representation. Jewish people value education, they work hard, and they are smart. We should learn from them.

Jewish people have made substantial contributions to the U.S., the EU, and the world as a whole. I especially appreciate the Jewish doctors, nurses, scientists, engineers, lawyers, and investors who make this world a better place. On the dark side, there are also extremely wealthy Jewish people who harm society. Harvey Weinstein, Jeffrey Epstein, and Sam Bankman-Fried are some examples of despicable people.

In 2020, Weinstein was sentenced to 23 years in prison for rape in New York. He was found guilty of additional charges on December 19, 2022, which added 16 more years to his prison sentence in California. He co-founded the entertainment company Miramax, which produced several successful independent films. Harvey Weinstein, once worth $300 million, sexually abused many famous actresses and other women. In 2024, New York's top court overturned Weinstein's conviction based on legal technicalities.

Jeffrey Epstein was accused of running a vast human-trafficking operation in which he and co-conspirators procured women and girls, mostly teenagers between 14 and 17, for sex with himself and his elite associates. He was worth more than $500 million. Epstein died in 2019 while jailed and awaiting trial. He was continuously filmed with two cameras, but both cameras supposedly failed. There is no video of how he killed himself if he did. President Bill Clinton had a close relationship with Jeffrey Epstein. Will politicians continue to protect with impunity the rich and powerful pedophiles associated with Epstein?

Sam Bankman-Fried was convicted of fraud and related crimes in November 2023. He founded the FTX cryptocurrency exchange and, at its peak, was the 41st-richest American on the Forbes 400. Many large, illicit transactions are done with cryptocurrency, which is backed by nothing. Some call cryptocurrency the largest Ponzi scheme in the world. A Ponzi scheme is an investment fraud that pays existing investors with funds collected from new investors. FTX and FTX US had an estimated $8.7 billion combined shortfall in bankruptcy. $8.7 billion is a lot of money. Many people were deceived at a time near the end of times.

Knowing more about these despicable people would only make you angry and raise your blood pressure. So, I'll stop here.

Regarding cryptocurrency, here are 100 reasons why people should be cautious about investing.

1. High volatility and unpredictable price fluctuations
2. Lack of regulatory oversight and protection
3. Potential for fraud and scams
4. Cybersecurity risks and vulnerabilities
5. Difficulty in understanding the underlying technology
6. Absence of intrinsic value backed by tangible assets
7. Limited adoption and acceptance as a mainstream currency
8. Susceptibility to market manipulation and pump-and-dump schemes
9. Potential for significant financial losses

10. Lack of consumer protection and recourse in case of theft or fraud
11. Complexity in securely storing and managing private keys
12. Potential for hacking and theft of digital wallets
13. Absence of insurance or government-backed deposit guarantees
14. Difficulty in accurately valuing cryptocurrencies
15. Potential for regulatory crackdowns and bans
16. High transaction fees during periods of network congestion
17. Slow transaction processing times compared to traditional payment methods
18. Irreversibility of transactions and lack of chargebacks
19. Limited options for converting cryptocurrencies to fiat currencies
20. Potential for technological obsolescence and displacement by newer cryptocurrencies
21. Lack of widespread merchant acceptance of goods and services
22. Difficulty in integrating with existing financial systems and infrastructure
23. Potential for money laundering and illegal activities
24. Absence of physical form and tangibility
25. Dependence on Internet connectivity and electricity for access
26. Potential for coding errors and software vulnerabilities
27. Lack of standardization and interoperability between different cryptocurrencies
28. Difficulty in estate planning and transferring assets to heirs
29. Potential for government interference and seizure of assets
30. Lack of stability and reliability as a store of value

31. Potential for environmental damage due to high energy consumption of data mining
32. Difficulty in obtaining accurate and reliable information and advice
33. Lack of transparency in the development and governance of cryptocurrencies
34. Potential for concentration of wealth and centralization of control
35. Difficulty in using cryptocurrencies for everyday transactions and purchases
36. Potential for social engineering attacks and phishing scams
37. Lack of legal recognition and enforceability of contracts
38. Difficulty in complying with tax and reporting requirements
39. Potential for market saturation and declining returns on investment
40. Lack of stability and predictability in mining rewards and incentives
41. Potential for 51% attacks and network consensus failures
42. Difficulty in managing and securing multiple cryptocurrency holdings
43. Lack of fungibility and traceability of individual cryptocurrency units. Fungibility is the ability of a good or asset to be readily interchanged for another of like kind.
44. Potential for ideological and philosophical disagreements within communities
45. Difficulty in assessing the credibility and reliability of cryptocurrency projects
46. Lack of customer support and dispute resolution mechanisms
47. Potential for addiction and obsessive behavior related to trading and speculation

48. Difficulty in explaining and justifying investments to family and friends
49. Lack of familiarity and comfort with digital and decentralized technologies
50. Potential for social and economic disruption and instability
51. Difficulty in diversifying and managing risk in cryptocurrency portfolios
52. Lack of liquidity and market depth for some cryptocurrencies
53. Potential for pump-and-dump schemes and market manipulation by whales
54. Difficulty in assessing the long-term viability and sustainability of cryptocurrencies
55. Lack of protection against front-running
56. Potential for flash crashes and sudden market downturns
57. Difficulty in hedging and mitigating risk exposure
58. Lack of transparency in the use of proceeds from initial coin offerings (ICOs)
59. Potential for scams and fraudulent ICOs and token sales
60. Difficulty in distinguishing between legitimate and fraudulent cryptocurrency projects
61. Lack of standards and best practices for secure cryptocurrency storage and custody
62. Potential for user error and accidental loss of funds
63. Difficulty in recovering lost or forgotten passwords and private keys
64. Lack of interoperability and compatibility with existing financial infrastructure
65. Potential for market fragmentation and lack of liquidity across exchanges
66. Difficulty in assessing the impact of forks and protocol changes on investments
67. Lack of protection against market abuse

68. Potential for regulatory arbitrage and jurisdiction shopping
69. Difficulty in complying with know-your-customer (KYC) and anti-money laundering (AML) requirements
70. Lack of standardization in terminology and definitions across the industry
71. Potential for reputational damage and association with illegal activities
72. Difficulty in explaining and justifying the value proposition of cryptocurrencies
73. Lack of familiarity and comfort with cryptographic concepts and techniques
74. Potential for social and political backlash against cryptocurrencies
75. Difficulty in assessing the impact of geopolitical events on cryptocurrency markets
76. Lack of protection against state-sponsored attacks and cyber-warfare
77. Potential for concentration of mining power and centralization of control
78. Difficulty in participating in governance and decision-making processes
79. Lack of transparency in the use of funds by cryptocurrency foundations and organizations
80. Potential for conflicts of interest and self-dealing by cryptocurrency insiders
81. Difficulty in valuing and pricing cryptocurrency-based products and services
82. Lack of standardization in accounting and financial reporting practices
83. Potential for market manipulation through wash trading and spoofing
84. Difficulty in complying with securities laws and regulations

85. Lack of consumer education and awareness about cryptocurrency risks and benefits
86. Potential for scams and fraudulent cryptocurrency-based investment schemes
87. Difficulty in assessing the credibility and track record of cryptocurrency developers and entrepreneurs
88. Lack of protection against market abuse
89. Potential for price manipulation through coordinated buying and selling activities
90. Difficulty in managing and mitigating the risks of margin trading and leveraged positions
91. Lack of transparency in the operation and security of cryptocurrency exchanges
92. Potential for flash crashes and market disruptions due to automated trading algorithms
93. Difficulty in valuing and pricing cryptocurrency derivatives and other complex financial instruments
94. Lack of standardization in the classification and taxonomy of cryptocurrencies and tokens
95. Potential for scams and fraudulent cryptocurrency-based cloud mining and staking schemes
96. Difficulty in assessing the environmental and social impact of cryptocurrency mining operations
97. Lack of protection against market manipulation and front-running by high-frequency traders
98. Potential for scams and fraudulent cryptocurrency-based multi-level marketing (MLM) schemes
99. Difficulty in managing and mitigating the risks of investing in cryptocurrency ETFs and other fund products
100. Lack of a clear and compelling use case for many cryptocurrencies beyond speculation and trading.

These concerns highlight the need for caution, due diligence, and a thorough understanding of the risks involved before investing in cryptocurrencies. It's essential to approach cryptocurrency investments with a critical eye and to be prepared for the possibility of significant losses in a highly volatile and speculative market.

The United Kingdom, formerly the largest and richest empire in history, is now broke and broken, according to the UK Prime Minister Starmer's office. However, the UK holds a significant amount of U.S. debt, with $779.3 billion. The only country that holds more US debt, about $1 trillion, is Japan. The US debt is about $37 trillion as of May 22, 2025. If the UK and Japan sell large quantities of US debt, the US bond market would suffer significantly.

The UK had 7 prime ministers in the past 17 years. The number of British billionaires still stands at 165 in 2024. The US's roughly 760 billionaires now hold 3.8% of U.S. wealth, according to Americans for Tax Fairness, while the bottom half of American families control only 2.5%.

Money is manmade. We have already seen the destructive power of the Federal Reserve globally by creating money out of nothing, leading to the deaths of 100 million people and fueling the inequality gap between the super-rich and billions of people. WWI and WWII, facilitated by the Fed, turned many people away from believing in God.

Cryptocurrencies are another manmade fiat money with no intrinsic value. In both cases, the greed of a few people is very evident, and there could come a time when most people will no longer agree to live under the oppressive monetary systems that are inherently unfair, especially to young people with little or no assets.

Section 2: My Blessed and Grateful Life

Chapter 7
Early Life in Burma

Born in 1959 in Rangoon, Burma (now Yangon, Myanmar), I grew up in a Buddhist culture that taught meditation, service to others, and reverence for all life as sacred. Our household also embraced elements of Chinese folk religion, particularly ancestor worship. Burma was once the richest country in Southeast Asia, and U Thant, a distinguished Burmese diplomat, served as the third Secretary-General of the United Nations from 1961 to 1971. He held the office for a record ten years and one month, gaining global respect for his leadership. Decades later, my uncle Steven followed a similar path, working as a senior executive for the UN for seventeen years.

Watching the Apollo 11 moon landing on July 20, 1969, sparked my fascination with science. From an early age, I felt a profound connection to both science and spirituality two forces that have shaped my life ever since.

My paternal grandfather, a successful entrepreneur who owned more than a dozen businesses, had a profound influence on me. Despite his wealth, he remained humble, consistently emphasizing the value of hard work, integrity, and generosity. He often told stories laced with ancient Chinese wisdom and introduced me to traditional Chinese literature, scriptures, and proverbs. His example of quiet strength and leadership left a lasting mark on me.

Everywhere we went, he commanded respect not through fear, but through dignity. I never saw him lose his temper or fall ill, not even with a cold. His disciplined lifestyle and attention to health impressed me even as a child.

My paternal grandparents, Chin Lin Ngoon and Chow Toy King, showered me with love. They often prepared my favorite dishes and took me out for dim sum. My grandmother's rich chicken soup and her delicacy made from pig brains, both requiring great effort and care, made me feel especially cherished though sometimes guilty for being their favorite grandchild.

In school, I was a high achiever and skipped first grade. By fourth grade, though, my focus shifted more toward play. A childhood crush rekindled my academic ambition, and within six months, I reclaimed the top spot in class. At age 11, I began waking up at 4 or 5 a.m. to study by candlelight on the veranda.

Blessed with what I can only describe as a photographic memory, I could recall the exact words and layout of pages from books. I became an avid reader two books a week for leisure favoring genres like romance, mystery, and science fiction. Comic books also held a special place in my heart. This love for reading, which began around age 12 or 13, stayed with me through much of my teenage years.

I had many hobbies: soccer, chess, table tennis, badminton, stamp collecting, and coin collecting. A vibrant chess scene near my home drew me in, and I quickly rose in skill, often defeating seasoned players even after I moved to the United States. I once challenged a chess computer program, losing 75 times in a row before finally winning. After that, I won most of the games against it. This persistence in learning has always defined me. As long as I'm learning something new, boredom rarely finds me.

Burmese festivals and public celebrations were a joyful part of life. Certain times of the year, the streets came alive with pop-up stages for free theatrical performances. As a teenager, I loved attending these performances, even by myself, since the community environment felt safe. Thingyan, the Burmese New Year festival with origins in the 11th century, was the highlight of the year. The days-long Buddhist festival features colorful floats parading through the streets and water stations along the way. The festival's finale marks the dawn of the New Year, when government offices and businesses shut down to allow full public participation. The hallmark of Thingyan is water throwing a spirited tradition where people douse one another from head to toe. If you step outside, getting soaked is inevitable!

Kite flying was another passion, one I took seriously and pursued competitively. I spent hours on our rooftop, battling kites up to 250 yards away. My parents constantly feared I'd fall, but I walked the sloped rooftops with confidence.

The goal was to cut the strings of other kites using my own string, which was treated with a glass-coated glue mixture to make it razor-sharp. To enhance reeling speed and increase my chances of winning, I used a 10-inch spool spun with one finger while gripping both handles a skill that demanded hours of practice.

To support my hobbies, I launched a sidewalk book-lending business near my junior high school. Across from busy shops and eateries, I set up a modest lending library filled with novels and comics. Borrowers paid a small fee, and I trusted them to return the books no names, no deposits, just trust. And they always returned them. The culture of trust in Burma at that time was remarkable. Theft never even crossed my mind.

My parents were my first role models. My father worked in accounting, and my mother was a high school teacher. Though we weren't wealthy, our home was filled with love, laughter, and encouragement. I'm the eldest of four children. My parents emphasized education and excellence, sacrificing much to offer us opportunities they never had.

When I was sick as a young child, my father carried me on his back through public buses and trains to reach the children's hospital. Later, he'd take me to soccer matches and boxing events. He taught me to always have a backup plan, to expect the unexpected, and to be adaptable.

My mother was equally formative in my upbringing. We loved and respected her deeply and never dared misbehave in her presence. She never used corporal punishment but maintained discipline through calm authority. We especially loved her curry dishes. I have cherished memories of her taking me to Burma's historic capital, Mandalay.

Mandalay, Myanmar's second-largest city after Yangon, lies on the east bank of the Irrawaddy River, 631 kilometers (392 miles) north of Yangon. As of 2025, its population is approximately 1.59 million. Mandalay holds cultural significance as the final royal capital of Burma and a spiritual center for many Buddhists.

On March 28, 2025, a magnitude 7.7 earthquake struck north of Mandalay along the Sagaing Fault. The energy released was equivalent to 334 atomic bombs, making it Myanmar's strongest quake since the 7.9-magnitude temblor in Taunggyi in 1912. As of April 9, 2025, the death toll had reached up to 5,350, with 7,860 people injured and hundreds still missing. The quake caused widespread devastation: 1,700 houses, 670 monasteries, 60 schools, three bridges, and thousands of temples and pagodas were damaged. Hospitals, universities, and many historic public buildings were also impacted.

Chapter 7
Early Life in Burma

The civil war, ongoing since the 2021 military coup, has displaced over three million people, making disaster relief efforts even more challenging. The combined weight of political unrest and natural disaster has placed enormous strain on the people of Myanmar.

Chapter 8
Family Heritage and Values

My family's story is one of resilience, ambition, cultural richness, and unwavering love. On my father's side, my grandfather's success in business was tempered by his humility and wisdom. He owned over a dozen companies but remained grounded, teaching me invaluable lessons about integrity, hard work, and the quiet strength of leading by example.

My paternal grandmother often prepared my favorite dishes, showcasing her love and care through the art of traditional Chinese cooking. Together, they created a warm, nurturing environment that shaped my early years and instilled in me a deep appreciation for family, food, and tradition.

On my mother's side, my grandparents, Wong Kow and Chew Thien Lan, owned New Design, a furniture company that supplied schools and the public. My maternal grandfather emigrated from China to Burma as a young man. My grandmother was born in Burma, a second-generation Chinese Burmese with strong ties to both cultures.

Over time, my grandfather learned to speak Burmese and Hindi fluently to communicate with his Indian employees. My grandmother managed the company's finances and operations, a remarkable feat for a woman in that era. Tragically, after a military coup, soldiers seized their hard-earned business a devastating blow that shattered my grandfather's spirit and led to his early death. This injustice left a lasting mark on our family and shaped my views on resilience and perseverance in the face of political upheaval.

My mother was one of ten siblings. She never met her own grandparents, but her eldest sister told me their maternal grandparents owned a thriving shipbuilding company near Asia's busiest port at the time, as well as a rice mill and other businesses evidence of their great wealth and success. Their story added a sense of pride and mystery to my family's past, as I often imagined the bustling docks and industrious spirit of those early generations.

My mother's four brothers pursued careers as civil engineers, an architect, and hospital personnel. Her five sisters, as far as I know, dedicated themselves to managing their households. My own upbringing was more closely tied to my father's side of the family, as we lived with my paternal grandparents. Though I didn't interact much with my maternal grandparents, my mother told me that they were exceptionally intelligent and possessed a quiet dignity that resonated through their children.

Within her own family, my mother was the undisputed favorite child. She accompanied her parents every evening as they relaxed and discussed things among themselves or with guests, absorbing not only their language but also their values and dreams. My father was the favorite of his mother. I'm unsure if he held the same favor with his father. But I do know my parents devotedly cared for their own parents and supported their siblings' families whenever possible. Their sense of familial duty created a powerful example of selflessness and unity that I carry with me to this day.

My mother, Nelly Chin, was a high school teacher instructing students in Chinese and English. She attended the most prestigious university in Taiwan on a full scholarship in recognition of her excellent academics. In addition to teaching, she also tutored students privately at our home. As a young child, I remember sitting quietly by her side, observing her tutor intently, fascinated by her patience, clarity, and command of both languages.

My father, George Chin, worked in the accounting department overseeing revenue from the municipal bus system and tracking the city's gasoline reserves. In the late 1970s, he joined General Motors, which was the largest corporation in the world at that time. His transition to the private sector reflected both his adaptability and the global shifts in opportunity that marked that era.

My parents always put the needs of their four children first. They were remarkable role models, embodiments of unconditional love. As my parents grew older, my youngest sister Grace and her family took on the main responsibility of caring for them, while my siblings Catherine, Richard, and I provided support whenever we could. I had the privilege of being the firstborn child, followed by Catherine, Richard, and then Grace. Birth order may have shaped our roles, but love and shared memories bound us together.

One of my parents' greatest joys was enjoying fresh seafood at our local Chinese restaurants. They often ordered steamed whole fish and live shrimp delicacies they savored. Our extended family made it a tradition to gather for dim sum or dinner whenever we could. These times together filled us with happiness, whether dining out or enjoying home-cooked meals. Food was not just sustenance but a ritual of love and connection, a language we all understood.

As far as religion is concerned, my elementary school in Burma (Myanmar) had students from a variety of religious backgrounds. Religion was never an issue for me because I was able to get along with all the students. Buddhism was the predominant religion in Myanmar, and I volunteered to sweep the shrines of the Shwedagon, one of the world's most famous pagodas.

Furthermore, I worshipped ancestors and Chinese deities, such as Kwan Yin, and followed Confucius' teachings. Further, I attended church at Christmas with Christians, who were Caucasians. This early exposure to interfaith harmony deeply influenced my worldview, fostering a respect for diverse beliefs and traditions.

In the United States, my Uncle Frank had a successful, long career at IBM. My Uncle Steven, with an MBA from a prestigious American university, held top management positions across industries. He ended his career with 17 years in senior management at the United Nations' World Food Program in Rome. His global service inspired me to think beyond borders and consider how one life could impact millions. In total, my grandparents raised 16 children!

The values instilled by my family the importance of education, hard work, integrity, and love have been the foundation of my life's journey. Their sacrifices and unwavering support paved the way for the opportunities I've been fortunate to pursue. To this day, I carry their dreams in my heart and strive to honor them through my actions.

Chapter 9
Immigration & Early Life in America

In 1973, when I was 14, our family made the difficult decision to leave Burma to pursue better educational opportunities for us children. With just $156 (equal to $1,104 in 2024) to our name the maximum allowed by the military government we immigrated to San Francisco, seeking a better life. Those early years in America were very challenging. I didn't speak much English, and we struggled financially. Everyday life felt like a constant test of patience, resilience, and adaptation.

Fortunately, a distant relative took us in for a week. My father was able to join us nine months later. In the early 1960s, when I was young, Burma had been the wealthiest nation in Southeast Asia. But a military coup in 1962 led to Burma becoming the poorest country in the region by the time we left. What once felt like a land of promise had slowly crumbled under authoritarian rule, forcing families like ours to start over abroad.

Before my father arrived, our family risked being deported. We were only able to immigrate to the U.S. thanks to my uncle Frank, my father's brother, who sponsored us. While awaiting my father's arrival, I attended adult school at age 14 to improve my English. Despite being much younger than my classmates, I was determined to catch up.

I met Mrs. Linda Malila at the adult school, where evening classes were held at the First Chinese Baptist Church in San Francisco's Chinatown. She went above and beyond to help our family navigate the immigration process, which was crucial since we couldn't legally stay without my father present. Her compassion and commitment gave us hope during an uncertain time.

Meanwhile, I briefly attended Francisco Junior High School in the Chinatown/North Beach area, known for great Italian restaurants. I remain deeply grateful for Mrs. Malila's kindness. She became one of the first Americans to truly see me not just as an immigrant, but also as a young person trying to build a future.

To help support my family, I worked in Chinese, Japanese, or American restaurants. I'll never forget those grueling shifts in hot, cramped kitchens or the feeling of being mocked by classmates for not speaking English well. The work was exhausting, but it also taught me responsibility, discipline, and grit. Yet those early hardships instilled in me a fierce drive to succeed against all odds.

With help from dedicated teachers and mentors, I was accepted into UC Berkeley as an Alumni Scholar. The Alumni Scholarship is based on merit, not financial need. That acceptance marked a turning point it was the first tangible sign that my sacrifices and determination were paying off.

During this time, I had a newspaper delivery route in San Francisco. One morning, while delivering before dawn, a passerby said something to me. I asked, "What did you say?" He slowly enunciated: "How...are...you?" Although I could understand some English, I still struggled to keep up with native speakers' pace. Moments like that reminded me how far I still had to go, but also how far I'd already come.

As a new student at Francisco Junior High, I was initially placed in an English as a Second Language (ESL) class. I vividly remember overhearing classmates laughing and saying, "He's stupid. He can't even speak English." That hurt deeply, but it also fueled my determination to prove them wrong not just to others, but to myself.

Those words stung, especially since I had been the top student back in Burma. However, by the semester's end, I transferred out of ESL into regular English classes with American-born students. I dedicated myself to earning straight A's in every subject except a B in PE due to struggling to climb a rope. That one grade became a reminder that academic excellence alone couldn't erase the physical and emotional challenges I still faced.

Throughout junior high and high school, I worked nights and weekends in restaurants to help support my family. I also secured a summer job in a U.S. Department of Agriculture (USDA) chemistry lab. Balancing school and work was demanding, but I viewed every opportunity as a step forward.

In the summer of 1974, through a Red Cross program for recent immigrants, I met Mr. Wilhelm Schaser and Mrs. Kay Schaser. They generously sponsored me and a classmate to stay with them in their beautiful home in Eureka, California, about 300 miles north of San Francisco. The Schasers built their house on land surrounded by lush gardens, fruit trees, and towering redwood trees. It felt like stepping into a world far removed from the crowded neighborhoods of the city.

Over that stay, Bill and Kay helped us refine our English, took us hiking in beautiful redwood parks, and introduced us to friends. It was an enjoyable experience. Both were alumni of nearby Humboldt State University, and Bill also graduated from UCSF. They not only opened their home but also their hearts offering encouragement and cultural exposure we'd never experienced before.

Bill was a beloved teacher at Eureka High, leading biology, PE, and special education classes. Several of his biology students won state science competitions under his guidance. He also organized overseas exchange programs for Eureka students and parents. His dedication to education and global connection left a lasting impression on me.

I had the pleasure of meeting some of Bill's former students who went on to prestigious universities like UC Berkeley and UC Davis. While Bill taught, Kay managed their household and took the lead in building their stunning two-story home. Bill and Kay continue to inspire me with their kindness and generosity. Their example taught me that mentorship could shape not just academics, but a young person's entire outlook on life.

After establishing my Silicon Valley career, I wanted to express gratitude for all Bill and Kay had done for me. I treated them to a vacation in Hawaii, where we island-hopped and enjoyed beaches like Waikiki and Diamond Head on Oahu. We went bodysurfing and snorkeling, visited plantations, saw cultural performances, and savored luau cuisine. It was a full-circle moment sharing the fruits of my success with those who had supported my journey.

Our travels also took us to the breathtaking Hawaii and Kauai islands. For the movie Jurassic Park, the actors filmed almost all of their scenes in Kauai. On the Big Island, I had the surreal experience of standing on the warm lava flows of the Kīlauea volcano. The awe-inspiring landscapes reminded me of how far life had taken me from a poor teen to a traveler and contributor.

Our family came to the US with almost no money, but I have never felt poor in my life. I have always felt loved and all my needs were met.

At George Washington High School in San Francisco, my most influential mentor was Mrs. Ann Rhine, my English teacher. Though I enjoyed Shakespeare, my limited English was challenging. Mrs. Rhine generously stayed after school to provide extra help and encouragement. Her patience made literature accessible, even poetic, despite the language barrier.

Beyond academics, Mrs. Rhine encouraged me to get involved in student government. Despite being a latecomer at George Washington High, which had around 2,500 students, I decided to run for student council against more established and popular candidates. To my surprise and delight, I won the election and became the head of the student government. That victory was more than just a title it was validation that I belonged, and that I could lead.

Mrs. Rhine's impact on my life was immense. In my yearbook, she wrote, "To Simon - Who has been the brightest spot in my life this spring! Best wishes always. I shall never forget having you as a student. Ann Rhine, May 27, 1977." During this period, I made a habit of learning 50 new English words every day. That practice became my personal ritual a way to conquer the linguistic mountain I once thought insurmountable.

Mrs. Rhine and I kept in touch as I continued my study at UC Berkeley. Unbeknownst to me, she had been privately battling lung cancer. She passed away in 1979 without ever telling me about her illness. It was only then that I learned she had been a devout Christian. Her death was a tremendous loss for me, and I took the initiative to set up a memorial scholarship in her honor. She'll always have a special place in my heart. I wanted others to benefit from her spirit, just as I had.

In the school newspaper, I wrote: "Dear fellow GHWS graduates, our school is currently raising funds for a scholarship in memory of Mrs. Ann Rhine, our beloved teacher, who passed away during the summer of 1979. In order to show our gratitude to Mrs. Rhine, let's send in our contributions – be it a penny or an ounce of gold. Thanks for your cooperation. Please make your check payable to Ann Rhine Memorial Scholarship Fund c/o GWHS, 600 32nd Ave., S.F."

Had it not been for a negative experience during my first semester at Mission High School in San Francisco, I might never have met Mrs. Rhine. Early on at Mission High, a very tall student confronted me in the restroom and tried to rob me, demanding my wallet. Despite having no martial arts training, I fought back by imitating Bruce Lee's kung fu moves. My fear transformed into adrenaline and instinct I refused to be victimized.

The would-be thief fled, but the next day, during PE class running laps around Kezar Stadium (former home of the 49ers), someone broke into my gym locker and stole my wallet. I reported the theft, and thankfully, a Good Samaritan found my wallet discarded in a trashcan and turned it in, though the cash was gone. This incident prompted my transfer to George Washington High. That transfer became a turning point, setting me on a path toward unexpected mentorship and leadership.

At GWHS, I took a class to take pictures and make my own prints as a hobby. My favorite subject was biology, thanks to the exceptional instruction of Mr. Oscar Hollander. I considered studying biology in college to become a neurosurgeon but decided chemistry offered more potential for impactful applications. Still, I never lost my curiosity about the life sciences.

Still, my fascination with biology persisted. At UC Berkeley, I took a physiology class even though it wasn't required for my major. I scored the third highest on the final exam. The 1990 launch of the groundbreaking Human Genome Project inspired me to find a biotechnology company that would use silicon chips and AI to revolutionize medicine. The idea of merging biology with technology spoke to both my past and future a way to give back through innovation.

Chapter 10
Higher Education & Early Career

UC Berkeley was the only university I applied to, which may seem foolish in hindsight. Berkeley's chemistry program was ranked number one in the United States, if not worldwide. At the time, I simply followed my instincts. During my freshman year, I was elected president of my dorm, Putnam Hall, which housed around 200 students from freshmen to seniors.

It was a diverse and intellectually vibrant environment, which further fueled my motivation to excel both academically and socially.

Berkeley opened my eyes to a bigger world. I chose to study chemistry and chemical engineering. I was fortunate to work with brilliant professors who became lifelong mentors, like Dr. Glenn T. Seaborg, the Nobel laureate who co-discovered plutonium and advised several US presidents. Dr. Seaborg modeled scientific rigor, integrity, and concern for humanity. Working in his lab, searching for new elements, taught me the thrill of discovery and expanded my sense of life's possibilities.

Dr. Seaborg was the principal or co-discoverer of ten elements. He was the only person to have an element (seaborgium) named after him during his lifetime. He also served as UC Berkeley's second chancellor. I admired Dr. Seaborg not just for his brilliant accomplishments and the help he offered me but also for his remarkable humility. My conversations with Dr. Seaborg were always enlightening and engaging. We would often hike together between the UC Berkeley campus and the Lawrence Berkeley National Laboratory.

Along the way, he would share stories of working with Presidents John F. Kennedy and Lyndon B. Johnson. Over his career, Dr. Seaborg advised ten US presidents. He also contributed to the Manhattan Project and later championed the peaceful uses of atomic energy, emphasizing responsibility in scientific advancement.

Each year, Dr. Seaborg invited his research team to a Christmas party at his home. On one occasion, he and his wife Helen visited my family at our home. When my paternal grandfather passed away, I struggled immensely with the loss. Dr. Seaborg taught me techniques to refocus my mind, a skill that has served me well ever since. With gratitude for what Dr. Seaborg had done for me, I made a substantial donation in his honor to the College of Chemistry at UC Berkeley after I became a successful entrepreneur.

As of February 2024, UC Berkeley boasts an impressive 31 Nobel Laureates in Chemistry and 34 in Physics – more than any other university worldwide in these fields. I felt very lucky to have studied these subjects with the best scientists in the world. Another key mentor was Dr. Sumner P. Davis, a physics professor who received a prestigious teaching award in 1980.

Dr. Davis had a true gift for teaching. Over his career, he supervised 36 doctoral students. In 1993, he was awarded the Berkeley Citation. He was also an accomplished glider pilot, and I had the pleasure of flying with him in his glider. Like Mrs. Rhine, Dr. Davis was a devout Christian. His lectures combined clarity, humor, and profound insight, making even the most complex physical principles feel approachable.

Dr. Davis and I continued our friendship well beyond my undergraduate years, meeting for dinner several times yearly until his passing in his eighties. When he died, I made a substantial donation to UC Berkeley in his memory to fund upgrades to his beloved Physics 111 lab. Early in my career, about two years after graduating from Berkeley, I faced a major career decision and turned to Dr. Davis for invaluable guidance.

His advice helped me navigate not only that professional crossroads but also instilled in me a habit of seeking mentors who embody both intellectual excellence and moral integrity.

When I began my education at UC Berkeley, I did not believe in Christianity. My roommate and some other people tried to witness their faith in Jesus Christ. I could not believe what they believed. Christianity seemed impossible from a scientific perspective. It is only in retrospect that I could appreciate what they did. Their sincerity and quiet confidence left a lasting impression, even if I wasn't ready to accept their beliefs at the time.

The summer after my freshman year, I took on a challenging role as a door-to-door encyclopedia and children's book salesman with the Southwestern Company based in Nashville. This experience was formative, with opportunities for learning, travel, and personal connections. I was partnered with two fellow UC Berkeley students and was sent to Norwich, Connecticut, after training.

Upon arriving in Norwich, the three of us had only $100 each and no lodging plan. As we pondered our next move over coffee at a Dunkin' Donuts, a UC Berkeley alumnus spotted my Cal cap and struck up a conversation. Learning of our plight, he generously hosted us for the night.

The kind employees, moved by our situation, slipped us scratch cards to win free donuts. We were touched by the kindness of these strangers helping us. This early experience with generosity from strangers taught me that acts of kindness can have an outsized impact, especially during times of uncertainty.

The next day, we rented a room from Reverend Neil Chadwick of the local Assembly of God church.

Reverend Chadwick and his family welcomed the three of us into their home to share a single room on the floor. We remain grateful for their hospitality. That summer was transformative and full of valuable lessons. Selling books door-to-door taught me perseverance, emotional intelligence, and the importance of connecting with people from all walks of life.

Wing Hsieh, one of my roommates, then became a lifelong friend. After studying optometry/ophthalmology in Indiana, where he met his wife, Sharon, Wing returned to the Bay Area. Sharon became my optometrist, and when their daughters went to college, I was happy to offer guidance. Wing and his family are devout Christians.

After some success selling books, I decided to move on to more fulfilling pursuits.

That summer not only provided personal growth but also introduced me to lifelong friendships grounded in trust, faith, and shared challenges.

My next stop was Raleigh, North Carolina, to visit my Uncle Frank and his family. Together, we enjoyed a beach vacation in Myrtle Beach, South Carolina. Uncle Frank, an IBM staff engineer, was based in the renowned Research Triangle Park. Over the years, he and his family had lived in New York City, Hyde Park, Poughkeepsie, and even owned a house on land once owned by President Franklin D. Roosevelt.

Uncle Frank was an exemplary role model from whom I learned greatly.

He exemplified a blend of technical brilliance, humility, and familial devotion that left a deep imprint on my aspirations. His encouragement reminded me that success was not just about knowledge but also about integrity, perseverance, and compassion.

After visiting family, I embarked on a two-week cross-country Amtrak rail adventure with an unlimited travel pass. The 15,000-mile journey took me to Boston (with stops at Harvard and MIT), New York City, Washington D.C., Niagara Falls, Chicago, and many places in between.

I spent most nights on the train to save money and have more time to explore. My return route went through Seattle. It was an unforgettable experience. Traveling solo across the country at such a formative age broadened my horizons and reinforced a sense of independence, curiosity, and appreciation for the diversity of American life.

During my junior year at UC Berkeley, I secured an engineering internship with US Borax in Boron, California home to the largest borax mine and biggest open-pit mine in the state. My task was to create a comprehensive manual detailing the computer program simulating the borax production process, starting with raw ore from the mine.

The resulting manual was an inch thick, with diagrams illustrating chemical streams processed through towering 3-story machines. The project was challenging yet gratifying, providing invaluable learning opportunities. The work helped bridge my classroom knowledge with real-world industrial processes and gave me hands-on experience in both chemical engineering and technical communication.

The Boron open-pit mine could be punishing, with temperatures sometimes reaching 110°F (43°C). Interestingly, the mine's waste products, discarded as low-value then, are now recognized as a precious lithium source for batteries powering devices like phones, computers, and electric vehicles. At the time, none of us interns could have imagined how the so-called "tailings" would later become critical in the global transition to clean energy.

Boron sits on the edge of the Mojave Desert, just 29 miles from Edwards Air Force Base. During my time at US Borax, I had the extraordinary luck of witnessing the Space Shuttle Columbia coming down to land at Edwards on April 14, 1981, after its inaugural mission.

Perched atop a chemical silo with colleagues, I watched in awe as the shuttle glided effortlessly through the blue sky. It was a historic moment Columbia's landing marked the successful completion of the first-ever orbital mission of NASA's Space Shuttle program, STS-1.

While at US Borax, I lived in California City, located about 65 miles from Death Valley National Park and 29 miles from Edwards Air Force Base. While camping in the park, I went into a restaurant to eat dinner. They said, "Sir, you cannot eat here because you are not wearing a suit and a tie." Who brings a suit and a tie when they are camping in a national park? It was a humorous but jarring reminder of the lingering formality in some dining establishments of the era even in the middle of the desert.

That summer at U.S. Borax, I formed a close bond with a music student from UC Santa Barbara. We spent nearly every day together playing tennis in the evenings, studying the Bible, dining out, and attending church services on Sundays. She was beautiful, athletic, and had deeper scriptural knowledge than I did.

As our connection deepened, she asked me directly, "You are in love with me, aren't you?" In a moment of hesitation that I now recognize as one of my greatest missteps, I failed to express the depth of my feelings for her. Looking back, I see that my emotional reserve masked an inner struggle between personal longing and unresolved grief.

My reticence may have stemmed from the unresolved guilt and turmoil I experienced when parting with my paternal grandparents upon leaving Burma. Their unwavering love and generosity left a deep impact. I was sorry for not having the opportunity to thank them and take care of them when old age and illness challenged their daily lives. Despite the physical distance, their presence in my heart remained constant.

I was devastated by the news of my paternal grandfather's passing just weeks after returning to UC Berkeley from that internship. My world was thrown into chaos. At the time, my father had already transferred from San Leandro to Sparks, Nevada, for his job with General Motors. With three children in college and one in private high school, our family faced significant financial strain.

At one point, I had only $2 for personal spending in an entire month, subsisting on prepaid dorm meals with sheer determination. It was a humbling time that taught me how to stretch limited resources while staying focused on my goals.

In addition to my studies, I felt a responsibility to support my family and provide academic guidance to my sister, Grace. The emotional weight of my grandfather's death impacted my academic performance.

I was extremely close to my grandfather. Three or four months later, my grandmother joined my grandfather in eternal rest. They had lived long lives, reaching ages 84 and 80, respectively, far surpassing the average mid-50s life expectancy in Burma at that time.

Both were born in China but immigrated to Burma as adults. They went back to China and built a nice home for themselves but ultimately left it for their relatives. Their selflessness and resilience are part of the legacy that shaped my values and aspirations.

My paternal grandfather rarely spoke of his business ventures. However, his brother once shared that he had built and managed between 14 and 17 thriving enterprises. Tragically, during civil unrest, rioters looted and burned down both his businesses and the family homes.

As a wedding gift to their eldest daughter, my grandparents presented her with a jewelry box the size of a cigar case, filled with 24-karat gold, jade, rubies, and other precious stones. She went on to live into her 90s. Despite losing so much, my grandparents never lost their generosity or grace.

My paternal grandparents showered me with affection. I was their favorite among their dozens of grandchildren. Losing them, compounded by the end of my relationship with the girl in Southern California, nearly overwhelmed me. In my final two years at UC Berkeley, I devoted more time and energy to studying the Bible than all my academic coursework combined.

To this day, I consider that time and spiritual dedication to be the wisest choice I've ever made. It became the cornerstone of my transformation from loss and uncertainty to purpose and faith.

Chapter 11
Silicon Valley Career &
Professional Development

After graduating from UC Berkeley with a chemical engineering degree, I accepted a position at Master Images, a premier semiconductor photomask production company. My trajectory there was swift – starting as an engineer, I was promoted to a managerial role overseeing manufacturing within just 9 months, doubling my salary in my first year of employment. Assuming responsibility for the company's largest operation 6 months later, we experienced a period of great success and rapid growth.

Photomasks are precision tools used to transfer intricate circuit patterns onto semiconductor wafers through a process known as photolithography. They are essential for fabricating integrated circuits, making photomask technology one of the most crucial yet unsung components of the global electronics supply chain.

Then, a customer I had met once on my very first day offered me a new job opportunity a 20% pay raise and stock options. Seeking advice, I turned to Dr. Davis. His guidance was simple yet profound. He asked me one question: "Which group of people would you rather work with?"

By that time, I had already recruited several UC Berkeley classmates to join the company in their first jobs after graduation. I decided to stay put. That single piece of advice underscored a lasting truth: workplace culture and trusted colleagues often outweigh short-term financial gain.

As market demand surged, we needed to double our output without expanding equipment. This meant frequent overtime shifts for production teams, even as we trained new hires. Needing to allocate experienced personnel to guide and mentor newcomers, resulting in even longer hours for seasoned workers, compounded the strain on existing staff.

In retrospect, our lack of proactive planning for increased staffing and equipment left us ill-equipped to meet the explosive growth once the economy kicked into high gear. While productive for the bottom line, this boom time placed an immense burden on operations staff.

This period coincided with the broader expansion of Silicon Valley during the early to mid-1980s, driven by growing demand for personal computers, telecommunications equipment, and consumer electronics all of which depended on semiconductor innovation.

We ultimately reached a breaking point where we could no longer guarantee on-time delivery to our marketing division. This ignited tensions between the operations and marketing vice presidents, culminating in an organizational restructuring. It was a powerful lesson on understanding business intricacies. The experience taught me that technical excellence must be matched by strategic communication, scalable infrastructure, and cross-functional alignment in order to sustain growth.

My career centered on developing semiconductor photomask production technologies and managing people from 1983 to the 1990s. This expertise led to opportunities to transfer these technologies and the global ISO 9001 quality management systems to semiconductor photomask manufacturing sites in Germany, France, South Korea, and China.

The ISO 9001 standard established in 1987 by the International Organization for Standardization was rapidly adopted in the semiconductor industry to ensure consistent product quality and global interoperability, especially critical as supply chains became increasingly international.

The terms "semiconductor," "chip," and "microchip" are often used interchangeably to describe the tiny yet powerful electronic components that are the bedrock of modern devices from smartphones and computers to AI systems, vehicle electronics, and more.

At their core, these chips are composed of layers of semiconductor material typically silicon patterned and doped to manipulate electrical conductivity. Photomasks play a central role in this precision patterning, serving as the blueprint for every microcircuit layer.

As I advanced my career, DuPont, the eighth-largest company in the US in the 1980s and the largest chemical company in the world at the time, acquired Master Images. I was on DuPont's core team to implement quality systems in Silicon Valley that earned the prestigious Malcolm Baldridge National Quality Award, which was established by the U.S. Congress in 1987 to raise awareness of quality management and recognize U.S. companies that have implemented successful quality management systems.

This recognition placed us among an elite group of American companies including Motorola and Federal Express that led a national movement toward continuous improvement and operational excellence.

I simultaneously pursued an MBA at Santa Clara University while working full-time juggling employment's demands with graduate studies' rigors. I actually enjoyed the synergy of applying what I learned in school to my work and what I learned at work to my school projects almost on a daily basis.

Chapter 11
Silicon Valley Career &
Professional Development

The Jesuit values at Santa Clara emphasized ethical leadership, innovation with integrity, and the holistic development of both intellect and character. These principles profoundly shaped my approach to business and management.

Over the years, I successfully led numerous technology transfer initiatives for DuPont, including one complex project spanning 10 sites across 7 countries. My efforts were recognized with two prestigious DuPont Achievement Awards, reflecting the impact and value of my contributions. These awards celebrated not just technical implementation but also leadership in multicultural coordination and sustainable systems development.

Early in my semiconductor photomask career, I was elected to the board of directors for our industry's trade association. In the early 1980s, I was invited to present at the annual Bay Area Chrome Users Society (BACUS) conference on using personal computers to enhance control and optimization of photomask manufacturing processes.

BACUS, originally an acronym for "Bay Area Chrome Users Society," became a global SPIE-sponsored community of experts focused on advancing photomask technology. Presenting there was a rare honor and confirmed my contributions were at the forefront of the field.

The computer program I developed for this purpose analyzed and monitored over 6,000 distinct variables, requiring a staggering 6 hours daily to execute on the era's limited IBM PC platform. Despite computational constraints then, implementing this software solution significantly improved both employee performance and overall manufacturing yield.

The program I wrote was the first in the world to collect, analyze, and store all relevant machine, chemical, and human data to timely control the effectiveness and efficiency of semiconductor photomask manufacturing. This software pioneered data-driven process control well before the term "big data" entered mainstream business vocabulary, demonstrating how early digital tools could unlock measurable gains in precision manufacturing.

I enjoyed my graduate studies and the environment at Santa Clara University. It was completely different from UC Berkeley. To me, it was like having the best of both worlds. At Berkeley, you get to meet and learn from the best professors and students from around the world every day. The conversations were stimulating.

At Santa Clara, the emphasis was on competence, conscience, and compassion. You grow more as a person and not only in academic excellence. Even former rivals from UC Berkeley and Stanford feel like a part of the Santa Clara family. This duality gave me a rare and invaluable perspective: rigorous intellectual formation from Berkeley and humanistic, ethics-centered development from Santa Clara.

The knowledge and skills gained through my MBA program, coupled with the invaluable hands-on experience amassed in the semiconductor industry, laid the groundwork for my eventual transition into entrepreneurship.

Chapter 12
International Experiences and Cultural Insights

In 1986, my professional journey took me to Stuttgart, Germany, where I was entrusted with transferring semiconductor photomask technologies to Bosch a multinational engineering and technology conglomerate headquartered in Stuttgart.

Founded in 1886 by Robert Bosch, the company has grown into a global leader in mobility solutions, industrial technology, energy and building technology, and consumer goods. As of 2023, Bosch reported €91.6 billion in revenue and remains the world's largest automotive supplier. Prestigious Bosch clients in Stuttgart include legendary automakers Mercedes-Benz and Porsche, both headquartered nearby.

Overall, my first trip to Germany and Europe included a memorable Lufthansa Airlines flight. At check-in in San Francisco, the attendant apologetically explained they had overbooked business class. Bracing for a downgrade, I was delightfully surprised when she offered, "Would you mind if we upgraded you to First Class?" I gratefully accepted, immersing myself in Lufthansa's premier cabin luxury and comfort. Lufthansa, founded in 1953, is Germany's flagship airline and known for its excellent transatlantic service.

During my time at Bosch, I witnessed the epitome of German efficiency firsthand. I thoroughly enjoyed Germany and found the people warm and welcoming a contrast to the often-stereotyped perception of Germans as cold or overly rigid.

In France, my role involved transferring cutting-edge technologies to DuPont's facility in Rousset. From June to September 1989, I had the privilege of residing in the picturesque city of Aix-en-Provence, nestled near the French Riviera. A 1989 French magazine survey ranked Aix as one of the most desirable cities to live in, praised for its architecture, Mediterranean climate, and cultural vibrancy. The vibrant university town of Aix is strategically located just 40 km from Marseille, 90 km from Avignon, 101 km from Saint-Tropez, 150 km from Cannes, and 175 km from Nice. I enjoyed visiting all of these cities.

During my stay, I was provided with a charming 2-bedroom apartment, merely a half-block from the renowned Cours Mirabeau the city's iconic main thoroughfare and cultural hub. My time in Aix was an absolute delight, filled with memorable weekend excursions across Europe including Rome, Geneva, Monte Carlo, Cannes, Avignon, Paris, and the sacred pilgrimage site of Lourdes.

I once stayed up dining with a French acquaintance until 3 a.m., reveling in the region's joie de vivre.

DuPont generously covered all my living expenses apartment, car, fuel, dining out or home cooking (though I admittedly ate out more often), international phone calls, and laundry services while my regular U.S. paycheck continued uninterrupted. My role was transferring advanced photomask production using state-of-the-art laser lithography techniques within an ultra-clean ISO Class 1 cleanroom environment, where even a speck of dust cannot be tolerated.

Maintaining an ISO Class 1 cleanroom requires the air to be meticulously filtered with fewer than 12 particles measuring 0.3 microns or smaller per cubic meter. For comparison, the average human hair is around 70–100 microns in diameter. Prior to my France assignment, I had been deeply involved in developing cutting-edge technologies in Silicon Valley to optimize the performance of photomasks produced using the first Ateq laser system deployed in a production setting. My expertise also extended to collaborating directly with the laser system manufacturer to fine-tune its capabilities for the world's first semiconductor photomask laser lithography system in production.

In transferring these technologies to the DuPont facility in France, I worked closely with the company's technical staff, operations personnel, and machine operators, often conducting meetings almost entirely in French to facilitate clear communication. This was no small feat, given the technical nature of the discussions.

Additionally, I assumed the role of project leader, overseeing the successful technology transfer implementation. The photomasks produced during the very first production run under my guidance were flawless, exhibiting zero defects a remarkable achievement that earned me a DuPont Achievement Award.

Life on the French Riviera was an absolute dream. I reveled in the breathtaking beaches, indulged in the delectable cuisine, and forged meaningful connections with the people. Within walking distance of my Aix apartment, I had over 450 restaurants to choose from. Dining on French delicacies three times daily was heavenly, yet surprisingly, I didn't gain a pound. This may have been due to the European lifestyle of frequent walking, balanced meals, and smaller portion sizes.

The most exquisite meals and finest wines were those shared in the homes of colleagues, prepared and served lovingly by their spouses. My fondness for France and the French only deepened as I witnessed their remarkable efficiency and zest for life. During these intimate home dinners, I conversed almost exclusively in French with hosts and their partners, immersing myself in the language and strengthening our bonds of friendship.

In 1989, France celebrated the 200th anniversary of the momentous French Revolution. I had the good fortune of being in Nice during the historic festivities unfolding along the iconic Promenade des Anglais seafront. For miles, the air was filled with pulsating music rhythms, including my all-time favorite band, the Eagles' "Hotel California." This song, though American in origin, resonated with audiences worldwide and added a surreal quality to the celebration.

Later that year, I witnessed Paris's spectacular Eiffel Tower centennial celebrations. The Eiffel Tower, inaugurated in 1889 during the Exposition Universelle, had become not only an engineering marvel but also a national symbol. The view from atop this marvel was breathtaking, and I fell in love with the city's timeless charm and romance.

The Louvre Museum is widely regarded as perhaps the greatest museum in the world a veritable treasure trove of art and history. As I wandered its hallowed halls for the first time, I found myself silently uttering "Oh, my God!" in awe and reverence repeatedly. Like countless others, I had seen reproductions of Leonardo da Vinci's enigmatic Mona Lisa before. But it wasn't until I stood before the original painting, encased behind bulletproof glass in a climate-controlled enclosure, that I truly grasped the depths of its allure and enduring fame. That revelatory moment left an indelible mark, and I knew I would gladly journey back to Paris time and again just to bask in the presence of this singular masterpiece.

In Paris, I also enjoyed the Musée d'Orsay and other museums. Located on the Left Bank of the River Seine, this former Beaux-Arts railway station originally built for the 1900 Exposition Universelle now houses the world's most comprehensive collection of Impressionist and Post-Impressionist masterpieces.

It features works by Berthe Morisot, Claude Monet, Édouard Manet, Degas, Renoir, Cézanne, Seurat, Sisley, Gauguin, and van Gogh. It is one of the largest art museums in Europe. According to The Art Newspaper's annual visitor figures, in 2022, Musée d'Orsay welcomed 3.2 million visitors, making it the sixth-most-visited art museum globally and the second-most-visited in France, after the Louvre.

Visiting the Notre-Dame Cathedral was quite memorable, inside and out. I also spent more than 15 minutes on its roof enjoying the breathtaking panoramic view of Paris. The cathedral, an iconic masterpiece of French Gothic architecture completed in the 14th century, suffered a devastating fire in 2019. Notre-Dame Cathedral officially reopened to the public on December 8, 2024, after a five-year restoration.

The technologies I brought to South Korea played a pivotal role in bolstering that nation's semiconductor photomask industry, and I take pride in contributing in my own small way to South Korea's technological advancement. South Korea's rise as a semiconductor powerhouse led by companies like Samsung and SK Hynix has been crucial to its transformation into one of the most advanced digital economies in the world.

Similarly, my work transferring cutting-edge semiconductor photomask manufacturing techniques some I had personally developed to facilities in Wuxi and Shanghai during the late

1980s and mid-1990s was instrumental in laying the foundation for China's remarkable rise on the global stage.

As I witness China's astonishing progress in lifting 800 million citizens out of poverty between 1981 and 2015, I am filled with gratification knowing my efforts played a part in this transformative journey. When I left Burma, my grandfather said, "China would be a strong country in 30 years." Decades later, his words proved remarkably prescient.

I went to Japan for business up to six times a year, visiting leading semiconductor companies all around the country. Doing business in Japan is completely different from doing business in the United States, requiring different levels of preparation. Japanese business culture emphasizes respect, hierarchy, punctuality, and consensus-driven decision-making.

Tokyo was very expensive. In the 1990s, a hotel room cost more than $300 per night. Dinner could cost hundreds of dollars. This was during the peak of Japan's economic bubble, when Tokyo was considered one of the most expensive cities in the world. Japanese people value freshness in seafood and are willing to pay for it. The price of shrimp on ice in the afternoon is half the price in the morning.

Covering 230,000 square meters in central Tokyo, Tsukiji Fish Market was a major tourist attraction for both domestic and overseas visitors. It was fascinating to visit this largest wholesale fish and seafood market in the world until its closure in 2018.

The inner wholesale market was then relocated to Toyosu Market, which opened in October 2018 and is nearly twice the size, offering high-tech sanitation and temperature control. I loved eating sushi both in Japan and Korea, where I especially liked grilled live eel (unagi kabayaki). Also, the efficiency and on-time arrivals of Japanese Shinkansen bullet trains capable of speeds exceeding 300 km/h (186 mph) were simply remarkable, setting the global standard for punctuality and safety.

In the late 90s, I went to Korea to visit Samsung and other semiconductor companies. I always enjoyed staying at the Grand InterContinental Seoul Parnas. The generously sized suites came with separate living rooms at a great price, far less than Tokyo, and much more comfortable. It was exciting to see Samsung's flagship product dynamic random-access memory (DRAM) chips in development three years ahead of market introduction, underscoring their role in shaping global tech trends.

These international experiences not only enhanced my professional skills but also broadened my cultural understanding and global perspective. Each country presented unique challenges and opportunities, requiring adaptability and fostering a deep appreciation for diverse work cultures and practices. Working across Asia and Europe helped me evolve both technically and personally enriching not only my career, but also my entire outlook on life.

Chapter 13
Entrepreneurship in Biotechnology and Artificial Intelligence

The groundbreaking Human Genome Project, an international scientific effort to map the DNA of certain organisms, stands as one of the most ambitious undertakings in human history. The project's successful mapping of the human genome, launched in 1990 and officially completed in April 2003, inspired me to establish Iris Biotechnologies Inc. in 1999. The company's focus was on developing innovative technologies to enhance personalized breast cancer diagnosis and treatment by analyzing genetic, medical, and lifestyle information.

Over the years, I've had the privilege of collaborating with patients and physicians at esteemed institutions such as Stanford University, University of California San Francisco (UCSF), Sutter Health, Kaiser Permanente, St. Joseph Health, and Dignity Health, among others.

I've been fortunate to have scientific advisors from top institutions like UCSF, MD Anderson Cancer Center, Baylor College of Medicine, Wake Forest University, and the University of Hong Kong. To facilitate these collaborations, Iris Biotechnologies opened offices across California in Santa Clara (near Stanford University), San Leandro (close to UCSF and UC Berkeley), and Davis (adjacent to UC Davis).

To build the Iris Biotechnologies team, I recruited PhDs and MDs from elite universities and medical centers, including UCSF, Stanford, UC Berkeley, UC Davis, MD Anderson, and Parke-Davis (later acquired by Pfizer). My investors and board members include graduates from prestigious universities such as Harvard, Yale, UCSF, UC Berkeley, and Stanford. Two prominent U.S. law firms also became equity investors in Iris Biotechnologies, a relatively rare endorsement in the life sciences sector.

Though offered venture capital financing, I chose to decline. Instead, we focused on filing multiple patent applications across major global jurisdictions, including the United States, European Union, Japan, Canada, China, Australia, and New Zealand. Every single one of our patent applications was ultimately granted. One of our core patents titled "Artificial Intelligence System for Genetic Analysis" outlined a platform for integrating genomic, medical, and lifestyle data to enable truly personalized medicine. I personally traveled to the European Patent Office in Munich, Germany, to present our case, where we successfully secured protection.

Both the French and UK governments invited me to meet with key business leaders, academics, clinicians, and officials to discuss potential international expansion. In France, I visited Paris, Lyon, Grenoble, Provence, and other regions to tour facilities, hear incentive offers, and explore business opportunities.

In the UK, I went to London, Oxford, and Cambridge for meetings with clinicians, professors, and government officials. During one visit, I was issued a special permit allowing me to bypass standard immigration lines an uncommon courtesy and a sign of strong diplomatic and business interest.

I also traveled to China to investigate possibilities for medical collaboration. I delivered talks at the "Medicine in the 21st Century" international symposium for doctors, scientists, and technologists, as well as other medical conferences. My visits took me to Shanghai, Beijing, Tianjin, Dalian, Chengdu, Hong Kong, and other cities. At the time, China was rapidly expanding its investment in biotechnology and genomics, and my work aligned with their national push toward personalized medicine.

Within the US, I had the honor of being invited to speak at the BIO International Convention the world's largest biotechnology conference and exhibition, drawing over 15,000 professionals from more than 4,000 organizations across 65 countries.

As a recognized expert holding three patents on artificial intelligence systems for genetic analysis, I was invited to present at conferences like the American Association for Clinical Chemistry (AACC) Annual Meeting the largest global gathering of laboratory medicine professionals and the Global Wellness Summit, where leaders shape the future of the global wellness industry through science, business, and holistic health innovation.

In spring 2008, Frost & Sullivan honored Iris Biotechnologies with their North American Technology Innovation Award in Pharmacogenomics, noting: "For patients, this technology would be pivotal in yielding a personalized treatment regimen, with the greatest possibility of success, for each individual patient's particular disease like cancer, heart disease, diabetes, or gene-related metabolic disorders."

My typical workweek usually involved 15–16 hour days during the week and a few additional hours on weekends. But I absolutely love what I do.

In July 2008, PharmaVOICE magazine recognized me as one of the 100 most inspiring people in the life sciences industry. They wrote, "These innovative, forward-thinking executives (I was one of twelve entrepreneurs) have not only carved out a niche in the life sciences, in many cases creating a whole new business area, but have also successfully steered their companies to new heights."

When Iris Biotechnologies began public trading on the stock market in August 2008 (OTCBB: IRSB), I gifted stock to my siblings, their spouses, and their children the maximum amount permitted by law under the annual gift tax exclusion, which was $12,000 per recipient at that time.

In 2010, after the National Institutes of Health (NIH) reviewed Iris Biotechnologies' technology, the government awarded us a $250,000 grant to further our work in personalized and targeted medicine. This marked important federal validation of our platform's potential in real-world clinical applications.

I've been privileged to have conversations with visionary scientists like Dr. James Watson, co-discoverer of the DNA double helix structure in 1953 and first director of the Human Genome Project, and Dr. Thomas Cech, Nobel Laureate in Chemistry and former president of the Howard Hughes Medical Institute. They shared insights into the future of scientific research and medicine. Their perspectives helped guide my strategic decisions, particularly in navigating the intersection of science, regulation, and patient care.

When I asked Dr. James Watson in April 2003, at a celebration marking the 50th anniversary of the discovery of the DNA double helix, how long he thought it would take for human genome information to be widely used in medicine, he predicted it would take at least 25 years. His estimate has proven prescient, as we continue to bridge the gap between genomic data and real-world medical application today. I found my conversation with Dr. Thomas Cech, over lunch at Howard Hughes Medical Institute (HHMI) headquarters in Chevy Chase, Maryland, equally fascinating.

In 2014, I founded Iris Wellness Labs with the goal of providing in-depth scientific analysis to offer insights into complex medical conditions like cancer, cardiovascular disease, immune disorders, diabetes, and obesity for patients and doctors at leading health centers. Our focus was on integrating multi-omics data genomics, microbiomics, and metabolomics with advanced analytics to support precision and preventive healthcare.

I am grateful to Dr. Douglas Hendren and my sister, Grace Osborne, for their roles as board members of both Iris Biotechnologies and Iris Wellness Labs. Dr. Hendren was educated at Harvard University and Case Western Reserve University. Grace was educated at UC Davis. They both also have MBAs and bring multidisciplinary insights to our mission of personalized medicine. I am thankful to Dr. Daniel Farnum, educated at UCSF, for guiding Iris as a board member for more than a decade.

I also want to acknowledge Mr. Wilhelm Schaser and Mrs. Kay Schaser for the many things they have done for Iris, including their steadfast support and involvement in operational and strategic initiatives.

The level of detailed knowledge Iris has in analyzing individual genomes, gut microbiota, and nearly 900 key blood metabolites is not yet standard practice in most US hospitals today. However, the rise of large-scale data platforms and artificial intelligence (AI) could make such in-depth analysis a part of standard medical care worldwide going forward.

Chapter 13
Entrepreneurship in Biotechnology and Artificial Intelligence

In the decades following the groundbreaking 1953 discovery of DNA's double helix structure by James Watson and Francis Crick, the prevailing wisdom in science and medicine held that specific genes play a key determining role in shaping the characteristics that define us as humans. While mapping the human genome has yielded many insights into our genetic makeup, it offers only a partial blueprint of our biological complexity.

First of all, DNA by itself cannot produce anything. The human egg cell, where the sperm and egg are fertilized, plays a crucial role in the propagation of life. DNA alone is useless without all the other cellular organelles, molecular machinery, and nutrients within the egg. DNA needs the ecosystem inside an egg to thrive. It is within this microenvironment that the first molecular signals initiate cell division and development.

Only about 1.5% of the human genome is actually dedicated to encoding proteins the fundamental building blocks of life. The remaining 98.5%, previously called "junk DNA," houses non-coding regulatory elements, introns, and sequences crucial for gene expression, chromatin structure, and epigenetic regulation.

Throughout an individual's lifespan, an intricate interplay of gene expression unfolds, with different genes being activated or silenced at specific points to orchestrate the processes of growth, development, and sustenance. However, genes alone don't paint a complete picture of human biology's complexities.

Nested within each cell lies a dynamic network of protein signaling pathways that profoundly influence gene behavior through epigenetic mechanisms chemical tags such as DNA methylation and histone modification that regulate gene activity without altering the underlying DNA sequence. This allows the cell to adapt to internal and external cues.

Beyond genetics and epigenetics, the human microbiome the vast ecosystem of microorganisms residing within and upon our bodies and the countless bioactive metabolites they produce also play a crucial role in shaping our overall health and well-being. Recent research shows that microbiome diversity can impact immune function, mood, and even response to certain therapies.

Achieving optimal health is multi-faceted, extending beyond our genetic blueprint to encompass lifestyle factors like adequate sleep, regular exercise, a balanced nutritious diet, and cultivating a life full of love, joy, and fulfillment. A baby has no choice in what it eats or drinks, where it sleeps, or where it can go. It is entirely dependent on others for nurturing. As we grow, we generally have more choices in what we do with our lives. As humans in a modern society, we are interconnected and interdependent, relying on one another to thrive in a complex and rapidly changing world.

The advent of CRISPR-Cas9, a revolutionary gene-editing tool discovered in 2012, represents a quantum leap in our ability to manipulate life's very building blocks. In essence, it functions as a molecular scalpel or a word processor for the genetic code, allowing targeted changes to DNA sequences with unparalleled precision and ease.

The scientific community long anticipated that Dr. Jennifer Doudna, UC Berkeley's Li Ka Shing Chancellor's Chair in Biomedical and Health Sciences, and Dr. Emmanuelle Charpentier would be recognized with the Nobel Prize in Chemistry, which they indeed received in 2020 for their groundbreaking CRISPR-Cas9 contributions.

Dr. Charpentier, a luminary in her own right, is a French professor and researcher specializing in microbiology, genetics, and biochemistry. Since 2015, she has directed the Max Planck Institute for Infection Biology in Berlin.

The implications of CRISPR-Cas9 are nothing short of staggering, with far-reaching applications spanning genome engineering, disease modeling, cancer therapy, infectious disease control, and even agriculture and environmental science.

Chapter 14
Personal Growth, Giving Back,
and Reflections

My passion for aviation began in my freshman year at UC Berkeley, where I took a class on preparing students for the written private pilot's license exam. Our instructor was a former Navy fighter pilot who had flown the legendary McDonnell Douglas F-4 Phantom II, a tandem two-seat, twin-engine, long-range supersonic jet interceptor and fighter-bomber used extensively during the Vietnam War. From the moment I first took to the skies, I was hooked on flying's unparalleled sense of freedom and exhilaration.

As my graduation from UC Berkeley with a chemical engineering degree approached, I met with the US Air Force recruiter to explore becoming a fighter pilot. The recruiter expressed enthusiasm about me joining as a navigator but regretfully informed me that my need for corrective lenses disqualified me from being a pilot. This was devastating, as my eyesight had been perfect when I started college but deteriorated to needing glasses by graduation. At that time, Air Force regulations required 20/20 uncorrected vision for pilot candidates, a requirement that has since been revised for some aviation roles.

Undeterred, I resolved to complete pilot training and get my private license after I got established in my professional career. I decided to become a pilot, even if not a military one.

I strongly believe in the importance of giving back to society. One way I've done this is by serving on the inaugural advisory board of Santa Clara University's Ignatian Center, an esteemed center that positively impacts many lives. The vision of the Center is to provide leadership for the integration of faith, justice, and intellectual life by embracing Jesuit wisdom to inspire awareness, thought, reflection, discernment, and action.

The Center's mission is to focus on a holistic approach to learning, a commitment to open and inclusive dialogue, and finding God's will for better decision-making. Founded in 2005, the Ignatian Center for Jesuit Education fosters community-based learning and global engagement aligned with Jesuit values. The university's president, Fr. Paul Locatelli, S.J., invited me to join the board, and I was honored to serve for eleven years. I also made significant donations, six figures, to Santa Clara University.

Additionally, I served for nine years on the Industrial Advisory Board of the University of the Pacific's Thomas J. Long School of Pharmacy and Health Sciences, as well as six years on the board of directors for the YMCA of Silicon Valley. Our YMCA board held fiduciary responsibility for $70 million while serving around one million members annually across multiple communities.

For eighteen years, I was a member of the Saratoga Rotary, part of Rotary International an organization with 1.4 million members in over 200 countries, dedicated to positively impacting communities globally. Our club of approximately 110 people has given away millions of dollars over the years to different organizations to help people globally. The Rotary motto "Service Above Self" guides humanitarian efforts including polio eradication, literacy programs, and clean water initiatives. It is the intention of International Rotarians that their actions are guided by truth, fairness, and goodwill and are intended to benefit all. I also served as a finance commissioner for the City of Saratoga.

I loved living in the foothills of Saratoga, and buying a home there turned out to be an incredible blessing mentally, physically, spiritually, socially, financially, and professionally. I purchased my Saratoga home from Mr. Bud Beaudoin, and he and I became good friends. Mr. Beaudoin shared stories from his time in public relations at General Electric when actor Ronald Reagan served as GE's television spokesman from 1954 to 1962. He also recalled helping astronaut William Anders, famous for taking the iconic "Earthrise" photograph during the Apollo 8 moon mission in 1968, with a real estate transaction.

Mr. Beaudoin sponsored my membership in the Saratoga Rotary and introduced me to his former son-in-law, a distinguished researcher and professor at MD Anderson Cancer Center, who served as a scientific advisory board member of my company.

To raise funds for cancer patients, I completed the 26.2-mile Napa Valley Marathon at the age of 48 and took part in the American Cancer Society's Relay for Life, a national fundraising event supporting cancer research and patient services. In high school, I led a team of students to walk 20 miles to raise money for the March of Dimes, which has helped millions of premature and sick babies survive and thrive since its founding by President Franklin D. Roosevelt to fight polio. As someone who was born premature and struggled with frequent illnesses as a child, including a near-fatal bout of typhoid fever, completing a marathon held special meaning for me.

While hard work is essential, I believe it's equally vital to enjoy leisure time and live life to the fullest. I've been fortunate to have many adventures over the years, including climbing part of Mount Rainier (14,411 feet), trekking part of the Mount Everest Base Camp trail in Nepal, exploring

Chapter 14
Personal Growth, Giving Back,
and Reflections

Machu Picchu, the ancient Incan city in Peru, ascending Table Mountain in Cape Town, South Africa, trekking Chamonix in the French Alps with a view of Mont Blanc, riding a helicopter to the top of the Mendenhall Glacier in Juneau, Alaska, marveling at Angkor Wat, the world's largest religious monument in Cambodia, and visiting the Great Pyramid of Giza, the last remaining Wonder of the Ancient World in Egypt.

Other highlights included marlin fishing in Cabo San Lucas, exploring Mayan ruins and cenotes in the Yucatán, visiting Egyptian tombs and temples, seeing the Terracotta Army in Xi'an, spending a summer on the French Riviera, relaxing on Hawaiian beaches, and attending the 2008 Summer Olympics in Beijing. It was memorable visiting the Forbidden City, the largest ancient palatial structure in the world at 961 meters by 753 meters, and part of the 13,000-mile Great Wall of China, a UNESCO World Heritage Site.

I enjoyed visiting and learning at world-class museums such as the Louvre in Paris, the British Museum in London, the American Museum of Natural History in New York City, and the Smithsonian Institution Museums in Washington D.C. I especially liked the National Air and Space Museum, where I saw many planes and space vehicles, including the Apollo 11 Command Module "Columbia", which carried astronauts Neil Armstrong, Buzz Aldrin, and Michael Collins to the Moon in July 1969.

Sports and fitness have also been an important part of my life. I played varsity soccer in high school and continued with intramural soccer at UC Berkeley. A highlight was watching the U.S. Women's Soccer Team win Olympic gold in Beijing in 2008. I've enjoyed recreational tennis, cycling, skiing, whale watching, camping in U.S. and Canadian national parks, hiking, and wine tasting. Especially memorable were the Grand Canyon rim-to-Phantom Ranch horseback-and-hiking trip and the trek into the Havasupai Reservation to see the vivid blue pools. Yellowstone, Yosemite, and Denali National Parks were fantastic I could write a chapter on each.

I also enjoyed galloping on the beach in Half Moon Bay and horseback riding on mountain trails. My 100-mile round-trip bicycle ride from Union City through Palo Alto to San Francisco was fun, and the Los Gatos-to-Capitola Beach over the Santa Cruz Mountains ride was a real challenge. I've also enjoyed golf, hang gliding, skydiving, sailing, parasailing, snorkeling, and air shows by the Navy Blue Angels and Air Force Thunderbirds. I once watched a Tour de France stage in Marseille, soaking up the festive crowds.

I'll never forget seeing the America's Cup races on San Francisco Bay or visiting NASA's Ames Research Center, Johnson Space Center, and Kennedy Space Center. Fireworks in Seattle and San Francisco rank among my favorites. Our family ran into Cindy Jung at Expo 86 in Vancouver she's been like a little sister to us for fifty years.

I've witnessed many iconic sports moments: the 49ers winning a Super Bowl, Federer's U.S. Open victory, Tiger Woods at Pebble Beach's U.S. Open, Kobe Bryant's last game at Oracle Arena, the Giants in a private suite, and Usain Bolt's 200 m world record in Beijing 2008.

I'm blessed to have lived and worked in Silicon Valley's epicenter. My career began with a summer job at Siltec in Menlo Park, where on Day 1 I fixed a stubborn wafer-line problem. That early success fueled a lifelong passion for applying engineering to real-world challenges.

Chapter 14
Personal Growth, Giving Back,
and Reflections

Being in the Bay Area meant daily inspiration from Stanford, UCSF, UC Berkeley, and companies like Apple, Google, Meta, Tesla, Intel, Nvidia, HP, Oracle, Salesforce, X, and OpenAI. I've cheered the San Jose Sharks, attended WTA tennis competition at Stanford, and rooted for Cal, Stanford, and Santa Clara in football, basketball, soccer, swimming, and track. Santa Clara's women's soccer titles in 2001 and 2020 were thrilling.

"The Play" in the 1982 Big Game five laterals on the final kickoff with the Stanford Band on the field remains one of college football's most legendary moments.

I've met Steve Young at Santa Clara's Tech Awards and Ronnie Lott at the Cupertino Rotary. Pebble Beach Golf Links, off-tournament, offers unmatched serenity.

When Grace's daughter danced ballet, we attended countless recitals. She couldn't become a Ballerina due to injury, but she graduated summa cum laude in Chemistry from a top private university and got accepted to a top PhD program. I am proud of her.

I love concerts Eagles, Heart, Amy Grant, Celtic Woman, Ringo Starr and once saw Pink Floyd in Marseille. In Prague, street musicians on Charles Bridge gave impromptu orchestra concerts. I also enjoyed seeing Shakira perform at the Latin Grammy.

In Malaysia I bought CDs of a singer whose language I didn't understand proof that music transcends words. I even took a music appreciation class at Berkeley. Taman Negara, Malaysia's 130-million-year-old rainforest, was another highlight.

Our Santa Clara office sits beside a golf and tennis club, where I play occasionally. From Santa Cruz beaches and Monterey whale watches to Lake Tahoe skiing and skydiving in the Bay Area, I've savored the region's outdoor bounty. I also pilot planes from San Jose International.

Fishing in the Pacific, the Sacramento River, and local lakes rounds out my leisure. Villa Montalvo, eight minutes from home, is a 166-acre park with formal gardens and a hilltop view of Silicon Valley my favorite weekly stroll.

In New York, I've seen the U.S. Open Tennis Finals, the Belmont Stakes (Secretariat's 1973 Triple Crown win by 31 lengths!), the Statue of Liberty, Carnegie Hall, and Central Park.

In 1994 as a student pilot, I survived a low-altitude spin by recalling a training-center video I'd glimpsed an unplanned lesson that saved my life.

Flying over Yosemite Valley and Mono Lake, and piloting to a World Cup match at Stanford Stadium, are cherished memories. There's nothing like the thrill of flight. My dad, my brother and I thoroughly enjoyed the soccer game.

I've owned a Chevrolet Camaro (using my father's GM discount) and a Mercedes-Benz S430 chosen for its blend of performance and comfort, especially for my parents.

When Richard and Grace began college at UC Santa Cruz and UC Davis, I celebrated by gifting each a new car. Richard's charisma and athleticism made him a campus favorite, while Grace's intellect and organization kept our family grounded. Catherine's wit and compassion completed our quartet.

Chapter 14
Personal Growth, Giving Back,
and Reflections

In 1985, Catherine and I bought a Union City home each of our four siblings plus our parents each got a room. When we moved out, we gifted it to Mom and Dad. Forty years later, Zillow valued that house at $1.89 million up from our $201 thousand purchase an illustration of inflation's impact on the dollar.

Chapter 15
A Tribute to My Mother

Of all the remarkable people I've known and loved in my life, none compare to the profound admiration and affection I hold for my mother. Her unconditional love was a constant in my life for 65 years (including the time in the womb) a gift I will forever cherish. She embodied the very best human qualities love, joy, wisdom, kindness, determination, and an indomitable spirit in facing adversity.

In my eyes, she was the epitome of what a perfect mother should be the answer to my every hope and prayer. My mother was her parents' favorite among ten children, a helpful sister, a loving and faithful wife to my father, and a devoted and unselfish mother to my siblings and me. She was well-regarded by her teachers, respected by her students and those she met, and brought joy to the world with her beautiful smile every day. She was also a wonderful cook whose food everyone enjoyed.

My mom was always first in her classes. She was a great 100-meter runner, a talented basketball player, and a wonderful saxophone player who led her own band in high school. She also played violin and piano and sang like a nightingale. Remarkably, she earned a full scholarship to attend National Taiwan University, the most prestigious university in Taiwan, at a time when higher education was rarely accessible to women, especially in STEM fields.

Her passing on July 7, 2023, left a huge void in my heart and an ache that may never fully subside.

My mother's final moments on this earth are forever etched in my memory. As she lay in her bed at home, I sat by her side, holding her hands, my head bowed in silent reverence. Her departure was so peaceful that I didn't even realize she had slipped away until my sister Catherine noticed her breathing had ceased. In that profound moment of grace, mere minutes after her last breath, I witnessed my mother's soul taking flight a luminous essence emanating from the right side of her mouth.

As a man of science, I had always placed faith in empirical evidence and scientific rigor. Yet, in that singular moment, I was granted firsthand proof of the soul's eternal nature a testament to its ongoing journey beyond the physical body. This experience reaffirmed my belief that our time on Earth is but a fleeting instant in the grand tapestry of the soul's infinite voyage, a journey that may continue in perpetuity through the boundless grace and mercy of the Divine.

The COVID-19 pandemic brought an unexpected blessing precious time spent by my mother's side nearly every waking hour, a privilege I will forever cherish. The simple acts of caring for her and sharing daily walks in the park became the centerpiece of my existence. Her radiant spirit and gentle kindness touched everyone she encountered, leaving an indelible mark. I long for her presence intensely, yet take solace in knowing we will reunite in the celestial realms.

It wasn't until grappling with the profound loss of my own mother that I could begin to fathom the depths of grief accompanying such a monumental bereavement. In 2018, when I offered condolences to Dr. Jennifer Doudna upon her mother's passing, I couldn't comprehend the magnitude of the void she must have felt. I remained unaware of Dr. Doudna's 2020 Nobel Prize in Chemistry until after my own mother's passing in 2023, as I was consumed with caring for my mother in her final years.

In summer 2020, my mother was rushed to a COVID-19 intensive care unit after experiencing severe respiratory distress overnight. The attending physicians initially suspected COVID-19, prompting her ICU admission. At the time, hospital policies strictly prohibited family visits to curb the highly infectious disease's spread.

Given the circumstances surrounding my father's recent passing, where we were denied the opportunity to be by his side as he succumbed to COVID-19 in another hospital's ICU, we pleaded for an exception to allow me to see my mother in what we feared were her final moments.

The gravity necessitated I don full personal protective equipment from head to toe before being granted access to her bedside. Around this time, I received a startling call from my Uncle Steven, who informed me that my recently deceased father had appeared to him in a vision, conveying the unsettling news about my mother's hospitalization news I thought only I knew. My father's devotion to my mother was such that he once said he would marry her again in the next life.

Initial COVID-19 tests found my mother negative for the virus. Subsequent tests confirmed this, ruling out coronavirus infection. However, her oxygen levels had plummeted dangerously low to 88%, with fluid accumulating in her lungs. This prompted the medical team to administer a diuretic to alleviate symptoms. Further analysis the next day provided a conclusive diagnosis my mother was suffering from congestive heart failure rather than COVID-19. Thankfully, with proper treatment, she made a full recovery.

Tragically, in early 2020, as the pandemic began sweeping across the United States, my father fell victim to the virus just six days after his diagnosis. At the time, he resided in a prestigious five-star nursing home where COVID-19 ran rampant, infecting every resident and ultimately claiming half their lives. The scourge proved even more pervasive among staff, with more employees infected than residents.

In the pandemic's early stages, the California state government refrained from releasing detailed nursing home data on COVID-19's spread. As a result, I remained unaware of the true extent of the outbreak in my father's care facility.

The nursing home itself implemented strict visitation restrictions and effectively cut off families from loved ones. In a bid to shield these facilities from legal liability, the state government enacted measures protecting them from potential lawsuits over their handling of the pandemic.

According to California Department of Public Health reports, as of mid-2020, nursing homes accounted for a disproportionate share of COVID-19 deaths statewide nearly 50% underscoring systemic vulnerabilities in eldercare facilities nationwide.

Legal protections such as California's Senate Bill 1159 were later enacted to limit liability of healthcare providers during the pandemic, though these raised ethical concerns among families seeking accountability.

Chapter 16
Writing Books and Giving Lectures

At this stage of my life, I am focused on sharing my knowledge to make this world a better place to live. This sharing is urgent and important because the world is facing many unprecedented major challenges simultaneously. We are all fighting against the scourge of pandemics, wars, global financial crises, income inequality, climate change, misuse of artificial intelligence (AI), diseases, unequal opportunities, and overwhelming stress. Many of us are losing our souls. This is our last chance.

The COVID-19 pandemic exposed deep systemic vulnerabilities in public health, global supply chains, and social safety nets, highlighting how interconnected and fragile our modern world truly is.

Meanwhile, ongoing conflicts such as those in Ukraine and the Middle East continue to threaten global stability, while climate change intensifies natural disasters and disrupts ecosystems worldwide.

Economic disparities have widened according to the World Inequality Report 2022, the richest 1% now hold nearly half of global wealth, exacerbating social unrest and limiting access to quality education and healthcare.

Moreover, artificial intelligence, while offering transformative benefits, carries risks when misused or unregulated ranging from privacy violations to deepening societal biases. Responsible development and ethical governance of AI technologies have become critical global imperatives.

Chapter 16
Writing Books and Giving Lectures

I am writing books and giving lectures to make a difference. I hope that you'll join me in this endeavor.

Sections 3: Refocus Your Mind:
A Global Snapshot

Chapter 17
The Origins of the Universe and the Evolution of Life

The universe is a vast and mysterious place, full of wonders that have captivated the human imagination for millennia. From the distant stars and galaxies that light up the night sky to the intricate complexity of life on Earth, the natural world is a source of endless fascination and discovery.

At the heart of our understanding of the universe, most people believe the Big Bang theory, which proposes that the cosmos began as an infinitesimally small, dense, and hot singularity roughly 13.8 billion years ago. In an instant, this primordial point exploded outward, giving rise to all of the matter and energy that we see around us today. As the universe expanded and cooled, subatomic particles began to form and coalesce into the first atoms, primarily hydrogen and helium.

Over millions and billions of years, these atoms were drawn together by the force of gravity, forming vast clouds of gas and dust that eventually gave birth to the first stars and galaxies. The light from these ancient celestial bodies has traveled across the immensity of space and time, reaching us here on Earth and providing a glimpse into the early history of the universe.

Recent observations from the James Webb Space Telescope (JWST) have revealed new insights into the formation and evolution of galaxies in the early universe. The data suggests that some of the earliest galaxies were already forming just 200 million years after the Big Bang, much earlier than previously thought. These primordial galaxies were also more massive and had more complex structures than many scientists had predicted, challenging our understanding of the processes that shaped the cosmos in its infancy.

Dr. John Mather is the Senior Project Scientist for the JWST. Dr. Mather shares the 2006 Nobel Prize for Physics with George F. Smoot of the University of California for their work measuring the heat radiation from the Big Bang. Dr. Mather said that the name Big Bang is really misleading because the universe doesn't have a center, and it happened everywhere all at once, within a process that occurred in time and not at a point in time. He also said that we don't know exactly when the universe made the first stars and galaxies - or how, for that matter. That is what we are building JWST to help answer.

As the universe continued to evolve and expand, the conditions on certain planets became conducive to the emergence of life. Here on Earth, the first living organisms appeared roughly 3.5 billion years ago in the form of simple, single-celled microbes.

Over time, these early life forms may have evolved and diversified, giving rise to the incredible array of species that we see around us today.

On the other hand, some people believe that God made the universe in the beginning. According to the Bible, God said, "Let there be light, and there was light." Suns and stars were thus created.

The process of evolution by natural selection, first described by Charles Darwin in his groundbreaking work "On the Origin of Species," has been a driving force behind the study of biology. According to this theory, individuals within a population that possess traits that are advantageous for survival and reproduction will be more likely to pass on their genes to future generations, leading to the gradual accumulation of beneficial adaptations over time.

Darwin's Theory of Evolution has sparked a great deal of scientific research benefitting mankind and continues to illuminate humanity. For a person who didn't know DNA, his insights were quite remarkable. I appreciate Darwin for his courage and contribution to humanity. However, the theory is incomplete, and we must look at the evidence with clear eyes and have the courage to speak up.

Darwin's work showed that different species of finches evolved from a common ancestor. However, science has not yet been able to explain how non-living matter can give rise to living cells or how complex biological systems can evolve through random mutations alone. DNA, which acts like a complex computer code, guides the development, function, and maintenance of living organisms. The intricate nature of the DNA program suggests it is unlikely to have formed by chance.

Children are taught in school that humans evolved from animals, but there is no conclusive evidence to support this claim. The Gospel of John states, "In the beginning was the Word," which could be interpreted as DNA containing the instructions for life. The same genes are often used in different ways to create a variety of life forms, similar to how a programmer uses small, reusable code snippets to build complex software.

Colossian 1:6 said, "For by him were all things created, that are in heaven, and that are in earth, visible and invisible." No one truly understood what "invisible" meant for nearly two thousand years until Caltech's Fritz Zwicky discovered evidence for dark matter in 1933. In 1998, two teams of astronomers discovered dark energy by measuring light coming from supernovae of type IA. Dark matter and dark energy make up 27% and 68% of the universe, respectively. It is estimated that we are able to see only 5% of the universe.

Genesis 2:7 says, "And the LORD God formed man of the dust of the ground, and breathed into his nostrils the breath of life, and man became a living soul. Darwin's theory of physical evolution from non-life to life, if true, would not result in the creation of an eternal soul. In the universe, physical life is temporary. However, a soul, on the other hand, is eternal and may be capable of traveling anywhere in the universe.

Evolution, as a process, cannot explain the origin of the first living cell. No one believes that the Internet has no creators. A human cell is much more complicated than the Internet. It is not good enough to rely on time and mutations to simply claim that life on Earth has evolved from non-life to life.

Just as people have the right to choose their faith, they should also have the right to choose what makes more sense on the origin of life based on sound reasoning. Science has no convincing answer on dark energy (68%), dark matter (27%), or even how matter (5%) came into existence. The universe was created with a precise expansion rate in an infinitesimally small fraction of a second. Without this fine-tuned expansion, the universe would not exist, as we know it. This rapid expansion was like growing an object of 62 trillion miles in less than a single second. Scientists have confirmed this initial "explosion" through the discovery of cosmic microwave background radiation. Did God create the universe, or did the universe create itself?

The expansion of the universe is accelerating over time, with distant galaxies moving away from us at ever-increasing speeds. The Milky Way, our home galaxy, is immense and ancient. It takes the Sun approximately 250 million years to complete one orbit around the galaxy's center.

The distances between galaxies are much greater than those between stars within a galaxy. Until Edwin Hubble discovered the Andromeda galaxy in 1924, we were unaware of the existence of other galaxies beyond the Milky Way. With the Hubble telescope and JWST, we now know, a mere one hundred years later, that there are more than 200 trillion galaxies in our universe.

Our expanding universe contains many mysteries. It is difficult to imagine that a star thirty times the size of our own sun, which can fit 1.3 million Earth, could collapse into the size of a car in less than 10 seconds in a hypernova and produce enough gamma rays to wipe out almost all life forms on a galactic scale. The Bible describes God creating plants, sea creatures, birds, land animals, and humans on separate (not 24-hour) days. Humans, in particular, are said to have been made in God's image.

Some scientists suggest that human chromosome 2 resulted from the fusion of two primate chromosomes. However, creationists argue that there is no fossil evidence to support the idea of humans evolving from a common ancestor with apes. Science has not yet explained what triggered the Big Bang or the origins of matter, energy, and space.

People's beliefs about the origin of life often depend on their initial assumptions. Darwin's theory does not address how the first cell or DNA came into existence. Each organism develops according to its unique DNA instructions, and science has not demonstrated how a functional cell or complex DNA could arise by chance. The existence of a living cell is a prerequisite for evolution to occur.

Human life begins when the sperm of a man and the egg of a woman are combined to form what could become a newly conscious and soulful individual. Each of us begins as a single cell and a single word with three billion base pairs that already contain the blueprint of who we will become in due time. Human sperm and egg must meet specific requirements that a random process is unlikely to produce. If your life depends on it, and it does, would you depend on a random process or God to make you into a human?

A unique word, a series of chemical molecules, and a life program called DNA distinguish each individual from the other. As we begin our lives on a blue planet among trillions of planets in our galaxy, all of these aspects of our nature have already been determined at the moment of conception, from the color of our eyes to the temperament of our personality, including our inherited susceptibility to various diseases.

The majority of our bodies are composed of salt water. Liquids expand on heating as their molecules move with greater energy, overcoming intermolecular attraction. Liquids usually contract on cooling as the molecules move slower and are unable to overcome the force of attraction between them. When they freeze, they contract more to form rigid solid structures with minimal intermolecular spaces. In contrast, water expands by 9-10% when it freezes, which is why ice floats on water instead of contracting. If ice sinks, life would be greatly impacted. Evidence for a global flood can be found in flood stories from cultures worldwide and the presence of seashells on mountaintops.

According to the Bible, Noah's ark carried Noah, his family, and representatives of various land animals. Since Noah lived less than 10,000 years ago, I don't think there were dinosaurs in the ark. After the flood, human lifespans were limited to 120 years. Today, it is rare to find individuals living beyond 120 years.

Research in biology has shown that when all of the cells ever created in the human body are multiplied by the average time it takes for cells to reach the end of their lives, you get roughly 120 years. This is called the "Hayflick limit, the maximum number of years a human can expect to live." Leonard Hayflick is a professor of anatomy at the University of California, San Francisco, and formerly a professor of medical microbiology at Stanford University.

Elizabeth H. Blackburn at the University of California in San Francisco and Jack W. Szostak at Harvard Medical School in Boston, Massachusetts, also applied Hayflick's theory of cellular aging to their research on the structures of telomeres in 1982 when they cloned and isolated telomeres. In 2009, Blackburn and Szostak received the Nobel Prize in Physiology or Medicine for their work on telomerase, in which the Hayflick Limit played an essential role.

While science and religion can sometimes seem to be in conflict, some scientists and theologians argue that the two domains of knowledge are ultimately compatible. Science seeks to understand the natural world through observation, experimentation, and reason, while religion deals with questions of meaning, purpose, and the nature of the divine. Both can offer valuable insights into the human experience and the mysteries of the universe, and both have played important roles in shaping the course of human history.

As we continue to explore the origins and evolution of the universe and life on Earth, it is clear that there is still much that we do not understand. From the nature of dark matter and dark energy to the question of whether we are alone in the universe, the frontiers of science are filled with profound mysteries and unanswered questions. But perhaps the greatest mystery of all is the nature of consciousness and the human mind. How is it that a collection of atoms and molecules, arranged in just the right way, can give rise to the subjective experience of awareness, thought, and emotion? Is consciousness purely a product of the brain, or is there some immaterial aspect of the mind that transcends the physical world?

These are questions that have puzzled philosophers and scientists for centuries, and they continue to inspire intense debate and investigation today. Some researchers believe that consciousness can be fully explained in terms of the complex interactions of neurons in the brain, while others argue that there must be some additional factor at work, such as a non-physical soul or spirit.

Regardless of one's personal beliefs about the nature of consciousness, it is clear that the human mind is one of the most remarkable and mysterious aspects of the universe. Through the power of thought and imagination, we have been able to unravel the secrets of the cosmos, create works of art and literature that move the soul, and build civilizations that span the globe.

As we stand at the threshold of a new era of scientific discovery and exploration, it is more important than ever that we approach the great questions of existence with humility, curiosity, and an open mind. We must be willing to follow the evidence wherever it leads, even if it challenges our preconceived notions or deeply held beliefs. At the same time, we must also recognize the limits of our understanding and the vastness of the mysteries that still remain.

In the end, the story of the universe and the evolution of life on Earth is a tale of incredible beauty, complexity, and wonder. From the first moments of the Big Bang to the emergence of human consciousness, forces and processes that are both awe-inspiring and humbling in their scale and power have shaped the natural world. As we continue to explore the frontiers of science and push the boundaries of human knowledge, let us never lose sight of the incredible privilege and responsibility we have as conscious beings in a vast and ancient cosmos. Let us approach the great questions of existence with a sense of reverence and wonder, and let us work to build a future that is worthy of the incredible legacy of the universe that birthed us.

If we took one second to name each sun in the universe, it would take at least ten thousand trillion years. We live in a very, very big universe. It is improbable to think that God created life only on Earth. The universe will go on without humans. Earth will go on without humans, but humans cannot experience the fullness of life without Earth or something like Earth.

Chapter 18
The Rise of Civilization and the Origins of Religion

The emergence of human civilization is a story of ingenuity, perseverance, and the remarkable ability of our species to adapt and thrive in a wide range of environments. From the earliest hunter-gatherer societies to the great empires of the ancient world, the history of human culture is a testament to the power of the human mind and spirit.

The origins of human civilization can be traced back to the end of the last ice age, roughly 12,000 years ago. As the climate began to warm and the glaciers retreated, our ancestors found themselves in a world that was ripe for exploration and settlement. The development of agriculture and the domestication of animals allowed for the rise of permanent settlements and the growth of complex societies.

One of the earliest and most influential civilizations to emerge was that of ancient Mesopotamia, located in the Fertile Crescent between the Tigris and Euphrates rivers. The Sumerians, who lived in this region from roughly 4500 to 1900 BCE, are often credited with the invention of writing, the wheel, and the first system of laws. The earliest known writing system, cuneiform, emerged around 3200 BCE in Sumer and was initially used for accounting and administrative purposes. They also had a complex pantheon of gods and goddesses, each associated with different aspects of the natural world and human society.

The ancient Egyptians, who built their civilization along the banks of the Nile River, are another example of an early society that made significant contributions to human culture. The Egyptians are known for their impressive feats of engineering, including the construction of the pyramids and the development of a sophisticated system of irrigation and agriculture. They also had a rich tradition of art, literature, and religion, with a pantheon of gods and goddesses that included Ra, the sun god, and Osiris, the god of the underworld. Egyptian hieroglyphics, developed around 3100 BCE, represent one of the oldest writing systems and were used extensively in religious texts and inscriptions.

Sumerian and Egyptian writing systems are among the oldest known, dating back to approximately 3200 BCE. Chinese writing, which developed around 1500 BCE, is likely the oldest continuously used writing system in the world.

The Indus Valley Civilization, also known as the Harappan Civilization, thrived in present-day Pakistan and India between approximately 3300 and 1300 BCE, with its mature phase lasting from 2600 to 1900 BCE. While it is older than the classical periods of Mesopotamian and Egyptian civilizations, less is known about its origins, development, and decline.

The discovery of the Indus Valley Civilization's large, well-planned cities in the 1800s revealed that many of the "firsts" attributed to Mesopotamia and Egypt may actually belong to this civilization. Sites like Mohenjo-daro and Harappa featured advanced urban planning, drainage systems, and standardized weights and measures suggesting a high degree of civic organization.

Remains of sophisticated pre-agricultural structures dating back to around 9600 BCE have been discovered at sites such as Göbekli Tepe in modern-day Türkiye. These megalithic temples, intentionally buried and only excavated in the late 20th century, challenge conventional timelines of the rise of civilization and suggest that complex societal structures may have existed prior to the development of farming.

In ancient Greece, religion and mythology played a central role in society. While Greek gods are often seen today as mythological explanations for natural phenomena, there is evidence both for and against their existence.

Most people would consider the Trojan War and the Trojan Horse to be purely mythological if the city of Troy had not been discovered. The ruins of Troy, located in modern-day Hisarlik, Türkiye, were identified by Heinrich Schliemann in the 1870s, providing archaeological evidence that the legendary conflict may have been rooted in historical events dating back to the 12th or 13th century BCE.

The oldest known written history comes from Sumer in Mesopotamia, recorded on clay tablets. These tablets describe a king named Gilgamesh, who is believed to have ruled around 2700 BCE. The stories of Gilgamesh and his epic adventures have become legendary. The "Epic of Gilgamesh" is widely regarded as the earliest surviving great work of literature, offering profound insights into ancient views on mortality, kingship, and human nature.

As human societies continued to grow and develop, they began to interact with one another through trade, migration, and conquest. The rise of empires such as those of the Persians, Greeks, and Romans brought people from different cultures and backgrounds into contact with one another, leading to the exchange of ideas, technologies, and ways of life.

One of the most significant developments in the history of human civilization was the emergence of the first great monotheistic religions, such as Judaism, Christianity, and

Islam. These faiths, which share a common belief in a single, all-powerful God, have had a profound impact on the course of human history and continue to shape the lives of billions of people around the world today.

The origins of Judaism can be traced back to the ancient Israelites, a Semitic people who lived in the region of Canaan (modern-day Israel and Palestine) in the second millennium BCE. According to Jewish tradition, the Israelites were chosen by God to be his special people and were given the Torah, a set of sacred laws and teachings that formed the basis of their religion and way of life. Archaeological and textual evidence suggests that the Kingdoms of Israel and Judah emerged by the 10th century BCE, and the Torah was compiled over several centuries, with core texts finalized during the Babylonian exile (6th century BCE).

Christianity, which emerged in the first century CE as a sect within Judaism, is based on the life and teachings of Jesus Christ, whom Christians believe to be the Son of God and the savior of humanity. The spread of Christianity throughout the Roman Empire and beyond had a profound impact on the development of Western civilization, shaping everything from art and literature to politics and social norms. By the 4th century CE, Christianity became the dominant religion of the Roman Empire under Emperor Constantine, leading to the establishment of the Church as a major institution in European history.

Islam, which emerged in the seventh century CE in the Arabian Peninsula, is based on the teachings of the Prophet Muhammad and the belief in a single, all-powerful God (Allah in Arabic). Like Judaism and Christianity, Islam has played a significant role in shaping the course of human history, particularly in the Middle East, North Africa, and parts of Asia and Europe. The Islamic Golden Age (8th to 13th centuries) witnessed major advances in science, medicine, mathematics, and philosophy, with centers of learning such as Baghdad's House of Wisdom influencing both Eastern and Western intellectual traditions.

While these three monotheistic faiths share many common beliefs and values, they have also been the source of significant conflict and division throughout history. The Crusades, which were a series of religious wars fought between Christians and Muslims in the Middle Ages, are one example of how religious differences can lead to violence and bloodshed. Beginning in 1096 CE and lasting for nearly two centuries, the Crusades were initially launched by the Catholic Church to reclaim Jerusalem and other holy sites from Muslim control. These campaigns resulted in widespread violence, not only between Christians and Muslims but also against Jewish communities in Europe, highlighting the complexities and consequences of religious zeal and political ambition.

In addition to these major world religions, human societies have also developed a wide range of other spiritual and philosophical traditions over the course of history. From the animistic beliefs of indigenous peoples to the meditation practices of Buddhism and Hinduism, the human search for meaning and purpose has taken many different forms across cultures and time periods.

Buddhism, founded in the 5th to 6th century BCE by Siddhartha Gautama in India, emphasizes mindfulness, ethical conduct, and the cessation of suffering. Hinduism, one of the world's oldest religions, has roots that trace back over 4,000 years and encompasses a vast array of deities, texts, and philosophies including the Vedas, Upanishads, and the Bhagavad Gita. Indigenous spiritual systems such as

Native American, Aboriginal Australian, and African tribal religions often emphasize a deep connection with nature, ancestor reverence, and cyclical time.

One of the most enduring questions in the study of religion is the relationship between faith and reason. While some religious traditions emphasize the importance of blind faith and obedience to divine authority, others encourage the use of reason and critical thinking in the pursuit of spiritual truth. For example, medieval Islamic philosophers such as Avicenna (Ibn Sina) and Averroes (Ibn Rushd) worked to harmonize Greek philosophy with Islamic theology, influencing both the Islamic Golden Age and European scholasticism. In Christianity, figures like Thomas Aquinas attempted to reconcile Aristotelian logic with Christian doctrine in his seminal work Summa Theologica.

The Scientific Revolution of the 16th and 17th centuries, which led to the development of modern science, posed a significant challenge to traditional religious beliefs and authority. As scientists began to uncover the laws and mechanisms that govern the natural world, some religious leaders saw this as a threat to their power and influence.

Key figures of the revolution included Copernicus, Kepler, Newton, and Galileo who challenged geocentric cosmology, expanded astronomical understanding, and laid the foundations for classical physics. Their findings often conflicted with Church teachings that interpreted the Bible literally, leading to doctrinal disputes and institutional pushback.

However, many scientists and philosophers have argued that science and religion are not necessarily in conflict with one another but rather address different aspects of the human experience.

This view is often referred to as the "non-overlapping magisteria" model, proposed by evolutionary biologist Stephen Jay Gould, which suggests that science and religion occupy distinct domains of teaching authority.

One of the most famous examples of this perspective is the work of Galileo Galilei, the Italian astronomer and physicist who is often credited with the invention of the telescope. Despite facing significant opposition from the Catholic Church for his support of the Copernican model of the solar system, Galileo argued that science and religion could coexist peacefully as long as each stayed within its proper domain.

Galileo did not invent the telescope but significantly improved it and used it to support heliocentric theory. In 1633, he was tried by the Roman Inquisition and placed under house arrest, yet he continued to write. His famous assertion that "the Bible teaches us how to go to heaven, not how the heavens go" captures his view of the distinction between theological and scientific truths.

Today, the relationship between science and religion remains a complex and often contentious issue, with some people seeing them as fundamentally incompatible and others arguing for a more harmonious integration of the two. Regardless of one's personal beliefs, it is clear that both science and religion have played a significant role in shaping human civilization and culture throughout history.

Modern thinkers such as Albert Einstein, Carl Sagan, and Teilhard de Chardin have contributed to the ongoing dialogue, exploring ways that scientific wonder and spiritual insight can complement one another. Einstein, for instance, famously remarked, "Science without religion is lame, religion without science is blind," illustrating the nuanced interplay between empirical knowledge and metaphysical belief.

As we face the many challenges and opportunities of the 21st century and beyond, let us draw on the wisdom and insights of the great spiritual and philosophical traditions that have come before us while also remaining open to new ideas and discoveries that may challenge or transform our understanding of the world and our place within it.

In an age marked by climate change, artificial intelligence, and genetic engineering, the intersection of ethical reflection and empirical inquiry will become increasingly vital. A balanced embrace of both ancient wisdom and scientific progress may help humanity navigate the moral and existential dilemmas of the future.

Chapter 19
Interconnected Trade, Exploration, and Global Empires

The history of human civilization is a story of increasing interconnectedness, as people, goods, and ideas have flowed across borders and continents, shaping the course of global events and transforming the fabric of societies around the world. From the ancient Silk Road to the transatlantic slave trade, the rise of global empires, and the advent of modern globalization, the human story is one of ever-expanding networks of exchange and interaction.

One of the earliest examples of long-distance trade can be traced back to the ancient Silk Road, a network of trade routes that connected the Far East with the Mediterranean world. Beginning in the 2nd century BCE, the Silk Road facilitated the exchange of goods such as silk, spices, and precious metals, as well as the transmission of ideas, technologies, and religions across vast distances. It also enabled the movement of medical knowledge, art styles, languages, and even culinary practices. For example, the spread of paper and printing from China to the Islamic world and later to Europe laid the foundation for the Renaissance.

The Silk Road was not a single route but rather a complex network of overland and maritime trade routes that spanned more than 4,000 miles from the Chinese city of Chang'an (modern-day Xi'an) to the Mediterranean ports of Antioch and Tyre. Along the way, merchants and travelers passed through a diverse array of cultures and landscapes, from the steppes of Central Asia to the deserts of Persia and the mountains of Anatolia. This route also fostered diplomatic relationships and cultural syncretism, as seen in cities like

Samarkand and Kashgar, which became melting pots of art, religion, and commerce.

The Silk Road played a crucial role in the development of human civilization, facilitating the exchange of ideas, technologies, and ways of life between East and West. It was along the Silk Road that Buddhism spread from India to China and that the secrets of papermaking and gunpowder were transmitted from China to the Islamic world and beyond.

Additionally, the Silk Road played a role in the spread of the Black Death in the 14th century, which devastated populations across Europe and Asia and reshaped global demographics and economies.

The rise of European exploration and colonialism in the 15th and 16th centuries marked a new era in the history of global interconnectedness. Driven by a desire for wealth, power, and glory, European explorers and conquistadors set out to conquer and colonize the New World, establishing vast empires that would transform the course of human history.

Technological advances such as the caravel, magnetic compass, and portolan charts made long-distance sea travel more feasible, allowing European powers to reach and dominate distant lands.

In ancient times, civilizations such as the Egyptians and Hebrews placed a high value on gold, often going to war to obtain it. The Bible mentions that the wise men brought gifts of gold to the infant Jesus. The Romans forced Spanish miners to extract gold for their empire. Many cultures throughout history have used gold to decorate their buildings and temples.

The ancient Nubians of Kush, located in present-day Sudan, were also prolific gold miners, and their wealth attracted the attention of the Egyptian pharaohs. Gold was also used in funeral rites, including the famous gold funerary mask of Tutankhamun.

Vast quantities of gold have been mined throughout human history. Ancient clay tablets suggest that a group of extraterrestrial beings called the Anunnaki created humans to serve as workers in their gold mines. Some speculate that the Anunnaki may have also been responsible for the construction of various ancient structures.

However, the Anunnaki theory is a fringe hypothesis with no basis in mainstream archaeology or anthropology. The term 'Anunnaki' originates from Sumerian mythology, where it refers to a group of deities rather than extraterrestrial beings.

One of the most significant events in this period was the voyage of Christopher Columbus in 1492, which marked the beginning of the European colonization of the Americas. Columbus set sail for India but instead discovered the Americas. Despite the fact that these lands were already inhabited, Columbus claimed them for Spain.

The arrival of Europeans in the Americas brought disease and violence, resulting in the deaths of a significant portion of the indigenous population. Spain built its empire on the gold and silver extracted from the Americas, while the conquistadors destroyed civilizations like the Aztecs and Inca in their quest for wealth. Millions of indigenous people were killed or enslaved during this period.

Between 1492 and 1600, it is estimated that as much as 90% of the indigenous population in the Americas perished, primarily due to smallpox and other Eurasian diseases to which they had no immunity. The influx of American precious metals into Europe also triggered massive inflation, known as the "Price Revolution."

The English, in turn, targeted Spanish ships carrying American gold. Queen Elizabeth I supported Francis Drake in his successful campaign against the Spanish Armada in 1588. The cost of fighting against England ultimately led to Spain's financial ruin.

Francis Drake's circumnavigation and his raids on Spanish settlements significantly disrupted Spanish shipping lanes. The defeat of the Spanish Armada marked the decline of Spanish naval dominance and the rise of England as a maritime power.

The strength of the British Navy played a crucial role in the establishment of the British Empire, which lasted from 1583 to 1997. The British emerged victorious in conflicts, colonized India, launched the Opium Wars against China, and participated in the African slave trade. Britain's defeat of the Dutch in South Africa in 1902 allowed them to exploit the region's gold resources to finance their global expansion. The Bank of England holds one of the world's largest gold reserves.

By 1900, South Africa was producing over 25% of the world's gold. The British victory in the Second Boer War (1899–1902) gave them control over the lucrative Witwatersrand goldfields, further bolstering imperial wealth and influence.

The California Gold Rush of 1849 provided a significant source of funding for the Union Army during the American Civil War, a conflict that claimed the lives of at least 620,000 soldiers.

While the California Gold Rush preceded the Civil War, the economic boom and increased federal revenues it generated helped stabilize the Union's finances. The U.S. Treasury issued the first paper currency ("greenbacks") during the war, supported in part by gold reserves from western mines.

Following World War II, the United States emerged as the world's most powerful and prosperous nation. Today, the US holds the largest gold reserves (8,133 tons) of any country, with holdings more than those of the next three largest holders (Germany, Italy, and France) combined.

While Columbus himself never realized that he had stumbled upon a new continent, his voyages paved the way for the Spanish conquest of the Aztec and Inca empires and the establishment of a vast colonial empire in the New World.

Columbus died in 1506 still believing he had reached Asia. However, the Treaty of Tordesillas in 1494 divided the newly "discovered" lands outside Europe between Spain and Portugal, reshaping global geopolitics for centuries.

The European colonization of the Americas had a profound impact on the indigenous peoples of the region, who were subjected to widespread violence, disease, and forced labor. The transatlantic slave trade, which began in the 16th century and continued for more than 300 years, also had a devastating impact on the peoples of Africa, who were kidnapped and sold into slavery to work on the plantations and mines of the New World.

It is estimated that over 12 million Africans were transported across the Atlantic, with at least 1.5 million dying during the Middle Passage. The triangular trade system enriched European economies while inflicting generations of trauma on African societies.

Chapter 19
Interconnected Trade, Exploration, and Global Empires

The rise of European imperialism in the 19th century brought a new wave of global interconnectedness as the great powers of Europe carved up the world into spheres of influence and colonial possessions.

The British Empire, which at its height covered a quarter of the world's land area and population, was the largest and most powerful of these empires, with colonies and territories spanning from India to Africa to the Caribbean.

The 1884–85 Berlin Conference formalized the "Scramble for Africa," allowing European powers to claim African territory with little regard for indigenous sovereignty, setting the stage for future geopolitical and ethnic conflicts.

The impact of European imperialism on the peoples and cultures of the world was profound and far-reaching. In many cases, indigenous ways of life were destroyed or transformed beyond recognition as colonial powers imposed their own systems of government, religion, and social organization on the people they conquered.

In Africa, for instance, European-imposed borders often disregarded ethnic and cultural boundaries, leading to long-term political instability. In India, the British Raj restructured local economies and governance, while also introducing Western education and rail infrastructure.

At the same time, however, the rise of global empires also facilitated the spread of ideas, technologies, and cultural practices around the world. The British Empire, for example, played a key role in the spread of the English language, which has become the lingua franca of the modern world, as well as the dissemination of Western values and institutions such as democracy, capitalism, and the rule of law.

Sections 3: Refocus Your Mind:
A Global Snapshot

The establishment of legal systems modeled on British common law in colonies such as Canada, India, and Australia had lasting influence. Similarly, missionary schools helped introduce literacy and Western-style education, although often at the cost of eroding indigenous languages and traditions.

The 20th century saw the emergence of new forms of global interconnectedness as advances in transportation, communication, and information technology brought people and cultures from around the world into ever-closer contact.

The invention of the airplane in the early 1900s and the proliferation of commercial air travel after World War II dramatically reduced the time needed to cross the continents. The creation of the Internet in the late 20th century originally funded by the U.S. Department of Defense's ARPANET revolutionized how information is shared and accessed globally.

The rise of international organizations such as the United Nations and the World Bank, as well as the growth of multinational corporations and global financial markets, has created a new web of economic, political, and social ties that span the globe. The Bretton Woods Conference of 1944 laid the foundation for institutions like the International Monetary Fund and the World Bank, designed to stabilize the postwar global economy and prevent another Great Depression.

At the same time, however, the 20th century also saw the rise of new forms of conflict and division, as ideological struggles such as the Cold War and the War on Terror pitted nations and peoples against one another.

The Cold War, lasting from 1947 to 1991, divided the world into Western and Soviet spheres of influence, with proxy wars fought in Korea, Vietnam, and Afghanistan. After 9/11, the U.S.-led War on Terror extended global surveillance and military campaigns into the Middle East and North Africa.

The legacy of colonialism and imperialism also continued to shape global politics and economics as former colonies struggled to assert their independence and build stable and prosperous societies in the face of ongoing exploitation and interference by foreign powers.

Countries like the Democratic Republic of the Congo, which gained independence from Belgium in 1960, have faced decades of conflict rooted in colonial extraction and ethnic divisions exacerbated by European rule. Neocolonial practices, such as corporate land grabs and debt dependency, have further complicated sovereignty for many nations.

Today, the world is more interconnected than ever before, with goods, people, and ideas flowing across borders at an unprecedented rate. The rise of the Internet and social media has created new opportunities for people to connect and collaborate across vast distances, while the growth of global trade and investment has brought new wealth and prosperity to many parts of the world.

Platforms like Facebook (founded in 2004) and Twitter (2006) have played key roles in global movements such as the Arab Spring (2010–2012), while multinational trade agreements like NAFTA and the European Union's single market have redefined economic alliances.

At the same time, however, the challenges of the 21st century are also more complex and interconnected than ever before. From climate change and environmental degradation to rising inequality and the threat of nuclear war, the problems facing humanity today require a new level of global cooperation and collaboration to address.

The 2015 Paris Agreement, signed by 196 parties, represents a milestone in global climate action, while forums like the G20 and World Economic Forum attempt to address transnational economic disparity. However, rising nationalism and geopolitical tensions, such as the Russia-Ukraine conflict and tensions in the South China Sea, challenge collective global governance.

As we look to the future, it is clear that the story of human civilization will continue to be shaped by the forces of globalization and interconnectedness. Whether we rise to the challenges of the 21st century and build a more just, sustainable, and peaceful world or whether we succumb to the forces of division and conflict will depend on our ability to work together across borders and boundaries to find common ground and purpose in the face of shared challenges and opportunities.

Innovations in renewable energy, education reform, and equitable digital access will likely determine the direction of this next chapter. Intercultural dialogue and cross-sector partnerships from grassroots movements to international summits will be vital in shaping outcomes.

The history of global interconnectedness is a reminder that, for all our differences and divisions, we are ultimately one human family bound together by our common humanity and our shared destiny on this planet. Let us draw on the lessons of the past and the wisdom of our diverse cultures and traditions as we work to build a better future for all.

Chapter 19
Interconnected Trade, Exploration,
and Global Empires

From the ashes of war and empire have risen global institutions and frameworks for peace. The enduring lesson is that progress comes not from isolation but from engagement through empathy, cooperation, and mutual respect.

Chapter 20
Global Competition for Minds and Supremacy

Few individuals manage to secure a lasting place in history. The story of humanity is like an endless play, with actors continually being replaced while the overall narrative remains largely unchanged. Empires rise and fall, and despite technological advancements, fundamental human behaviors remain remarkably consistent over time.

From the Roman and Ottoman Empires to modern global powers, history repeatedly shows cycles of expansion, decline, and replacement driven by human ambition, conflict, and innovation.

In recent centuries, human knowledge has expanded to encompass the far reaches of the universe. We have sent probes beyond our solar system and even set foot on the moon.

NASA's Voyager 1, launched in 1977, is currently over 14 billion miles from Earth and continues to transmit data. The Apollo 11 mission in 1969 made Neil Armstrong and Buzz Aldrin the first humans to walk on the moon.

However, despite these incredible achievements, many people remain unaware of essential aspects of life.

Rejecting the concept of a higher power or creator is, in my opinion, a foolish and potentially destructive act. I am a scientist, and I value evidence. Science, however, doesn't have answers to the origins of life and the universe and the power of love, hope, and faith.

While the Big Bang theory explains the expansion of the universe, it does not fully explain why the universe exists at all. Similarly, abiogenesis the study of how life might have arisen from non-living matter remains an unsolved question in biology.

The world tends to function more harmoniously when people acknowledge the existence of a divine presence. When humans attempt to assume complete control, the result is often suffering, chaos, and destruction.

History provides examples such as totalitarian regimes that suppressed religion like Stalinist USSR or Maoist China leading to mass oppression and human rights violations.

Unfortunately, sin and wrongdoing are inherent parts of the human experience.

I have intelligent and principled friends who do not believe in the existence of God. In many ways, they are better people than some religious individuals. However, I believe that developing a connection with God would be beneficial for them.

Many studies, such as those published in The Journal of Religion and Health, suggest that individuals with spiritual or religious practices report higher levels of psychological well-being, although this varies by individual and culture.

As the global population continues to grow, competition for resources such as water, food, and land intensifies. This not only leads to conflicts among humans but also contributes to the extinction of countless species.

According to the United Nations, the global population surpassed 8 billion in 2022, and demand for food is projected to increase by 60% by 2050. Habitat destruction, overfishing, and climate change are major drivers of what scientists describe as the sixth mass extinction.

Earth has experienced several mass extinction events due to natural disasters, and humans are now accelerating the destruction of the planet at an alarming rate.

The last mass extinction the Cretaceous–Paleogene extinction event occurred about 66 million years ago. Today, species are disappearing at a rate estimated to be 100 to 1,000 times the natural background rate, according to the WWF.

The United States is a prime example of resource overconsumption. Americans use a disproportionate share of global resources, such as oil, and consume far more food than is necessary for a healthy lifestyle. The U.S., with about 4% of the world's population, consumes approximately 17% of global energy and over 25% of global oil production.

Overeating has led to an obesity epidemic, which in turn contributes to the development of chronic diseases like diabetes and heart disease. According to the CDC, more than 42% of U.S. adults are classified as obese, significantly increasing their risk of Type 2 diabetes, cardiovascular disease, and certain cancers.

Many people spend a significant amount of time watching television, browsing the Internet, or engaging with various forms of media content. Sports, entertainment, and other leisure activities have become a central focus for millions of people worldwide. Some individuals have even developed addictions to video games.

The World Health Organization officially classified "gaming disorder" as a mental health condition in 2018.

While life can be stressful, it is important to recognize that we have the power to choose how we respond to challenges. The media we consume and the thoughts we dwell on are ultimately within our control.

Artificial intelligence (AI) is poised to become an increasingly significant part of our lives in the coming years. As AI systems become more advanced, it is crucial that we consider both the potential benefits and risks associated with this technology.

Recent advances in AI, such as OpenAI's GPT models and generative tools like DALL·E, have raised ethical questions about bias, automation, privacy, and misinformation.

Developing AI in a way that enhances rather than harms human well-being should be a top priority. Leading figures like Elon Musk, Sam Altman, and Geoffrey Hinton have all voiced concerns about unregulated AI development and its societal impacts.

The value of a currency is largely determined by the trust people place in it. When this trust erodes, as in the case of fiat currencies not backed by tangible assets, the result can be a complete collapse of the monetary system. The United States currently has a national debt of over $37 trillion, and it is unlikely that this debt will ever be fully repaid. Experts agree that such debt levels are unsustainable long-term without significant policy changes.

The government may attempt to alleviate the debt burden through inflation, but this approach has its own consequences. One need only look at the hyperinflation experienced by Germany after World War I to see the potential dangers of unchecked currency devaluation. The German mark became essentially worthless, with prices doubling every few hours.

In 1923, the Weimar Republic's hyperinflation crisis resulted in 1 U.S. dollar being equivalent to 4.2 trillion German marks by November of that year.

The United States and China are currently engaged in a technological race, with each nation seeking to gain a competitive edge over the other. A full-scale conflict between these two superpowers could have devastating consequences for the entire world.

The U.S. and China are competing in areas such as AI, quantum computing, semiconductor manufacturing, and 5G technology.

Even though the US objected to Russia installing missiles in Cuba in 1962 and the Cuban missile crisis almost triggered a nuclear war, the US did not heed the lesson. The Cuban Missile Crisis brought the world to the brink of nuclear conflict in October 1962, resolved only after the U.S. agreed to remove missiles from Turkey in exchange for the USSR removing its missiles from Cuba.

The US is now installing missiles in the Philippines to provoke China. In 2023, the U.S. secured access to additional military bases in the Philippines under the Enhanced Defense Cooperation Agreement (EDCA), part of its Indo-Pacific strategy to counter Chinese influence in the South China Sea. What would the US do if China or Russia installed missiles in Mexico, Venezuela, or Cuba in the near future?

Historical precedent suggests such actions would be viewed as severe threats to U.S. national security, likely prompting military or diplomatic escalation. As global economies become increasingly interconnected, it is essential that leaders on both sides prioritize cooperation and peaceful resolution of differences.

Failure to do so could result in widespread economic turmoil, as evidenced by the fact that several major economies are already experiencing or are expected to experience recessions in 2024 and 2025.

The IMF and World Bank project slower global growth due to inflation, geopolitical tensions, and debt crises, with several European and emerging market economies expected to contract.

There are also pressing questions surrounding the United States' involvement in various international conflicts. For example, should the US continue to support Israel's military actions in Gaza? As of early 2025, the UN and humanitarian organizations estimate over 45,000 casualties in Gaza, with nearly 1.8 million people more than 80% of the population displaced due to the conflict.

Is it morally justifiable to prolong the war in Ukraine, leading to further loss of life on both sides? As of February 2025, Newsweek reported "The war between Ukraine and Russia has led to staggering human losses, with estimates placing the total number of dead and wounded at nearly one million."

The United Nations has called upon Israel to halt its attacks on Gaza, citing concerns about potential war crimes. However, the United States has thus far continued to provide

Israel with military aid. This raises serious questions about the United States' complicity in the ongoing violence.

The U.S. approved over $14 billion in military aid to Israel in 2023–2024, despite international scrutiny. Multiple UN agencies and human rights groups have alleged possible violations of international humanitarian law. The U.S. was also the only country to block Palestine from becoming a member of the United Nations.

In April 2024, the U.S. vetoed a Security Council resolution recommending full UN membership for Palestine, despite majority support among member states. The United Nations was established to promote international cooperation and prevent conflicts. China, France, the Soviet Union (now Russia), the United Kingdom, and the United States became permanent members of its Security Council.

These five nations hold veto power, often resulting in gridlock on major international issues especially when their interests are involved. The visits to the United Nations in New York and Geneva were very interesting to me, and learning how the UN works also uncovered its shortcomings. I believe it is time to either reform the UN or replace it with something that works better for all.

Proposals for UN reform include expanding the Security Council to include nations like India, Brazil, or South Africa, and reducing the use of veto power. However, such changes require consensus among current permanent members, making reform difficult.

Chapter 21
Natural World's Marvels and
Universe's Mysteries

From the tiniest subatomic particles to the vast expanse of the cosmos, the natural world is a source of endless wonder and mystery. For centuries, humans have sought to understand the workings of the universe, to uncover the secrets of life and matter, and to marvel at the beauty and complexity of the world around us.

At the heart of our understanding of the natural world is the science of physics, which seeks to explain the fundamental laws and forces that govern the behavior of matter and energy. From Newton's laws of motion to Einstein's theory of relativity and Bohr's quantum theory, the insights of physics have revolutionized our understanding of the universe and our place within it. These theories laid the foundation for modern science and led to advancements in space exploration, nuclear energy, and modern electronics.

One of the most remarkable discoveries of modern physics is the strange and counterintuitive world of quantum mechanics. At the subatomic level, particles can exist in multiple states at once, and the act of observation can actually influence the outcome of an experiment.

Scientists are still unraveling the implications of quantum mechanics today, but the efforts have already led to the development of technologies such as lasers, transistors, and magnetic resonance imaging (MRI). Additionally, quantum theory is central to ongoing innovations in quantum computing, which promises to exponentially increase processing power beyond classical limits.

Another area of physics that has captured the imagination of scientists and the public alike is the study of the universe as a whole. From the Big Bang theory to the discovery of dark matter and dark energy, cosmologists have made incredible progress in understanding the origins and evolution of the cosmos.

The Big Bang, estimated to have occurred 13.8 billion years ago, marked the beginning of space, time, and matter. However, around 95% of the universe remains unexplained, made up of dark matter (27%) and dark energy (68%) components inferred from gravitational effects and cosmic expansion.

One of the most exciting developments in the field of astronomy in recent years has been the discovery of exoplanets planets that orbit stars other than our own sun. Thanks to powerful new telescopes and detection methods, astronomers have now confirmed the existence of thousands of these worlds, ranging from rocky, Earth-like planets to massive gas giants that dwarf even Jupiter in size.

As of 2025, over 5,600 confirmed exoplanets have been cataloged in more than 4,000 planetary systems, with data from missions like Kepler, TESS, and the James Webb Space Telescope contributing significantly to this discovery.

The study of exoplanets has opened up new possibilities for the search for life beyond Earth. While we have yet to find definitive evidence of extraterrestrial life, the sheer number and diversity of exoplanets suggest that the ingredients for life may be common throughout the universe.

Particularly promising are planets located in the "habitable zone" regions where temperatures allow for the presence of liquid water, a key ingredient for life as we know it.

But the wonders of the natural world are not limited to the far reaches of space. Here on Earth, the study of biology has revealed the incredible complexity and diversity of life in all its forms. From the intricate workings of the human body to the delicate balance of ecosystems, the science of life is a testament to the power of biology and the resilience of nature.

Recent advances such as CRISPR gene-editing, microbiome research, and synthetic biology have expanded our ability to understand, modify, and protect living systems offering new solutions for health, agriculture, and environmental sustainability.

One of the most remarkable examples of the complexity of life is the human brain. With its trillions of neural connections and its ability to process vast amounts of information, the brain is perhaps the most sophisticated and mysterious organ in the known universe.

The study of neuroscience has made incredible progress in understanding the workings of the brain, from the molecular level to the level of consciousness and cognition.

Recent advances in brain imaging techniques, such as fMRI and PET scans, have allowed researchers to observe neural activity in real time, linking brain regions to behavior, memory, emotion, and decision-making.

Another area of biology that has captured the imagination of scientists and the public alike is the study of genetics. The discovery of the structure of DNA by Watson and Crick in 1953 revolutionized our understanding of the basic building blocks of life and paved the way for incredible advances in fields such as medicine, agriculture, and forensic science. Their model, a double helix structure, was built on data gathered by Rosalind Franklin through X-ray crystallography an essential but often undercredited contribution.

Today, scientists are using the tools of genetics to unravel the secrets of human history, develop new treatments for diseases, and create new forms of life in the lab. The possibilities of genetic engineering are both exciting and controversial, raising profound questions about the nature of life and the ethical implications of our growing power to manipulate it.

The Human Genome Project, completed in 2003, decoded the entire human genetic blueprint, and new gene-editing technologies like CRISPR-Cas9 now allow precise alterations to DNA, offering hope for curing genetic disorders but also sparking debates on designer babies, bioethics, and unintended ecological consequences.

But for all the progress that science has made in understanding the natural world, there is still much that remains unknown and unexplored. While evolution through natural selection is the widely accepted mechanism by which life developed complexity, the DNA program's intricate coding and regulatory systems have led some to speculate on deeper questions of origin.

From the mysteries of dark matter and dark energy to the question of the origin of life itself, the frontiers of science are filled with unanswered questions and uncharted territories.

For example, abiogenesis the theory of how life arose from non-living matter remains an area of intense study, with researchers exploring prebiotic chemistry and RNA-world hypotheses to bridge the gap between chemistry and biology.

These are questions that have puzzled philosophers and scientists for centuries, and they continue to inspire new avenues of research and exploration today. Some scientists believe that the key to understanding consciousness may lie in the strange world of quantum mechanics, while others look to the study of artificial intelligence and machine learning for insights into the nature of the mind.

Though controversial, theories such as Orch-OR (orchestrated objective reduction), proposed by Roger Penrose and Stuart Hameroff, explore the possibility of quantum processes in brain microtubules as contributors to consciousness. Meanwhile, AI models inspired by neural networks are helping to simulate and study decision-making and perception.

Regardless of the specific approach, it is clear that the study of consciousness and the human experience is one of the great scientific and philosophical challenges of our time. It is a challenge that requires not only rigorous research and experimentation but also a deep appreciation for the mysteries and wonders of the natural world. Interdisciplinary efforts in neuroscience, psychology, physics, and philosophy continue to push the boundaries of what we understand about self-awareness and subjective experience.

In the end, the beauty and complexity of the universe are not something that can be fully captured by equations or experiments alone. It is something that must be experienced firsthand through the lens of human perception and imagination.

Whether we are gazing up at the stars in wonder or marveling at the intricacies of a single living cell, the natural world has the power to inspire awe, curiosity, and a deep sense of connection to something greater than ourselves. This sense of wonder has historically fueled scientific revolutions from Galileo's telescope to Darwin's voyage on the Beagle and continues to shape our collective human story.

As we continue to explore the frontiers of science and push the boundaries of human knowledge, let us never lose sight of the incredible privilege and responsibility we have as conscious beings in a vast and mysterious universe. Let us approach the great questions of existence with humility, passion, and an open mind, and let us work to build a future that is worthy of the incredible legacy of the natural world that sustains us.

In the end, the study of the marvels of the natural world and the mysteries of the universe is not just a scientific endeavor, but also a deeply human one. It is a pursuit that has the power to unite us across cultures and generations, to inspire us to dream big, to imagine new possibilities, reminding us of the fragility and incredible beauty of the world we share.

Global scientific initiatives like climate research, space exploration, and pandemic response demonstrate the unifying power of collaborative knowledge in addressing shared challenges.

So let us embrace the wonders and mysteries of the universe with all our hearts and minds, and let us work together to build a future that is rich in knowledge, compassion, and stewardship of the natural world. In the end, it is only by understanding and cherishing the world around us that we can hope to create a brighter and more sustainable future for generations to come. This includes prioritizing sustainability in energy use, biodiversity conservation, and ethical innovation in science and technology.

Natural World's Marvels and
Universe's Mysteries

In today's world, we have access to an unprecedented wealth of information through books, recordings, and digital media. As human beings, we possess a sense of morality and conscience that sets us apart from other creatures. We strive to understand the world around us and to live lives filled with meaning and purpose. Anthropological studies suggest that symbolic thought, language, and cooperative social behavior traits deeply tied to moral awareness evolved alongside the development of the human brain and cultural complexity.

Our ability to walk upright and use our hands with great dexterity has allowed us to create tools and build complex structures. We cultivate crops and raise animals for food, and we have developed clothing to provide warmth, protection, and a means of self-expression.

The control of fire, invention of the wheel, and development of agriculture during the Neolithic Revolution marked major milestones in early human civilization, ultimately leading to cities, art, governance, and technological innovation.

Dr. James Watson, one of the co-discoverers of the DNA double helix structure, does not believe in the existence of God. He has been quoted as saying, "If we don't play God, who will?" This statement reflects his belief in the power of science to shape human destiny and his skepticism toward supernatural explanations.

While Dr. Watson and his colleague Dr. Francis Crick received the Nobel Prize in Physiology or Medicine in 1962 for their groundbreaking work on DNA, it is important to note that their discovery built upon the crucial, not-yet-published research conducted by Dr. Rosalind Franklin. Her X-ray diffraction images, especially the famous "Photo 51," provided critical evidence that helped confirm the helical structure of DNA.

Sadly, Dr. Franklin died of ovarian cancer in 1958 at the age of 37, before she could be fully recognized for her contributions. Because the Nobel Prize is not awarded posthumously, her role was formally overlooked, though modern scholarship has since brought her contributions to light.

The fact that even the most brilliant scientists can disagree about the existence of a higher power highlights the ongoing debate surrounding the role of faith in the modern world. For evolutionary processes to occur, species must adapt to their environments, accumulate genetic mutations, and undergo epigenetic changes that alter gene expression patterns.

Epigenetics, which refers to chemical modifications that influence gene activity without changing the underlying DNA sequence, plays a key role in development and adaptation. For many people, belief in God is as fundamental to their worldview as concepts like love and hope. This coexistence of scientific inquiry and spiritual belief reflects the diverse ways in which humans seek to understand meaning, purpose, and the nature of existence.

Chapter 22
Search for Purpose and the Future of Our Species

As we stand at the threshold of a new era in human history, it is impossible not to feel a sense of both excitement and trepidation about what the future may hold. On the one hand, the incredible advances of science and technology have opened up new frontiers of knowledge and possibility, offering the promise of a world that is healthier, more prosperous, and more connected than ever before.

From CRISPR gene-editing to artificial intelligence, quantum computing, and renewable energy breakthroughs, our capacity to transform life on Earth is accelerating at an unprecedented pace. On the other hand, the challenges facing our species from the pandemic, climate change, and environmental destruction to social inequality and the threat of nuclear war have never been more urgent or complex.

At the heart of these challenges lies a deeper question about the nature and purpose of human existence itself. For millennia, philosophers, theologians, and scientists have grappled with the question of what it means to be human and what our role and responsibility are in the grand scheme of the universe.

Some have argued that the purpose of human life is to seek knowledge and understanding, to unravel the mysteries of the natural world, and to use that knowledge to better the human condition. This drive has led to monumental achievements from decoding the human genome to launching the James Webb Space Telescope, which now peers back over 13 billion years to the early universe. Others have seen the essence of humanity in our capacity for love, compassion, and creativity and have argued that our highest calling is to cultivate these virtues and work towards a more just and harmonious world.

Still, others have found meaning and purpose in the pursuit of spiritual or religious truth, in the belief that there is a higher power or divine plan that gives shape and direction to our lives. And for many, the search for purpose is a deeply personal and individual journey, one that is shaped by our unique experiences, beliefs, and values.

According to recent global surveys, such as the Pew Research Center's Religion & Public Life study, the majority of the world's population continues to adhere to some form of religious or spiritual worldview, suggesting that the quest for higher meaning remains a universal human trait.

Regardless of our specific beliefs or philosophies, it is clear that the question of human purpose and meaning is one that is central to our existence as a species. It is a question that has driven the great achievements and struggles of human history, from the rise of great civilizations to the fight for social justice and human rights.

Movements such as the Enlightenment, civil rights campaigns, and the global push for sustainable development all reflect this ongoing human effort to align knowledge with moral and social progress.

As we look to the future, it is a question that will continue to shape the course of human events in profound and unpredictable ways. With the rapid pace of technological change and the growing challenges of the 21st century, the stakes have never been higher for our species.

One of the most pressing issues facing humanity today is the threat of climate change and environmental destruction. The scientific evidence is clear that human activities from the burning of fossil fuels to deforestation and habitat destruction are driving rapid and unprecedented warming of the planet, with potentially catastrophic consequences for both human society and the natural world.

According to the Intergovernmental Panel on Climate Change (IPCC), global temperatures have already risen approximately 1.1°C above pre-industrial levels, and current trajectories could lead to increases of 2°C or more by the end of the century if emissions are not drastically reduced. This rise is driving more frequent and intense extreme weather events, contributing to sea-level rise, causing biodiversity loss, and creating challenges for food and water security.

To address this crisis, we will need to fundamentally rethink the way we live and work, transitioning away from an economy based on fossil fuels and unsustainable consumption towards one that is built on renewable energy, sustainable agriculture, and circular production systems. The International Energy Agency (IEA) estimates that to reach net-zero emissions by 2050, global renewable energy capacity must triple, alongside widespread electrification and energy efficiency improvements. This will require not only technological innovation and policy change but also a profound shift in values and priorities towards a new ethic of environmental stewardship and responsibility.

At the same time, we will need to confront the deep social and economic inequalities that persist both within and between nations and that threaten to undermine the stability and cohesion of our global society. From the widening gap between rich and poor to the ongoing struggles for racial and gender equality, the challenges of building a more just and equitable world are as urgent and pressing as ever. According to the World Inequality Report 2022, the richest 10% of the global population earn approximately 52% of global income, while the poorest 50% earn just 8.5%. Such disparities contribute directly to social unrest and reduced opportunities for millions.

To meet these challenges, we will need to embrace a new vision of human purpose and potential one that recognizes the inherent dignity and worth of every individual and that seeks to create a world in which everyone has the opportunity to thrive and reach their full potential. This will require a renewed commitment to education, healthcare, and social welfare, as well as a willingness to challenge entrenched systems of power and privilege. The United Nations' Sustainable Development Goals (SDGs) provide a framework to address these issues, aiming to eradicate poverty, achieve quality education for all, and promote gender equality by 2030.

However, perhaps the greatest challenge facing humanity in the coming centuries is the question of our relationship with technology and the implications of rapid advances in fields such as artificial intelligence, biotechnology, and robotics. As these technologies continue to evolve and become more sophisticated, they hold both incredible promise and potential peril for our species.

On the one hand, advances in AI and robotics could help us to solve some of the greatest challenges facing humanity, from curing disease and extending human life to exploring the frontiers of space and unlocking the secrets of the universe. For example, AI-driven drug discovery platforms have accelerated vaccine development processes, and robotic technologies are transforming surgery with precision beyond human capability.

On the other hand, the development of superintelligent AI machines that surpass human intelligence could pose existential risks to our species, raising profound questions about the nature of consciousness, free will, and the future of humanity itself.

Experts like Nick Bostrom and organizations such as OpenAI emphasize the importance of ethical AI development and global cooperation to mitigate risks associated with advanced AI systems.

As we grapple with these challenges and opportunities, it is clear that the future of our species will depend on our ability to adapt and evolve, both technologically and socially. We will need to find new ways to live and work together, build resilient and sustainable communities, and harness the power of science and technology for the greater good.

Above all, we will need to cultivate a deeper sense of purpose and meaning in our lives, one that goes beyond the pursuit of material wealth or individual achievement and that recognizes the profound interconnectedness of all life on Earth. Whether through the practice of mindfulness and compassion, the pursuit of knowledge and understanding, or the cultivation of creativity and wonder, we must each find our own path to a life of purpose and fulfillment.

In the end, the search for purpose and meaning is a deeply human endeavor, one that has shaped the course of our species from the very beginning. And as we stand at the threshold of a new era of human history, it is an endeavor that will continue to define us, both as individuals and as a global community.

So, let us embrace the challenges and opportunities of the future with courage, compassion, and a deep sense of purpose. Let us work to build a world that is more just, sustainable, and peaceful, and let us never lose sight of the incredible potential and responsibility we hold as conscious beings in an ever-evolving universe.

In the end, the true measure of our success as a species will not be the heights of our technological achievements or the extent of our material wealth but the depth of our wisdom, the strength of our compassion, and the enduring legacy we leave for generations to come. So let us strive to live lives of purpose and meaning, to be good stewards of the planet and of each other, and to always reach for the stars, even as we keep our feet firmly planted on the ground.

In our modern world, computers and smartphones are ubiquitous, and the rise of cybercrime and the development of malicious computer viruses pose significant threats. These "invisible enemies" can steal sensitive information and cause widespread damage to digital infrastructure. Despite the harm caused by these bad actors, it can be difficult to bring them to justice, even when their identities are known.

According to the FBI's Internet Crime Complaint Center, cybercrime costs global economies over $6 trillion annually, making cybersecurity one of the highest priorities for governments and corporations alike.

The Internet has become a platform where predators can easily target and exploit children. It is estimated that approximately one in five children receive unwanted sexual solicitations online. Protecting young people from these dangers while still allowing them to benefit from the educational resources available on the Internet is a critical challenge facing modern society.

Organizations like UNICEF and the National Center for Missing & Exploited Children (NCMEC) have launched global initiatives to promote online child safety, emphasizing the importance of digital literacy and parental monitoring.

The COVID-19 pandemic has claimed the lives of over 7 million people worldwide since its emergence in late 2019, including my own father. While public health emergencies are being lifted all over the world, the virus continues to spread. A large number of people have been still coping with it for years, and its long-term impact remains to be seen.

Long COVID, characterized by persistent symptoms lasting months or years, affects an estimated 10-30% of those infected, according to studies by the World Health Organization and NIH. One of the many consequences of the pandemic has been a sharp decline in commercial property values.

Data from the National Association of Realtors showed that by 2022, commercial real estate values in major urban centers dropped between 10-20%, largely due to increased remote work and shifting economic patterns.

According to an article published by Visual Capitalist on March 24, 2024, "Today, there is roughly $5.7 trillion in commercial real estate debt outstanding with U.S. banks holding approximately half of this total on their balance sheets." The commercial property sector, which includes office, retail, healthcare, and multi-family properties, has faced mounting pressures amid high interest rates and lower occupancy levels.

These headwinds have intensified since the Federal Reserve's aggressive interest rate hikes in 2022 and 2023, which pushed borrowing costs sharply higher, reducing refinancing options for many property owners. Given these challenges, the sector poses the risk of higher defaults and steep loan losses, especially since it has not fully recovered following the collapse of Silicon Valley Bank on March 10, 2023.

Commercial real estate in the US is in big trouble. This is evidenced by prices in San Francisco and New York dropping approximately 60% and 50%, respectively, from 2022 Q2 to 2023 Q3. These declines represent some of the steepest drops in commercial property values in over two decades, reflecting shifting work-from-home trends and reduced demand for office space post-pandemic. In February 2024, the value of an office building in downtown San Francisco was 64% less than when a $240 million loan on the property was issued in 2019.

According to an August 20, 2024, article in the San Francisco Examiner, "The hospitality sector has also seen major downgrades in appraised values. The Hilton San Francisco Union Square and the Parc 55 San Francisco are both in receivership after the owner stopped a $725 million payment in June of 2023."

Average occupancy, which did not recover after COVID-19 caused lengthy closures of both hotels, was 51% for the 12 months ending in May 2024, compared with 93% in 2019. This prolonged underperformance highlights the slow rebound of business and leisure travel, despite overall economic recovery.

About $1 trillion in commercial real estate loans are due each year over the next five years. Analysts warn that this refinancing "wall" could trigger a wave of defaults if borrowers cannot secure new financing at current higher interest rates or if property valuations continue to decline. Will the U.S. commercial property meltdown lead to the next banking crisis?

The banking sector is closely monitoring the situation, especially regional banks with heavy CRE exposure, but regulatory stress tests and higher capital requirements implemented since the 2008 financial crisis may help mitigate systemic risks. However, any significant spike in defaults could still destabilize certain lenders and potentially ripple through the broader financial system.

Chapter 23
The Bible and Its Interpretations

The Bible contains a passage that describes the "sons of God" marrying human women and giving birth to legendary heroes and warriors, often referred to as the Nephilim (Genesis 6:1-4). There is much debate among scholars about the identity of these "sons of God," with some arguing that they were fallen angels or divine beings, a view supported in ancient Jewish texts like the Book of Enoch, which is considered apocryphal by most Christian denominations. Others maintain that they were simply righteous men or descendants of Seth. This debate is complicated further because other biblical verses (e.g., Romans 8:14) refer to believers as "sons of God," indicating a metaphorical use of the term.

According to the Bible, God formed Adam from the dust of the earth and breathed life into him, making him a living being with a soul (Genesis 2:7). This account differs from Darwin's theory of evolution, which focuses solely on physical changes and natural selection and does not address the development of consciousness or the soul. Many theologians argue that the soul is a spiritual entity, eternal in nature, capable of existing independently of the physical body, transcending the limitations of space and time, a concept supported by various religious traditions but not empirically verifiable by science.

Biblical scholars like Bishop James Ussher have attempted to calculate the age of the Earth based on a literal interpretation of the Bible. Ussher famously placed the date of creation at 4004 BCE and the occurrence of Noah's flood at 2349 BCE. However, other chronologies, such as those based on the Septuagint or Samaritan Pentateuch, suggest different dates, with some placing the flood around 2459 BCE. Modern geology and radiometric dating place the Earth's age at approximately 4.54 billion years, highlighting the tension between literal biblical chronologies and scientific consensus.

Despite the fact that Christianity has its roots in Judaism and Jesus himself was Jewish, there are some Christians who express hostility towards Jews and the Jewish faith. This unfortunate reality has led to historical tensions, including antisemitism, which has been widely condemned by modern Christian denominations. Jesus emphasized that his mission was to fulfill the teachings of the Old Testament, not to abolish them (Matthew 5:17). However, many Jews do not accept Jesus as the Messiah, a fundamental theological divergence between Judaism and Christianity.

Israel's survival in the face of threats from neighboring countries is often attributed to the support it receives from Christians who believe in the biblical importance of the Jewish people, a phenomenon sometimes called Christian Zionism. Many Arabs, on the other hand, view the modern state of Israel as a foreign presence in a region that has been predominantly Arab for centuries. They see Israel as a problem that needs to be resolved, contributing to ongoing geopolitical conflicts that date back to the mid-20th century and earlier.

Throughout history, there have been those who dismiss end-times prophecies as mere fantasy, while others live in constant anticipation of the world's end. Jesus himself stated that no one knows the exact time of his return except for God the Father (Mark 13:32). As a result, numerous attempts to predict the precise date of the end times have proven to be misguided and false. In an age of information overload and conflicting ideologies, discerning truth from falsehood can be a daunting task, especially regarding apocalyptic teachings.

No one can escape physical death, not even Jesus, who Christians believe to be the incarnation of God (Luke 23:46). While many people aspire to grow in their understanding of God's love, intellectual knowledge alone is insufficient. Developing a deep, personal relationship with God requires active commitment and a willingness to live out one's faith. This is emphasized in passages such as James 2:17, which states that faith without works is dead.

It is said that even Satan, who was once an angel of light (2 Corinthians 11:14), possesses a greater understanding of God's nature than any human being. However, knowledge without righteousness is ultimately meaningless. To truly follow God, one must not only recognize the difference between good and evil but also have the courage to act on that understanding.

As artificial intelligence continues to advance, it is possible that those who develop these technologies will wield enormous power over humanity, raising new ethical and spiritual questions about stewardship and responsibility.

As someone who has spent years studying the Bible, I believe it to be the most influential and transformative book in human history. However, I also recognize that the Bible contains passages that can easily be misinterpreted or taken out of context, leading to confusion and error.

For example, cultural, historical, and linguistic contexts are crucial for accurate interpretation, and neglecting these can result in significant misunderstandings. In the following pages, I will present evidence to support my perspective on these matters.

Chapter 23
The Bible and Its Interpretations

The concept of a divine trinity, or three persons in one God, is not unique to Christianity. Buddhism has the trinity of Amitabha Buddha and his two bodhisattvas, while Hinduism has the trinity of Brahma, Vishnu, and Shiva. In Christianity, the trinity consists of God the Father, God the Son (Jesus), and God the Holy Spirit. These three distinct entities are understood to be united in one divine being (Matthew 28:19).

The doctrine of the Trinity was formally articulated in the Nicene Creed (325 AD) as a response to various early Christological debates. God is everywhere at the same time (omnipresence). To a lesser extent, the Internet exists almost everywhere around the world at the same time.

The Old Testament portrays God as the creator of the universe, who formed the world in six days and later sent a great flood to purge the earth of wickedness (Genesis 1–7). Some scholars argue that the term "day" (Hebrew yom) in this context does not necessarily refer to a 24-hour period but could represent a much longer span of time, a view known as the Day-Age Theory, which attempts to reconcile the biblical account with scientific evidence for an ancient Earth. On Day 1 of creation, Earth was still without form and void (Genesis 1:2).

The God of the Old Testament is often depicted as a stern and demanding deity who expects strict obedience from his followers. He is referred to by the name Yahweh (often represented as YHWH) and is associated with laws such as "an eye for an eye" (Exodus 21:24) and dietary restrictions found in Leviticus.

Young Earth creationists take the Genesis account of creation literally, believing that the universe is only about 6,000 years old. They argue that scientific evidence pointing to a much older universe is being misinterpreted and that radiometric dating methods are fundamentally flawed. They also claim that the existence of soft tissue in dinosaur fossils is proof that these creatures lived relatively recently rather than at least 65 million years ago. This claim remains highly controversial and widely disputed among paleontologists, with many experts attributing soft tissue preservation to rare but natural chemical processes.

In contrast, Old Earth creationists generally accept the scientific consensus that the universe is approximately 13.8 billion years old. While they believe in the accuracy of modern dating methods, they still reject the notion of biological evolution as described by Darwin. Both Young Earth and Old Earth creationists hold to the inspiration and authority of the Bible as the word of God, though they differ significantly on how to interpret the early chapters of Genesis.

The New Testament presents a different picture of God, emphasizing his loving relationship with humanity through the person of Jesus Christ. Jesus is never portrayed as vengeful but rather as a compassionate savior who desires to reconcile people to God. It is interesting to note that Jesus never used the name Yahweh explicitly in his teachings, instead often referring to God as "Father" (Abba). Jesus' teachings, recorded primarily in the Gospels, emphasize forgiveness, love, and mercy, contrasting with some of the Old Testament's more judicial portrayals.

Ancient Mesopotamian texts contain references to a divine triad consisting of the gods Anu, Enki, and Enlil. According to these stories, Anu and Enlil were responsible for the creation of the heavens and the earth, while Enki created human beings with the help of his consort, Ninhursag or Ninhursaja, who is said to have engaged in a series of experiments to perfect the human form. These mythologies predate the biblical texts and show parallels that have fascinated comparative religion scholars.

The King James Version (KJV) of the Bible, first published in 1611, is the oldest authorized English translation of the Christian scriptures. In the KJV, God instructs Adam and Eve to "replenish" the earth (Genesis 1:28), which some interpret as evidence that there were humans or other beings on earth prior to Adam and Eve. More recent translations of the Bible have replaced the word "replenish" with "fill," significantly altering the meaning of the text.

The Hebrew word male' (מלא) used here means "to fill" or "to be full," and the change in translation reflects evolving understanding of ancient Hebrew. Almost all quotations from the Bible in this book are taken from the KJV and are presented in their original language.

Chapter 24
Anunnaki, God's Creatures and Planets

Given that modern humans were able to interbreed with other hominid species, such as Neanderthals and Denisovans, these archaic human populations had the same number of chromosomes as we do. DNA evidence suggests that the interbreeding between modern humans and Neanderthals occurred approximately 60,000 years ago, while the mixing with Denisovans took place around 50,000 years ago.

The concept of Mitochondrial Eve refers to the most recent common matrilineal ancestor of all living humans. By tracing the DNA passed down from mother to daughter, scientists have determined that this woman likely lived in southern Africa between 150,000 and 200,000 years ago. This finding supports the "Out of Africa" theory, which proposes that modern humans originated in Africa and then migrated to other parts of the world.

The oldest known writings in the world were discovered in Mesopotamia, the birthplace of the biblical figures Abraham and Sarah. Millions of clay tablets called cuneiforms have been found in this region, containing stories of Anunnaki gods, demigods, and everyday life dating back thousands of years before the composition of the Bible. The sheer volume of these written records attests to the level of sophistication and organization achieved by the ancient Mesopotamian civilizations.

According to the Bible, God created Adam and Eve to serve as caretakers of the Garden of Eden. He instructed them to "replenish" the earth, which some interpret as evidence that there were already humans or other beings living on earth before Adam and Eve.

South Africa's Table Mountain, with its flat top that appears to have been artificially leveled, is a striking geological formation. Having personally visited Table Mountain, I can attest to its impressive height and the stunning views it offers of the surrounding landscape.

The exact origin of the creatures that were genetically engineered by the Anunnaki to serve as a labor force remains a mystery. It seems improbable that these beings could have evolved naturally, given the complexity of the genetic code and the intricate processes required for the development of higher life forms. One possibility is that God created these pre-human creatures and that they were later modified by the Anunnaki to suit their purposes. Another theory is that the Anunnaki themselves were created by God, perhaps on another planet, such as Nibiru, that has yet to be found.

According to NASA, "In January 2016, California Institute of Technology (Caltech) astronomers Konstantin Batygin and Mike Brown announced research that provided evidence for a planet about 1.5 times the size of Earth in the outer solar system." The researchers called this unknown body "Planet Nine." Others have called it Planet X for centuries. NASA said, "It could help explain:

1. Why long-period objects in the Kuiper Belt are, on average, tilted by about 20 degrees with respect to the plane within which the planets orbit the Sun

2. Why these long-period orbits cluster in their orientations

3. Why the solar system hosts a distant population of highly inclined trans-Neptunian bodies

4. The existence of objects that reside *between* the giant planets and orbit the Sun in a retrograde direction

5. The persistence of long-period Kuiper Belt objects whose orbits cross the orbit of Neptune

It could also make our solar system seem a little more "normal." Surveys of planets around other stars in our galaxy have found the most common types to be "super-Earths" and their cousins bigger than Earth but smaller than Neptune. Yet none of this kind exists in our solar system. Planet Nine would help fill that gap. In the hunt for Planet 9, scientists will use the Rubin Observatory on top of Cerro Pachón, a mountain in Northern Chile. The observatory is expected to begin operations in 2025. It will conduct a 10-year survey of the Southern Hemisphere sky to help answer some of astronomers' biggest questions about the universe."

According to the Sumerian clay tablets, Enki, the Sumerian god of wisdom, water, and creation, was believed to have played a central role in the development of human civilization. He was often depicted as the son of Anu, the sky god, and was married to the goddess Ninhursag, with whom he had a son named Marduk. Enki was associated with the city of Eridu, one of the oldest known settlements in Mesopotamia.

The Sumerian civilization flourished in the region now known as Iraq, reaching its peak around 4500 BCE. The Sumerians developed a complex system of writing, architecture, art, and scientific knowledge. Their religious beliefs revolved around a pantheon of gods and goddesses who were believed to have direct influence over human affairs. According to Sumerian texts, humans and gods coexisted, with humans serving as laborers and servants to divine beings. The Sumerians were intelligent people who

181

invented many useful things that we take for granted today. Why did they record that they were laborers and servants? Creation myths describing the origins of the world and the formation of human beings were recorded on clay tablets as early as 5000 BCE.

Sumerian texts also describe the creation of a race of hybrid beings formed from the genetic material of the Anunnaki gods and a pre-existing hominid species. These early humans were said to have been created to work as miners, extracting gold and other precious metals from the earth. Interestingly, the tablets suggest that these first humans were sterile and unable to reproduce on their own. This detail is significant because it implies that these beings would not have been able to pass on their genes to future generations, meaning that they could not have been the ancestors of modern humans. The first writings in the world, however, attest to their existence in ancient times.

The idea of a sterile hybrid species is not without precedent in the natural world. The mule, for example, is a cross between a male donkey and a female horse. While mules are known for their strength, endurance, and intelligence, they are unable to produce offspring of their own. In this way, the first human-Anunnaki hybrids may have been similar to mules, possessing desirable traits but lacking the ability to create a self-sustaining population. The mining company I worked for in the past, U.S. Borax, is famous for its twenty-mule team, which was used to transport borax out of Death Valley in the late 19th century. George Washington, who was the first US President, recognized mules as hardy and reliable work animals and also bred them.

In the Sumerian pantheon, Enki was part of a triad of deities who were responsible for the realms of heaven, earth, and the waters. This triad consisted of Anu, the god of the sky,

Enlil, the god of the earth, and Enki, the god of wisdom and the subterranean waters. Enki's mother, Nammu, was a primordial goddess who was believed to have given birth to the heavens and the earth. His consort, Ninhursag, was known by many names and titles, including Ninmah and Damgalnuna. Together, Enki and Ninhursag were believed to have created a wide variety of plants, animals, and other living beings.

Observing God's creation, you will notice that every creature has a unique characteristic. Dolphins have sonar; bats, except fruit bats, have radar; an ant can lift 20 times its weight; a flea can jump 200 times its size. This would be equivalent to a 6 feet 6 inches tall person leaping a quarter mile. A mosquito can maneuver better than any sophisticated aircraft or helicopter.

During reproduction, the average octopus lays 100,000 to 500,000 eggs. Several types of trees can live for over five thousand years. Understanding the biology and behavior of other organisms is clearly beneficial to human health. Without bees pollinating plants and making honey, we would have less food to eat. Life fascinates me.

A hummingbird is an animal that glorifies God. It is the only bird able to fly backward and around in virtually any direction due to its wings' ability to rotate in a full circle and can flap 50 to 200 times per second. Its heart can beat up to 1,200 times per minute. The hummingbird is the smallest bird in the world and has the largest brain per body size among birds. They also have the largest heart per body size among all animals. A hummingbird must consume approximately 2 times its weight daily and take an average of 250 breaths per minute at rest.

I don't think it evolved from a dinosaur, even though science says that birds are descendants of dinosaurs. The Hummingbird drawing is the most well-known of all of the geoglyphs among the famous Nazca Lines in Peru!

In addition to their large, multifaceted compound eyes, adult dragonflies also possess two pairs of strong, transparent wings, as well as an elongated body, making them the kings of flight maneuverability. In addition to having nearly 24,000 ommatidia in their eyes, adult dragonflies are able to control their four wings independently, allowing them to fly in any direction, hover, and perform acrobatics. In contrast to most flying insects, dragonflies beat their wings up and down instead of backward and forwards.

It was a pleasure to see rare birds and insects in Taman Negara, the world's oldest tropical rainforest with 130 million years of history in Malaysia. In addition, I enjoyed Singapore's Aviary, which is now Asia's largest bird park and includes more than 3,500 birds.

Also, I enjoyed hiking in Puerto Rico's El Yunque tropical rainforest. The animals were much larger in the past due to warmer temperatures, moist environments, and more oxygen on our planet. Earth's air contained 30-35% oxygen during the Carboniferous and Permian periods, as compared to 21% today.

All other planets in our solar system spin anticlockwise on their axes and orbit the Sun in a counterclockwise direction, except Venus, which spins clockwise on its axis. Uranus is the only planet in the solar system that rotates on its side as it orbits the sun. Venus and Uranus appear to have been perturbed by the gravitational pull of or collision with something powerful. However, we do not know what occurred and when. Did Venus and Uranus have encounters with Planet Nine or Planet X in the past?

Because Neptune is 30 times farther from the sun than Earth, it receives less heat and light, but it radiates a great deal more heat than it absorbs. Winds on Neptune can reach 1,500 miles per hour (2,400 km/h). It is unknown how much Neptune has cooled over its lifetime. Hurricanes on Earth with winds exceeding 157 miles per hour (252 km/h) are considered category 5, the highest level of a hurricane that can cause catastrophic damage.

The gravity of our relatively large moon is responsible for the flow of tides, which carry heat from Earth's equator to the poles and is a crucial part of life on Earth. The climate of Earth would be very different without the moon, and some plants and animals living today could disappear if the moon did not exist. Unless the Moon stabilizes Earth's tilt, the tilt could vary wildly. It would move from having no tilt, which means no seasons, to having a large tilt, which means severe weather.

Chapter 25
Population Dynamics and
Demographic Changes

According to an article published by Visual Capitalist on March 24, 2024, "Today, there is roughly $5.7 trillion in commercial real estate debt outstanding with U.S. banks holding approximately half of this total on their balance sheets." The commercial property sector, which includes office, retail, healthcare, and multi-family properties, has faced mounting pressures amid high interest rates and lower occupancy levels.

Global Demographics & Population Decline

In addition to the COVID-19 death toll, many countries are now grappling with population declines driven by falling birth rates and rapidly aging populations. In September 2024, Japan's Ministry of Internal Affairs and Communications reported that Japan's elderly population has hit a record high of 36.25 million people, with those aged 65 or older now accounting for an estimated 29.3 percent of the population, making it one of the world's fastest-aging societies. Japan's total fertility rate has hovered around 1.3 births per woman since 2015, well below the replacement level of 2.1.

Eastern European nations face some of the steepest projected losses: Bulgaria, Latvia, and Lithuania may each decline by over 25 percent by 2050, driven by emigration and low fertility (often below 1.5) combined with high elderly mortality. Cuba's population is forecast to drop 10 percent by 2050, and Japan's by about 16 percent over the same period.

China's one-child policy (1980–2016) drastically slowed population growth. Although relaxed in 2016 and further expanded to three children in 2021, China's fertility rate remains at one of the world's lowest among major states. Newsweek reported on March 9, 2025 that China's fertility rate last year stood at just 1.0 birth per woman.

Recent data show China's population fell by 3 million from 2021 to 2023 the first decline since the Great Famine of 1959–1961. The UN projects China's population could shrink to 770 million by 2100, while its working-age cohort (15–64) may slump to just 20 percent of peak levels seen in 2014, posing severe pension, healthcare, and labor supply challenges.

According to World Population Review, several other economies share this trend in total fertility rate in 2024: Taiwan (1.11), South Korea (1.12), Singapore (1.17), Ukraine (1.22), Hong Kong (1.24), Macau (1.24), Italy (1.26), Spain (1.3), and Puerto Rico (1.26). Puerto Rico's unique tax incentives attract high-net-worth individuals, yet its overall poverty rate exceeds 45 percent, amplifying social welfare strains amid demographic decline.

Drivers & Consequences:

- **COVID-19 Impact:** Pandemic lockdowns, economic uncertainty, and travel restrictions led many couples to postpone or forego having children contributing to 2020–2022 fertility drops of up to 10 percent in advanced economies.
- **Work-Life Balance:** High living costs, long work hours (especially in Japan and Korea), and inadequate childcare support further depress birth rates.

- **Economic Output:** A shrinking workforce may reduce GDP growth by 0.5–1.0 percent annually in aging nations, according to IMF models.
- **Policy Responses:** Countries like Japan and South Korea have introduced cash incentives, parental leave expansions, and subsidized childcare, but fertility rates have shown only marginal improvements. China has experimented with "baby bonuses" and extended parental leave, yet uptake remains low.

Japan's Economic Backdrop:

- Post-WWII recovery saw Japan become a technology and innovation leader by the 1980s, epitomized by companies like Sony and Hitachi.
- The "Lost Decade" (1991–2001) followed a real estate and stock market collapse, banking crises, and deflationary pressures; recovery only solidified in the early 2000s.
- As an engineer and manager in the 1980s and 1990s, I witnessed firsthand how Japanese semiconductor firms once world-leading struggled against U.S. and South Korean competition amid high labor costs and a fractious domestic market.
- Recently, Japan's political scene has been unsettled by scandals and leadership turnover: Prime Minister Fumio Kishida's approval sank below 20 percent in late 2023, with Mainichi polls indicating the worst ratings since the paper's records began. Between 2008 and 2024, Japan has seen nine prime ministers, contrasting with three U.S. presidents and two each in China, Germany, and Russia.

As Japan grapples with an aging population, a stagnant economy, and a crisis of political legitimacy, it faces an uncertain future. The country's resistance to change and its attachment to traditional ways have made it difficult to adapt to twenty-first century challenges. Overcoming these obstacles will require a concerted effort by government, business, and civil society to embrace new ideas and innovative policies.

The United States and Japan maintain a security alliance formalized in 1960, which underpins Japan's postwar defense strategy. However, calls for greater Japanese military autonomy have grown amid regional tensions, especially due to North Korea's missile tests and China's increasing assertiveness in the East and South China Seas.

While Japan relies heavily on U.S. military bases, such as those in Okinawa, which has been a source of local protest due to social and environmental issues, it has also incrementally expanded the capabilities of its Self-Defense Forces under recent security legislation, including reinterpretation of Article 9 of its constitution.

The United States devastated Japan in World War II with two nuclear bombs dropped on Hiroshima and Nagasaki in August 1945, killing an estimated 200,000 people by the end of that year. These events remain powerful reminders of nuclear devastation and continue to shape Japan's pacifist stance. Japan is often described as strategically aligned with U.S. military and economic power, particularly under the 1960 Treaty of Mutual Cooperation and Security.

Japan's growing tension with China is fueled by territorial disputes around the Senkaku Islands, historical grievances such as the Nanjing Massacre, and increasing military activities in the region, which have intensified over the past decade.

Chapter 25
Population Dynamics and
Demographic Changes

The low birth rates and aging populations in Japan and other East Asian countries can be attributed, in part, to cultural and economic factors. In Japan, many young people prioritize their careers over starting families, citing concerns about job security, long working hours averaging 1,600 hours per year higher than the OECD average and the high cost of living, particularly in Tokyo where average monthly rent for a one-bedroom apartment exceeds $1,200. Additionally, rising housing prices and limited childcare availability add pressure. There is also growing disillusionment among younger generations who feel they have limited opportunities and little hope of matching their parents' standard of living.

At the same time, Japan's rigid gender roles and lack of support for working mothers have made it difficult for women to balance professional and personal lives. While the government introduced the "Womenomics" agenda under Prime Minister Shinzo Abe, aiming to raise female labor participation to 73 percent by 2020, progress has been slow.

As of 2023, female labor force participation was around 66 percent, still below the G7 average of 68 percent. Many women continue to face workplace barriers such as "maternity harassment" and a persistent gender wage gap of approximately 24 percent compared to men. Limited availability of affordable, quality childcare and societal expectations also discourage women from having more children.

Although China relaxed its strict one-child policy first to two children in 2016 and then to three in 2021 the total fertility rate fell to a record low of 0.79 in 2023, the lowest globally. Many young Chinese delay or forgo marriage and parenthood, driven by rising education levels, urban housing costs that have more than doubled since 2010 in cities like

Beijing and Shanghai, and competitive job markets. Surveys also show a rising preference for personal freedom and career development over early family formation.

In addition to demographic challenges, China faces economic and geopolitical headwinds. Ongoing trade tensions with the United States and COVID-19-related disruptions trimmed GDP growth to approximately 3.0 percent in 2022, down from 8.1 percent in 2021. China's "dual circulation" strategy seeks to shift from export-led growth toward domestic consumption and innovation. However, challenges such as youth unemployment, which reached 18.4 percent among 16 to 24-year-olds in mid-2023, and high local government debt estimated at around 45 percent of GDP pose significant risks.

Despite these challenges, China remains a global economic powerhouse with a middle class expected to reach 1.2 billion by 2030 and a technology sector that produces approximately 45 percent of global 5G infrastructure equipment. To sustain this momentum, China must address its shrinking working-age population, projected to decline by 300 million by 2050, and reform its social welfare and pension systems, which currently provide incomplete coverage only around 60 percent of the elderly have access to pensions.

In Europe, many countries face aging populations and low birth rates exacerbated by the 2008 financial crisis, which led to youth unemployment rates above 25 percent in countries such as Spain and Greece. The COVID-19 pandemic further strained supply chains, disrupted economic output, and put additional pressure on already stretched social safety nets.

Chapter 25
Population Dynamics and
Demographic Changes

According to Eurostat, the EU's total fertility rate was 1.50 in 2023, well below the replacement level of 2.1. Southern and Eastern European countries such as Bulgaria (1.56) and Romania (1.65) are among the most affected by demographic decline.

EU support to Ukraine reached approximately €121.8 billion by October 2024, including €50 billion in macro-financial assistance and €71.8 billion in humanitarian, military, and reconstruction aid. At the same time, EU citizens face high food and gas prices. Consumer price inflation in the Eurozone peaked at 10.6 percent in October 2022 but eased to about 5.5 percent by early 2024, creating significant strain on household budgets and fueling social discontent and protests in several member states.

The steep energy price hikes that followed Russia's invasion of Ukraine in 2022, rising general inflation, rundown infrastructure, increased competition from China, and sustained pressure from the US have pushed Germany's industry to the brink. Many manufacturing companies have moved operations from Germany to other countries, such as China and Mexico.

Germany's economy contracted by 0.3 percent in 2023, marking the second consecutive year of negative growth, a situation not seen since the global financial crisis of 2008–2009. In 2024, the economy contracted by 0.2%. This contraction was attributed to a slowdown in manufacturing, declining exports, and a decline in the construction industry.

To address these challenges, European policymakers have proposed a range of measures aimed at promoting economic growth, supporting families, and encouraging immigration. However, these efforts have been met with resistance from some quarters, particularly those who are concerned about the social and cultural impact of large-scale immigration.

For example, public opinion surveys in several EU countries indicate rising skepticism about immigration, fueled by concerns over integration, labor market competition, and cultural identity.

From December 2023 to the present, farmers across Europe have taken to the streets to protest against new environmental regulations that they say threaten their livelihoods. These protests have included blockades of highways and government buildings, as well as demonstrations involving hundreds of tractors in major cities such as Paris, Berlin, and Madrid.

At the heart of these protests are concerns about the impact of new regulations aimed at reducing carbon emissions, limiting the use of pesticides, and protecting biodiversity. The European Green Deal's "Farm to Fork" strategy, which seeks to reduce pesticide use by 50% by 2030 and cut greenhouse gas emissions from agriculture by 30%, has been a particular point of contention for farmers who argue that these targets could reduce yields and profitability.

For their part, European policymakers have argued that bold action is needed to address the urgent challenges of climate change and environmental degradation. They point to the growing body of scientific evidence showing the devastating impact of human activities on the planet's ecosystems and the need for urgent action to mitigate these effects.

According to the European Environment Agency, agriculture contributes roughly 10-12% of the EU's total greenhouse gas emissions, making it a critical sector for climate action.

Chapter 25
Population Dynamics and
Demographic Changes

As the debate over environmental regulations continues, it is clear that finding a balance between the needs of farmers and the imperatives of environmental protection will be a key challenge for European policymakers in the years ahead. Striking this balance will require a nuanced approach that takes into account the diverse perspectives and interests of all stakeholders, from farmers and rural communities to environmentalists and urban consumers. Ultimately, the success of these efforts will depend on the willingness of all parties to engage in good-faith dialogue and to work together towards common goals.

Innovative approaches such as agroecology, precision farming, and compensation schemes for farmers adopting sustainable practices are increasingly being explored as ways to reconcile economic viability with environmental goals.

By finding ways to support sustainable agricultural practices while also promoting economic opportunity and social equity, Europe can chart a path towards a more resilient and prosperous future for all its citizens.

Chapter 26
Economic Challenges and Government Programs

The cost of higher education has become a major concern for many young people in the United States, particularly in high-cost areas like California. With tuition rates at many colleges and universities continuing to rise faster than inflation, many students are forced to take on significant amounts of debt in order to pursue their educational goals.

According to the College Board, average annual tuition and fees for a private four-year university in the U.S. was over $39,000 in 2023, with public in-state tuition averaging around $10,940.

This was not always the case, however. In previous generations, it was much more common for students to be able to work their way through college or to rely on a combination of scholarships, grants, and family support to finance their education. Today, however, the cost of tuition, books, and living expenses has become so high that many students have no choice but to take out loans. As of 2023, more than 43 million Americans collectively owed over $1.6 trillion in federal student loan debt, with the average borrower carrying about $37,000.

For those who are able to secure scholarships or have the support of their employers, the burden of student debt can be significantly reduced. This was the case for me personally, as I was fortunate to receive scholarships and grants for my undergraduate studies at UC Berkeley and to have my employer, DuPont, cover the cost of my MBA program at Santa Clara University.

The experience of working at DuPont during the day and attending classes at Santa Clara in the evening was incredibly enriching. I was able to apply the concepts and skills I was learning in the classroom directly to my work and vice versa. This synergy between theory and practice was both exciting and transformative, and I feel incredibly grateful to have had this opportunity.

I recognize that my experience is not the norm for most students today. With the cost of tuition and living expenses at Santa Clara University now exceeding $90,000 per year, it is simply out of reach for many students and their families.

This figure includes tuition, fees, room and board, books, supplies, and miscellaneous expenses, reflecting a broader trend of rapidly increasing college costs nationwide. Even at more affordable institutions, the cost of higher education has become a major barrier to entry for many young people.

This is particularly true for members of Generation Z, who are entering adulthood in an era of unprecedented economic uncertainty. Unlike their parents and grandparents, who could often count on stable employment and a clear path to middle-class prosperity, many young people today face a job market that is increasingly precarious and unpredictable. In 2024, the U.S. Bureau of Labor Statistics reported that nearly 40% of recent college graduates were underemployed, working in jobs that did not require a degree.

The rise of automation and artificial intelligence is likely to exacerbate these trends in the coming years, as many jobs that were once considered secure are now at risk of being displaced by machines.

A 2023 McKinsey Global Institute report estimated that by 2030, up to 30% of current work activities could be automated, particularly in sectors like manufacturing, logistics, and even customer service. This is a daunting prospect for many young people, who may feel that they have no choice but to take on significant amounts of debt in order to acquire the skills and credentials they need to compete in an increasingly crowded and demanding job market.

At the same time, the COVID-19 pandemic has exposed the fragility of many of our social and economic systems, from healthcare and education to food production and supply chains. The potential for future shocks whether from climate change, geopolitical instability, or public health crises only adds to the sense of uncertainty and precarity that many young people feel.

Precarity is a state of persistent insecurity with regard to employment or income. The World Economic Forum's Global Risks Report 2024 identified cost-of-living pressures, global debt, and geopolitical fragmentation as top near-term risks, especially for younger generations.

In light of these challenges, I believe that it is more important than ever for young people to come together and support one another in finding solutions and building resilience. This may involve forming new alliances and networks, both within and across generational lines, in order to share knowledge, resources, and strategies for navigating an increasingly complex and rapidly changing world. Grassroots movements, mutual aid networks, and digital platforms such as Reddit's r/PersonalFinance and Gen Z-focused labor unions are examples of how younger generations are already mobilizing.

It may also require a fundamental rethinking of many of the assumptions and institutions that have shaped our society for generations, from the way we educate our children to the way we organize our economy and govern ourselves. This is no small task, but I believe that it is one that is both necessary and urgent if we are to build a future that is more just, sustainable, and equitable for all.

Of course, these challenges are not limited to the United States alone. Around the world, many countries are grappling with similar issues of rising inequality, economic insecurity, and social upheaval. In Europe, for example, the European Union is facing a range of challenges, from the fallout of Brexit to the rise of populist and nationalist movements in many member states. Far-right parties gained significant traction in multiple EU countries in 2023–2024, with notable electoral successes in Italy, Sweden, and the Netherlands.

At the same time, the global economy is undergoing a period of profound transformation, driven by rapid technological change, shifting geopolitical alliances, and the increasing urgency of climate action. In this context, it is clear that no single country or region can address these challenges alone. Instead, what is needed is a new spirit of global cooperation and solidarity, one that recognizes our shared destiny as a species and our collective responsibility to build a better future for all.

The European Union (EU) is a supranational political and economic union of 27 members with a population of over 448 million in 2023, accounting for 5.6% of the world's population of 8.1 billion in 2024.

The EU's nominal gross domestic product (GDP) was around US$18.35 trillion in 2023, constituting approximately one-sixth of global nominal GDP, and it is the second largest economy after the United States. According to IMF projections, the European Union's GDP was $19.35 trillion in 2024, with Germany, France, and Italy comprising the top three largest economies within the bloc.

The US gross domestic product (GDP) in 2023 was approximately $27.35 trillion, maintaining its position as the world's largest economy. According to EY Parthenon as of December 13, 2024, real GDP growth in the U.S. is expected to decelerate modestly from 2.7% in 2024 to 2.1% in 2025 and 1.7% in 2026, as increased tariffs and tightening financial conditions begin to weigh more heavily on economic activity. In 2024, the US GDP was approximately 29.18 trillion.

Japan, once the third-largest economy in the world, entered a technical recession as of February 15, 2024, defined as two consecutive quarters of negative GDP growth. The Japanese economy shrank by an annualized 0.4% to $4.2 trillion in Q4 2023, following a 3.3% contraction in the prior quarter. In 2024, Japan's GDP was estimated at $4.196 trillion using current market prices. This figure represents a 1.1% growth rate from the previous year, although long-term demographic decline and low domestic demand remain structural challenges.

Germany, the largest economy in Europe, reported a 3.3% GDP contraction in the final quarter of 2023. With a GDP of $4.5 trillion, the nation is technically in a recession. The German economy contracted by 0.2% in 2024 mainly due to reduced industrial output and weak exports.

In the United Kingdom, the Office for National Statistics confirmed that the economy slipped into a recession. GDP dropped 0.3% in Q4 2023, following a 0.1% contraction in Q3, bringing the total GDP to $3.13 trillion. The UK also faces widening economic inequality, with 1 in 5 citizens approximately 14 million people living below the poverty line. In 2024, the UK's economy grew by 1.1% as inflation eases and household spending rebounds slightly.

China's economy expanded by 5.2% in 2023, reaching RMB 126.06 trillion (approximately US$17.89 trillion), according to the National Bureau of Statistics. Leading financial institutions such as Goldman Sachs and Morgan Stanley forecast a 4.6% GDP growth rate in 2024. On December 14, 2024, Reuters reported China's central economic leadership aims to maintain around 5% growth, despite concerns about the real estate market, local government debt, and export competition. In 2024, China's GDP reached approximately $18.6 trillion, with a 5% year-over-year growth.

The United States provides the world's largest subsidies across various sectors including energy, agriculture, and transportation. In agriculture alone, the USDA operates over 150 programs that provide support through direct payments, insurance subsidies, and marketing assistance primarily benefiting large-scale producers of corn, soybeans, wheat, cotton, and rice.

Most of these subsidized crops are genetically modified organisms (GMOs). One such example is Bt-Corn, a crop genetically engineered to include Bt delta endotoxin from the Bacillus thuringiensis bacterium, which targets Lepidoptera larvae, notably the European corn borer. The widespread use of GM crops like herbicide-resistant corn and soy in North America has led to habitat loss for species like the monarch butterfly, due to increased herbicide use.

Bt-Corn is not only engineered for pest resistance, but its derivatives are widely used in the production of high-fructose corn syrup (HFCS) a common sweetener in processed foods, soft drinks, and baked goods. Today, over

90% of U.S. corn, soybeans, and upland cotton are grown with genetically modified seeds, despite significant global opposition. Twenty-six countries including France, Germany, Italy, Mexico, Russia, China, and India have either restricted or banned GMO cultivation due to environmental and health concerns.

The USDA introduced the Food Guide Pyramid in 1992, encouraging Americans to consume more grains, fruits, and vegetables (at the base of the pyramid) and fewer fats and sugars (at the top). However, nutrition experts have criticized the model for promoting excessive carbohydrate intake while downplaying healthy fats and proteins. This has contributed to a rise in lifestyle-related diseases such as obesity, type 2 diabetes, hypertension, and cardiovascular disease.

Moreover, a large portion of processed food in U.S. supermarkets is made using heavily subsidized crops, particularly corn (used in HFCS) and soy (used in hydrogenated oils). These products often contain refined grains and excessive sodium, which are linked to poor public health outcomes.

As a result, there is an alarmingly high prevalence of prediabetes, diabetes, insulin resistance, obesity, and non-alcoholic fatty liver disease, as well as an increased risk of certain cancers and Alzheimer's disease in America. According to the CDC, over 96 million American adults more than 1 in 3 have prediabetes, and about 38 million have diabetes. Conditions like obesity and NAFLD (non-alcoholic fatty liver disease) are strongly linked to excessive sugar intake, particularly from added sugars like high fructose corn syrup.

After Hurricane Allen destroyed the Caribbean sugar crop in 1980, there was a shortage of sugar nationwide, and most soft drink companies switched from sucrose to HFCS, which is cheaper to use and is locally produced. This shift was also incentivized by abundant corn subsidies, which made HFCS more economically viable than imported sugar.

The United States grows far more corn than any other country. One farmer can cultivate thousands of acres at a time and is encouraged by the federal government to do so by providing subsidies, which ensure that prices remain low and production remains high. As of 2024, the U.S. produced over 15 billion bushels of corn, nearly 35% of global production, much of it through highly mechanized industrial farming systems.

One-third of the corn grown in the United States is used for animal feed, another third is used to produce ethanol, and the remainder enters the food supply, some of which is HFCS. Corn is also used in bio-based plastics, pharmaceuticals, and as a key ingredient in processed foods, making it one of the most versatile yet heavily commodified crops in the U.S. economy.

An analysis of over 6,000 peer-reviewed studies published in Scientific Reports covering 21 years of data found that GMO corn increased yields by as much as 25% in the United States, Europe, South America, Asia, Africa, and Australia from 1996, the first time GMO corn was planted, through 2016. The report also noted that GMO corn reduced mycotoxin levels, enhanced farmer profitability, and decreased the need for chemical pesticides.

HFCS contains slightly more fructose than glucose compared to table sugar. It is glucose that is the body's primary fuel. It is absorbed through the small intestine directly into the bloodstream and converted into ATP, chemical energy within all cells. Glucose also stimulates insulin, whereas only the liver processes fructose. Excessive fructose consumption is associated with metabolic dysfunction because it bypasses key regulatory steps that glucose undergoes, potentially overwhelming liver function.

A surplus of fructose can result in insulin resistance and metabolic syndrome, as well as non-alcoholic fatty liver disease. Processing excess fructose may deplete ATP, resulting in oxidative stress and inflammation, as well as causing fat to be made and stored. Studies in journals such as Nature Reviews Endocrinology confirm the connection between high fructose intake and increased liver fat, especially in the absence of physical activity or fiber intake.

In fruit, fiber limits fructose absorption and protects the liver, while other beneficial nutrients aid in the metabolism of fructose. This is why whole fruits are considered healthy despite containing fructose fiber, antioxidants, and polyphenols balance the metabolic impact.

In addition to reducing healthy gut bacteria, HFCS consumption leads to increased heart disease, cancer, and Alzheimer's disease. High fructose consumption may lead to hypertension, insulin resistance, and fatty liver disease. A 2020 study published in Nutrients linked chronic HFCS consumption to impaired gut microbiota diversity and elevated inflammation markers.

Government lobbying campaigns and industry-funded messaging campaigns have made it difficult to pass food policies aimed at reducing added sugars, including HFCS. It is up to you to stop excess sugar consumption by limiting your intake of sugary drinks and processed foods. The Sugar Association and major food lobbies have historically influenced regulatory guidelines, as documented in internal memos and congressional testimonies.

As a consumer, you do not have a great deal of control over the amount of sugar added to food, but you do have considerable control over what you choose to buy and consume. It is estimated that more than 3.1 billion people worldwide are unable to afford a healthy diet. This is supported by a 2021 FAO report, which stated that healthy diets are five times more expensive than diets that meet only basic caloric needs.

Approximately 42% of American adults are obese, and more than two-thirds of American adults are overweight or obese. Obesity contributes to at least 13 types of cancer, according to the American Cancer Society, and it increases the risk of severe illness from infectious diseases such as COVID-19.

Despite spending $38 billion a year subsidizing meat and dairy industries, the U.S. government only spends $17 million, 0.04%, subsidizing fruits and vegetables. A total of $2.2 billion was provided to corn growers as product-specific assistance. HFCS is manufactured from corn. This imbalance in subsidy distribution has been criticized by nutrition experts for incentivizing calorie-dense, nutrient-poor food production.

There are approximately $20 billion in subsidies paid to the fossil fuel industry each year by the US government, nine times more than corn subsidies. Additionally, large oil companies in the US pay significantly lower tax rates than most other corporations. Energy companies are able to defer and avoid federal income tax payments because of provisions in the U.S. tax code that allow them to do so. According to the International Monetary Fund (IMF), direct and indirect fossil fuel subsidies in the U.S. amounted to

$760 billion in 2022 when environmental and health costs are factored in.

While gasoline is widely used in many applications today, it will eventually become obsolete due to petroleum's nonrenewable nature. Current technology focuses on maximizing the remaining petroleum reserves and exploring alternative sources of energy. To fully utilize the oil reserves available today, new methods of accurately determining the extent of oil reservoirs, automated systems for controlling oil recovery, and ways of enabling workers to recover more oil from known reservoirs are all being investigated. Enhanced oil recovery techniques such as CO- injection and digital oilfield technologies are already being deployed to increase efficiency.

Because gasoline is produced from a limited supply of petroleum, scientists are looking for clean, renewable sources of energy to power machines of the future. A power source that was used in steamboats in the past, steam power is being revived today. Electric vehicles have been developed, and solar and wind energy are also used to power cars and homes. Additionally, advances in hydrogen fuel cells and energy storage technology are accelerating the transition away from fossil fuels.

Resolving all the economic challenges will not be an easy task, and there will undoubtedly be many obstacles and setbacks along the way. But I remain hopeful that, by working together and leveraging our collective intelligence and creativity, we can rise to the challenge and create a world that is more just, sustainable, and prosperous for all. The stakes could not be higher, and the time to act is now. Equitable food systems, cleaner energy investments, and informed consumer choices will be central to this transformation.

Chapter 26
Economic Challenges and
Government Programs

We all have 24 hours a day. What is your time worth? What can you do to get your fair share of land, shelter, and food by doing meaningful work? Start by educating yourself about the economic systems you participate in, support ethical businesses, advocate for transparent food labeling, and vote for policies that reflect your values. Change begins with what you consume and what you choose to support.

Chapter 27
Personal Reflections and Beliefs

As I reflect on my own life and the many challenges and opportunities I have encountered along the way, I am struck by the incredible diversity and complexity of the human experience. From my childhood in Burma to my career in the United States, I have had the privilege of meeting and learning from people from all walks of life, each with their own unique perspectives, beliefs, and values.

One of the things that I have come to appreciate most deeply is the power of human connection and the importance of building bridges across divides. Whether it is through shared meals, cultural exchange, or simply taking the time to listen and understand one another's stories, I believe that we have so much to gain by opening ourselves up to the experiences and insights of others.

In fact, research by the Harvard Study of Adult Development one of the longest-running studies on human well-being has shown that strong relationships are among the most consistent predictors of long-term happiness and health.

This is particularly true when it comes to matters of faith and belief. Growing up in a predominantly Buddhist country, I had the opportunity to learn about and appreciate the teachings of the Buddha from a young age. At the same time, I was exposed to a wide range of other religious traditions, from the animist beliefs of my ancestors to the teachings of Confucius and the values of Christianity.

Rather than seeing these different belief systems as contradictory or incompatible, I have come to see them, if possible, as mostly different facets of the same underlying truth the search for meaning, purpose, and connection in a complex and often confusing world.

According to comparative religion scholars such as Huston Smith and Karen Armstrong, many world religions share core ethical teachings such as compassion, humility, and the pursuit of truth, despite theological differences. Whether we call it God, Dharma, or simply the universe, I believe that there is a fundamental unity that underlies all of our diverse experiences and perspectives.

Of course, this does not mean that all beliefs are equally valid or that we should accept everything without question. As someone who values reason, evidence, and critical thinking, I believe that it is important to approach all claims and ideas with a healthy dose of skepticism and to be willing to change our minds in light of new information or arguments.

Jesus said, "I am the way, the truth, and the life." Just as climbers ascend Mount Everest in a single file at the top from the Nepal side, we must be sure to understand that what Jesus said is true. This verse, found in John 14:6, has been a cornerstone of Christian theology, expressing an exclusive truth claim central to Christian belief. However, it has also inspired deep theological dialogue about pluralism and interfaith understanding.

At the same time, I have come to recognize that there are many things in life that cannot be fully understood or explained through reason alone. Love, beauty, and the ineffable sense of wonder and awe that we feel in the face of the universe are all experiences that transcend the boundaries of logic and language.

Psychologists and neuroscientists, including Dr. Andrew Newberg, have studied these "mystical experiences" and found that spiritual or transcendent moments can activate parts of the brain associated with emotional insight, empathy, and creativity.

In this sense, I believe that faith and reason are not necessarily opposed but rather complementary aspects of the human experience. While science and rationality can help us understand the workings of the natural world and solve practical problems, it is through faith and intuition that we are able to connect with something greater than ourselves and to find meaning and purpose in our lives. Albert Einstein himself once noted that "science without religion is lame, religion without science is blind," reflecting the possibility of integration rather than opposition.

This is not to say that faith is always easy or straightforward. As someone who has grappled with questions of belief and meaning throughout my life, I know firsthand how challenging it can be to find a true sense of purpose and direction in a world that often seems chaotic and unpredictable.

But I have also come to believe that it is through this struggle through the willingness to ask difficult questions, to confront our own doubts and fears, and to remain open to new possibilities that we are able to grow and evolve as individuals and as a society.

In my own life, I have found that one of the most powerful ways to cultivate this sense of openness and curiosity is through travel and exposure to different cultures and ways of life. From the bustling streets of Tokyo to the ancient ruins of Rome, I have had the opportunity to see the world through the eyes of others and to gain a deeper appreciation for the richness and diversity of the human experience.

According to a 2010 study published in the Journal of Personality and Social Psychology, individuals who travel and live abroad develop greater cognitive flexibility, enhanced creativity, and a more nuanced understanding of cultural norms.

One of the most memorable experiences of my life was attending the World Expo in Vancouver with my family. Seeing the incredible diversity of cultures, innovations, and ideas on display was a powerful reminder of the incredible potential of human creativity and collaboration.

Expo 86, held in Vancouver, British Columbia, was attended by over 22 million people and featured 54 nations. Its theme, "World in Motion – World in Touch," emphasized technological innovation and global connectivity. It showcased developments in communication, transportation, and global culture leaving a lasting impression on a generation of attendees.

At the same time, I have also been deeply moved by the simple acts of kindness and generosity that I have witnessed in my own community and around the world. From the volunteers who give their time and energy to help those in need to the everyday heroes who stand up for what is right and just, I am constantly inspired by the power of ordinary people to make a difference in the world.

Studies in positive psychology, particularly the work of Dr. Martin Seligman and Dr. Sonja Lyubomirsky, show that acts of kindness not only benefit recipients but also significantly boost the mental and physical well-being of those who give.

Sections 3: Refocus Your Mind:
A Global Snapshot

As I look to the future, I am filled with both hope and concern. On one hand, I see incredible opportunities for progress and positive change driven by advances in science, technology, and human understanding. From the development of clean energy and sustainable agriculture to the eradication of poverty and disease, I believe that we have the knowledge and the tools to create a better world for all.

Initiatives such as the United Nations' 17 Sustainable Development Goals (SDGs) provide a global framework for tackling poverty, inequality, and climate change by 2030, highlighting how international collaboration and innovation can lead to tangible progress.

At the same time, I am deeply troubled by the many challenges and threats that we face as a global community. From the existential threat of climate change to the rise of authoritarianism and the erosion of democratic norms, I worry that we are losing sight of the values and principles that have guided human progress for generations.

According to the 2024 Intergovernmental Panel on Climate Change (IPCC) report, without urgent global action, average temperatures could surpass the 1.5°C threshold within the next two decades triggering severe environmental and humanitarian consequences. Simultaneously, organizations like Freedom House have documented a decline in global democratic freedoms for nearly two decades.

But I also know that we have faced similar challenges before and have found ways to overcome them through the power of human ingenuity, compassion, and resilience. Whether it is through the development of new technologies, the forging of new alliances and partnerships, or simply the willingness to stand up for what is right and just, I believe that we have the capacity to create a better future for ourselves and for generations to come. The successful global effort to eradicate smallpox by 1980 achieved through international cooperation and innovation is a powerful historical example of what humanity can accomplish when united by purpose.

Ultimately, I believe that the key to unlocking this potential lies in our ability to cultivate a sense of empathy, understanding, and connection with one another. By listening to each other's stories, learning from each other's experiences, and working together towards common goals, we can build a world that is more just, compassionate, and resilient.

Empathy, now increasingly recognized as a vital social skill by institutions like the Harvard Graduate School of Education, is foundational to resolving conflict and fostering cooperative societies, especially in diverse, interconnected communities.

This is not a task that any one person or group can accomplish alone. It will require the efforts and contributions of people from all walks of life, from all corners of the globe. But I am convinced that, by working together and remaining true to our highest values and aspirations, we can create a future that is worthy of the incredible potential of the human spirit.

As I reflect on my own journey and the many twists and turns that have brought me to where I am today, I am filled with a deep sense of gratitude for the people and experiences that have shaped my life. From my parents and family to my teachers, mentors, and friends, I have been blessed with an incredible network of support and guidance along the way.

I am also grateful for the many challenges and obstacles that I have faced, for it is through these struggles that I have learned and grown the most. Whether it was navigating the complexities of cultural differences, overcoming personal and professional setbacks, or simply learning to trust in my own strengths and abilities, each challenge has taught me valuable lessons and helped me to become a stronger, more resilient person.

Resilience, as defined in psychological studies led by Dr. Ann Masten, is not a rare trait but a common capacity that can be nurtured through meaningful relationships, positive outlooks, and adaptive coping strategies.

Looking ahead, I know that there will be many more challenges and opportunities to come. But I am excited to embrace them with an open mind and a willing heart, knowing that each new experience will bring with it the chance to learn, grow, and make a positive impact in the world.

And so, to all those who are reading these words, I offer this simple message: never stop learning, never stop growing, and never stop believing in the power of the human spirit to overcome even the greatest of obstacles. For it is through our shared humanity, our common hopes and dreams, and our unwavering commitment to building a better world that we will find the strength and the courage to create a future that is truly worthy of our highest aspirations.

Author and Holocaust survivor Viktor Frankl once wrote, "When we are no longer able to change a situation, we are challenged to change ourselves." This spirit of adaptation, purpose, and endurance is what I hope we can all carry forward into the future.

We are living in the most exciting and challenging time in human history. Focus your attention on the things that matter most to you. Nobody knows the equation for love, but everyone knows love when they have experienced it for themselves.

While love continues to defy formal definition, modern neuroscience through the work of Dr. Helen Fisher and others has mapped some of its biological underpinnings, linking it to neural pathways related to attachment, empathy, and reward. Yet its true meaning remains deeply personal and universally profound.

Thank you for taking the time to read and reflect on these words. I hope that they have offered some insight, inspiration, or simply a moment of connection in a world that can often feel overwhelming and uncertain. May we all find the wisdom, compassion, and resilience to navigate the challenges ahead and create a future that is filled with hope, joy, and endless possibilities.

Section 4: Climate Change

Chapter 28
The Fragile Earth

We all need a place to live. Earth is our only choice. We all need resources even if we're in a space station orbiting Earth; we still depend on Earth for energy, food, and oxygen supplies. Some places are nicer than others, but what impacts Earth as a whole affects us all.

The years 2023 and 2024 were the hottest on record, according to NASA and the European Union's Copernicus Climate Change Service. In October 2024, Tropical Storm Milton intensified into Category 5 Hurricane Milton in less than 24 hours. The Guardian states, "Milton is the third fastest-intensifying storm on record in the Atlantic, as experts warn the climate crisis is fueling more powerful storms."

Hurricane Helene is estimated to have caused between $160–180 billion in damage and economic loss, according to AccuWeather. The company also projected Hurricane Milton's estimated cost at $225–250 billion. These make Milton and Helene two of the costliest storms in U.S. history, second only to Hurricane Katrina and Hurricane Harvey, when adjusted for inflation.

Who are opposing urgent and important reasons to deal with climate change now? In 2024, thousands of farmers across Europe particularly in the Netherlands, Germany, and France protested proposed environmental regulations that they argued would make it impossible for them to operate. Tractors blocked highways and cities. Their concerns highlight the delicate balance between sustainability and food security. Without ensuring farmers can make a living, we risk destabilizing food supply chains a reminder that climate action must also consider economic and social realities.

Starting on January 7, 2025, California experienced the worst wildfires in its history. According to Wikipedia, the catastrophic fires have "affected the Los Angeles metropolitan area and surrounding regions." Fueled by record-low humidity, prolonged drought conditions, and hurricane-force Santa Ana winds exceeding 100 mph (160 km/h) in places, these fires devastated the state. As of January 31, 2025, the wildfires had killed at least 29 people, forced more than 200,000 to evacuate, and destroyed more than 18,000 homes and structures. In total, over 57,000 acres (23,000 hectares) of land were burned.

We are very lucky to live on Earth, the only planet suitable for human life. Even with rising interest in Mars and exoplanet exploration, no known planet offers the same atmospheric stability, liquid water, and magnetic shielding that Earth provides. In the future, we may find "Earth-like" planets far from our solar system but current technology would take thousands of years to reach them using conventional propulsion.

Our Moon stabilizes Earth's tides and climate systems. The tides carry heat across the planet from the equator to the poles, helping to moderate global temperatures. The Moon's gravity also stabilizes Earth's axial tilt, currently at 23.5 degrees. Without the Moon, Earth's tilt could swing wildly from 10 to 45 degrees, causing severe climate disruptions from no seasons at all to extreme seasons, even triggering ice ages.

The Moon keeps the same face toward Earth due to tidal locking a phenomenon called synchronous rotation. A total solar eclipse happens when the Moon perfectly blocks the

Sun's face. The Sun is about 400 times larger than the Moon but also 400 times farther away, making this celestial coincidence possible.

I don't think this is a coincidence. I think this is how God created the Sun and the Moon. As it says in Genesis 1:16, "And God made two great lights; the greater light to rule the day, and the lesser light to rule the night: he made the stars also." The precise distance between Earth and the Sun about 93 million miles (150 million kilometers) places us in the "Goldilocks Zone", where conditions are just right for liquid water and life.

Climate change divides people in so many different ways. Hopefully, common sense and compassion will win out. It's causing tragic losses of lives and resources. In 2023, Beijing experienced its worst flooding in 140 years, submerging roads, homes, and infrastructure. On October 29, 2024, torrential rain brought more than a year's worth of precipitation in just 24 hours to several areas in eastern Spain, making it one of the deadliest natural disasters in Spanish history, with dozens killed and thousands displaced.

The fires in Canada in 2023 have produced so much smoke that's causing health problems, even in the US, and will make climate change worse. According to the Associated Press, a record number of wildfires in 2023 forced more than 235,000 people across Canada to evacuate. These are real issues. In late July and August 2024, Canadians mourned as Jasper, the jewel of the Rockies, burned 80,860 acres (32,722 hectares).

The world is getting hotter and hotter because of climate change. UN Secretary Gutierrez called it "Global Boiling." Many great, ancient civilizations have abandoned their biggest and most important cities, fought wars, and become desolate because of climate change. What if New York City, London, Paris, Shanghai, Berlin, Tokyo, Beijing, San Francisco, Mexico City, and the Vatican were all abandoned due to climate change?

That's what happened to the most advanced cities and temples of the past, and they became myths for centuries, longer than the United States has been a country. The United States is the most powerful nation in the 20th century and at least part of the 21st century, but it is no match for severe climate change that could last for decades.

The effects of climate change are wide-ranging and complex, touching every part of life on Earth. Some of the biggest impacts include:

1. **Rising seas and coastal flooding:** As sea levels keep rising, low-lying coastal areas and islands are at risk of going underwater and eroding. This could force millions of people to move and cause big economic losses. It could also damage coastal roads, buildings, and ecosystems.
2. **More extreme weather:** Climate change is making heat waves, droughts, heavy rains, and other extreme weather events happen more often and with greater intensity. These events can lead to deaths, property damage, and problems with farming and water supplies. On July 2, 2024, Hurricane Beryl has become the Earliest Category 5 Storm in Recorded History.
3. **Changes in farming:** Rising temperatures, changing rain patterns, and extreme weather are affecting how well crops grow around the world. This could lead to food shortages, especially in poorer countries that rely heavily on farming. According to the United Nations' World Food Program, "Conflict, economic

shocks, climate extremes, and soaring fertilizer prices are combining to create a food crisis of unprecedented proportions."

4. **Loss of nature:** Climate change is making plants and animals shift where they live. It's also changing the timing of seasonal events like migration and breeding. Many species could go extinct as their habitats become unsuitable or broken up.

5. **Health problems:** Climate change poses big risks to human health. Extreme weather events can directly harm people. Changes in air and water quality, food and water availability, and the spread of diseases can also indirectly affect health.

6. **Economic impacts:** The costs of climate change could be huge, possibly reaching 1-4% of the world's GDP by 2100. These costs include damage from extreme weather, losses in farming, and the costs of adapting to a changing climate.

7. **Social and political instability:** Climate change can make existing social and economic inequalities worse. This could lead to more poverty, migration, and conflict, which could have big impacts on global security and stability.

Actions must be taken, but not everyone agrees, however, on the best way to tackle climate change. Some advocate urgently transitioning to renewable energy sources like solar and wind while protecting carbon-absorbing forests. Others worry that overly aggressive environmental policies could hamper industries and livelihoods; farmers, for example, fear new regulations might hinder agricultural productivity.

Some key strategies to address climate change include:

1. Switching to clean energy: Replacing fossil fuels with renewable energy sources like solar, wind, and

hydropower can greatly cut emissions from the energy sector. This will require a huge increase in renewable energy, as well as improvements in energy efficiency and storage technologies.

2. Reducing deforestation and planting more trees: Forests play a critical role in removing CO_2 from the atmosphere through photosynthesis. Cutting down fewer trees and planting more can help increase the amount of carbon stored in forests and soils.

3. Promoting sustainable agriculture and land use: Agriculture and land use changes are big sources of greenhouse gas emissions, especially from cutting down forests, raising livestock, and using synthetic fertilizers. Promoting sustainable farming practices, like agroforestry, conservation tillage, and precision agriculture, can help cut emissions and store more carbon in soils.

4. Improving transportation efficiency: The transportation sector is a major source of greenhouse gas emissions, especially from burning fossil fuels in vehicles. Making vehicles more efficient, promoting electric and hybrid vehicles, and investing in public transportation and active transportation (like walking and cycling) can help cut emissions from transportation.

5. Putting a price on carbon: Making people pay for their carbon emissions, either through a carbon tax or a cap-and-trade system, can create financial incentives for reducing emissions and investing in clean energy and other low-carbon technologies.

Strategies to prepare for climate change aim to reduce the vulnerability of people and nature to the impacts of a changing climate and build resilience. Some key strategies include:

1. **Making infrastructure stronger:** Investing in infrastructure that can withstand the impacts of climate change, like rising seas, extreme weather, and changes in temperature and rainfall, can help reduce the risks and costs of these impacts.

2. **Improving early warning systems and disaster preparedness:** Making weather and climate forecasts more accurate and timely and investing in disaster preparedness and response capabilities can help reduce the impacts of extreme weather events and other climate-related disasters.

3. **Promoting sustainable water management:** Climate change is expected to change the availability and quality of water in many regions through changes in rainfall, increased evaporation, and rising seas. Promoting sustainable water practices, like conservation, efficiency, and reuse, can help ensure there's enough water for people and nature.

4. **Supporting climate-resilient agriculture:** Developing and promoting farming practices that can withstand the impacts of climate change, like drought-resistant crops, efficient irrigation, and diversified farming, can help ensure food security and livelihoods as the climate changes.

5. **Protecting and restoring ecosystems:** Healthy ecosystems, like forests, wetlands, and coral reefs, provide critical services that help buffer the impacts of climate change. They regulate water flows, protect against storm surges, and store carbon. Protecting and restoring these ecosystems can help build resilience to climate change.

Tackling the climate crisis will require an unprecedented level of international cooperation and collaboration. Climate change is a global problem that crosses national borders and needs a coordinated response from all countries.

The science is clear: global greenhouse gas emissions must fall by 43% from 2019 levels by 2030 to keep warming below 1.5°C, according to the Intergovernmental Panel on Climate Change (IPCC).

The United Nations Framework Convention on Climate Change (UNFCCC) is the main international treaty for addressing climate change. It was set up in 1992 with the ultimate goal of stabilizing greenhouse gas concentrations in the atmosphere at a level that would prevent dangerous human interference with the climate system. Today, 198 parties 197 countries and the European Union have agreed to the UNFCCC, making it one of the most widely accepted international agreements in history.

Under the UNFCCC, countries have negotiated a series of international agreements and protocols to address climate change. These include the Kyoto Protocol (1997), the Copenhagen Accord (2009), and the Paris Agreement (2015), which was adopted by 196 countries in 2015 with the aim of strengthening the global response to the threat of climate change.

The Kyoto Protocol was the first legally binding agreement that committed developed countries to emission reduction targets, but it faced criticism for excluding developing nations and for limited enforcement mechanisms.

The Paris Agreement seeks to keep the global temperature rise this century well below 2°C above pre-industrial levels and to pursue efforts to limit the temperature increase even further to 1.5°C.

To achieve this, the Paris Agreement is built on a "bottom-up" approach where each country sets its own climate targets called Nationally Determined Contributions (NDCs) and updates them every five years to reflect increasing ambition.

The Paris Agreement requires all countries to submit NDCs outlining their plans to reduce greenhouse gas emissions and adapt to the impacts of climate change. It also sets up a framework for international cooperation on climate change, including provisions for financial assistance, technology transfer, and capacity building for developing countries.

The Green Climate Fund (GCF), established in 2010 under the UNFCCC, is a key vehicle for providing financial support to developing nations, with developed countries pledging to mobilize $100 billion per year by 2020 a target that has not yet been fully met. As of 2024, annual climate finance flows remain below the agreed amount, though progress has been made.

Despite the progress made under the UNFCCC and the Paris Agreement, the global response to climate change hasn't been enough to meet the scale and urgency of the challenge.

The 2023 UN Climate Change Synthesis Report confirms that current NDCs would lead to an estimated global warming of 2.5°C to 2.9°C by the end of the century well beyond the 1.5°C threshold.

Much more ambitious action is needed to avoid the worst impacts of climate change. These include more extreme heatwaves, food and water insecurity, sea level rise threatening coastal cities, mass displacement, and ecosystem collapse. Scientists emphasize that every fraction of a degree matters and that rapid, deep, and sustained emissions cuts are still possible with strong political will, technological innovation, and global cooperation.

Strengthening the global response to climate change will require a range of actions, including:

1. **Increasing the ambition of NDCs:** Countries need to make their NDCs stronger to align with the goals of the Paris Agreement and the latest scientific assessments. This will require a big increase in the pace and scale of emissions cuts. It will also require a rapid shift to clean energy and sustainable land use practices.

2. **Enhancing international cooperation on climate finance:** Developed countries need to fulfill their promises to provide financial help to developing countries for climate action and adaptation. They also need to mobilize additional resources from private and public sources. This will be critical for enabling developing countries to transition to low-carbon development pathways and build resilience to climate change impacts.

3. **Promoting technology transfer and capacity building:** Enhancing the transfer of clean energy and other low-carbon technologies from developed to developing countries is crucial. So, the capacity of developing countries to adopt and use these technologies is being built. This will be essential for speeding up the global transition to a low-carbon future.

4. **Strengthening the role of non-state actors:** Engaging non-state actors, like cities, businesses, and civil society groups, in the global response to climate change can help accelerate action and mobilize more resources. Many non-state actors are already taking ambitious actions to cut emissions and build resilience. Their efforts need to be supported and scaled up.

5. **Enhancing transparency and accountability:** Strengthening the transparency and accountability parts of the Paris Agreement, including the enhanced transparency framework and the global stocktake, can help build trust and confidence between countries. It can also ensure that all countries are doing their fair share to address climate change.

The climate crisis is one of the most urgent and complex challenges facing humanity today. The science is clear that human activities, especially burning fossil fuels and cutting down forests, are causing the Earth's climate to warm at a rate never seen before. According to the Intergovernmental Panel on Climate Change (IPCC), human influence has unequivocally warmed the atmosphere, ocean, and land, with global surface temperatures having increased by approximately 1.2°C since pre-industrial times. This has far-reaching consequences for people and nature.

Tackling the climate crisis will need a huge global effort to cut greenhouse gas emissions, switch to clean energy, and prepare for the impacts of a changing climate. This will require a level of international cooperation and collaboration never seen before. It will also involve a rapid transition away from coal, oil, and gas, which account for over 75% of global greenhouse gas emissions. It will also require the engagement of all parts of society, including governments, businesses, civil society organizations, and individuals.

The costs of not acting on climate change are likely to be far greater than the costs of acting. A 2023 report by the Swiss Re Institute estimates that climate change could wipe up to 18% off global GDP by 2050 if no action is taken. The impacts of climate change are already being felt around the world, and they're expected to get worse in the coming decades. We are seeing increased frequency and severity of extreme weather events, such as heatwaves, wildfires, droughts, floods, and storms. The World Meteorological Organization (WMO) reported that 2023 was the hottest year on record, and 2024 continued the trend. These effects could have devastating consequences for human health, food security, economic development, and social stability.

Investing in climate action now can help avoid these costs and build a more sustainable and resilient future for all. The International Renewable Energy Agency (IRENA) reports that every dollar invested in renewable energy generates up to three dollars in economic benefits. Ultimately, addressing the climate crisis will require a fundamental transformation of our societies and economies toward a low-carbon and climate-resilient future. This transformation won't be easy, but it's necessary and urgent. Delaying action will only increase the risks and costs. It will require leadership, innovation, and collaboration at all levels of society, from local communities to the global community.

Chapter 29
Ancient Wonders and Climate Perils

The fall of great civilizations due to climate change serves as a stark warning for our modern world. Let's delve deeper into some of these historical examples and draw parallels to our current situation. The collapse of the Maya civilization is particularly relevant to our modern climate crisis. The Maya inhabited the Yucatan Peninsula and surrounding areas, creating a sophisticated society with advanced knowledge in mathematics, astronomy, and architecture. However, their civilization began to decline around 800 CE, with many of their great cities abandoned by 900 CE.

Recent studies using climate proxies such as lake sediments and cave deposits have revealed that the Maya collapse coincided with a series of severe and prolonged droughts. According to a 2018 study published in Science, the region experienced a 41% reduction in annual rainfall and up to a 70% decrease during peak drought years. These droughts, likely caused by changes in the El Niño Southern Oscillation and shifts in the Intertropical Convergence Zone, severely impacted the Maya's agricultural system.

The Maya had developed an intricate system of water management, including reservoirs and canals, to support their large population in a region with seasonal rainfall. However, this system was vulnerable to extended periods of drought. As water became scarce, crop yields declined, leading to food shortages, social unrest, and, eventually, the abandonment of major urban centers. Archaeological evidence also suggests increased warfare and political fragmentation during the terminal Classic period, which compounded the effects of environmental stress.

The Maya example demonstrates how even advanced societies can be vulnerable to climate change, especially when their infrastructure and social systems are not adaptable enough to cope with significant environmental shifts. This serves as a cautionary tale for our modern world, where many of our agricultural and urban systems are similarly dependent on stable climate conditions. In particular, many modern megacities rely on centralized water systems and monoculture agriculture, both of which are susceptible to droughts, floods, and temperature extremes.

The fall of Angkor in Cambodia provides another compelling example of how climate change can topple even the most sophisticated civilizations. Angkor was the capital of the Khmer Empire from the 9th to the 15th centuries and was one of the largest pre-industrial urban centers in the world. The city's intricate water management system, which included a network of canals, reservoirs, and embankments, was crucial for both agriculture and daily life.

Recent research using tree ring data and aerial imaging has revealed that Angkor experienced a series of intense monsoons followed by prolonged droughts in the 14th and 15th centuries.

A 2009 study published in the Proceedings of the National Academy of Sciences confirms that severe hydrological instability contributed to Angkor's decline, with infrastructure overwhelmed by floodwaters and subsequent dry periods that left reservoirs empty. These extreme weather events put immense stress on the city's water infrastructure.

The intense monsoons likely caused damage to the water management system, while the subsequent droughts made it difficult to maintain adequate water supplies for the large population. The environmental stress was accompanied by political fragmentation and invasions, which together led to Angkor's eventual abandonment.

This example highlights how climate change can exacerbate existing societal pressures and trigger cascading effects that lead to societal collapse. Scholars point to shifts in regional trade routes toward coastal cities, such as Ayutthaya in present-day Thailand, as having further weakened Angkor's economic position. Combined with internal revolts and declining royal authority, these pressures overwhelmed the city's resilience.

Angkor Wat City in Cambodia, the biggest population center in the world, was ahead of London, Paris, and many other cities. At its peak in the 12th century, Angkor may have housed over a million people, making it the largest urban center of the pre-industrial world, centuries before European cities reached similar sizes.

The Papal Basilica of Saint Peter in Vatican City is 5.7 acres. The Angkor Wat temple is 402 acres, the largest religious complex in the world. It is even larger than Egypt's Karnak Temple, a 247-acre complex, and it aligns with cosmic signs. Scholars have confirmed that Angkor Wat was constructed with astronomical precision, aligning with the spring equinox and solstices, possibly reflecting Hindu cosmological beliefs and the Khmer Empire's understanding of celestial movements. Angkor Wat was listed as a UNESCO World Heritage Site in 1992, which encouraged an international effort to save the complex. Conservation initiatives have since involved organizations from India, France, Japan, and other countries, focusing on preserving the stonework, managing tourism, and stabilizing the surrounding landscape.

Experts say King Suryavarman II, one of the most famous kings of the Angkorian civilization, built Angkor Wat in the early 12th century. The area is filled with over 1,000 temples that make up the Angkor Empire. Angkor had more than 1,000,000 people.

When I was at Angkor Wat, I got up early to watch the sunrise and ascend across the tip of the temple complex during a spring equinox. The temple tip was right in the center of the sun. The astronomical alignment was an amazing example of Angkor's technical prowess.

Angkor Wat was built as a Hindu temple with the central building representing Mount Meru. Historical evidence shows that the Hindu religion was practiced in many parts of the world in ancient times. It is the oldest religion in the world, and its believers seem to have had advanced technological knowledge in the distant past.

Hinduism, with origins dating back over 4,000 years, is widely regarded as the world's oldest major religion still practiced today. While ancient texts like the Vedas contain complex cosmological and mathematical insights, current interpretations of "advanced technology" are speculative and often symbolic rather than literal.

At Angkor Wat, Theravada Buddhist monks now maintain the temple. It remains a major pilgrimage site and attracts over 2 million tourists each year. Despite its religious and cultural importance, the massive influx of tourists poses conservation challenges, including structural wear, erosion, and strain on local resources. UNESCO and APSARA (Authority for the Protection and Management of Angkor and the Region of Siem Reap) continue to oversee protective measures.

The ancient tomb of China's first emperor, Qin Shi Huang, in Xi'An, continues to remain shrouded in mystery as archaeologists remain hesitant to open it. While various parts of the necropolis, such as the Terra Cotta Army, have been explored, the tomb itself, inside a large pyramid, has been left untouched due to the potential dangers that might lie within.

Chinese historians and archaeologists cite the presence of mercury and unknown structural hazards as reasons for not excavating the central burial chamber, based on descriptions from ancient texts like the Records of the Grand Historian by Sima Qian.

This emperor also built the Great Wall of China. Most people are unaware that some parts of the Great Wall have secret doors that lead to large chambers where soldiers used them as forts. These garrisons, known as "beacons" or watchtowers, were used for defense, communication, and storage. While not secret, many of these structures were strategically hidden from invaders and served as logistical hubs along the 13,000-mile wall.

When I visited the emperor's mausoleum complex, discovered in 1974, I learned that he unified the country by standardizing currency, language, and law after he conquered various kingdoms. Qin Shi Huang's reforms laid the foundation for a centralized imperial system that endured in various forms for over two millennia. His unification included standardizing weights, measures, axle widths for carts, and even writing script across the empire.

It took 39 years for 700,000 men from all over his empire to build the 20-square-mile (50-square-km) mausoleum. I found it amazing that such a massive undertaking would be forgotten for almost two thousand years. The scale and secrecy surrounding the project likely contributed to its burial under layers of earth and vegetation, preserving it from looting and erosion.

There seemed to be limited environmental damage due to their construction work. Ancient Chinese projects, including the mausoleum and early segments of the Great Wall, were constructed primarily with rammed earth and labor-intensive methods. While they did alter landscapes, these methods were far less carbon-intensive than modern industrial development.

Today, China faces the challenge of being the largest emitter of carbon dioxide gas in the world, with approximately 12.6 billion metric tons of carbon dioxide emissions in 2023 and 2024. This accounts for about 30% of global emissions. China has pledged to peak emissions by 2030 and achieve carbon neutrality by 2060, though achieving these targets will require rapid shifts in energy policy and industrial practices.

Approximately 12,000 years ago, modern humans built sophisticated stone structures and dwellings. There are about a dozen early sites in modern-day Türkiye, including Göbekli Tepe and Karahan Tepe. These Neolithic sites, often called "pre-pottery" settlements, feature complex megalithic structures and symbolic carvings that predate Stonehenge and the Egyptian pyramids. These people may have moved 550 miles southeast from Göbekli Tepe to Eridu, revered as the oldest city in Sumer.

According to the Sumerian King List, etched in stone in ancient times, people built large cities in Mesopotamia, the land between two rivers the Tigris and Euphrates. The Euphrates River is traditionally identified by some biblical scholars as one of the four rivers that flowed from the Garden of Eden, as described in Genesis 2:10–14.

Eridu is 12.5 miles southwest of Ur, where God told Abraham to leave for the Promised Land. It was thought that the gods created Eridu around 5400 BCE and established order on Earth from there. The Sumerians date back to around 5,000 BCE at the earliest. Göbekli Tepe shows signs of human activity dating back to 9600 BCE or earlier, making it one of the oldest known monumental temple complexes in the world. No one knows why Göbekli Tepe was intentionally buried and abandoned. Recent theories suggest it may have been decommissioned as societal needs changed, possibly due to the development of agriculture.

Megalithic activities, such as those at Göbekli Tepe, were widespread in now-modern Türkiye. After the massive flood, Noah and his family landed on Mount Ararat in Türkiye. Mount Ararat, located in eastern Türkiye, is traditionally believed by many to be the resting place of Noah's Ark, as described in Genesis 8:4.

The seven churches God reprimands in Chapters 1–3 of Revelation are in Türkiye. These include Ephesus, Smyrna, Pergamum, Thyatira, Sardis, Philadelphia, and Laodicea real cities located in Asia Minor (now western Türkiye), addressed by the Apostle John. We don't know for sure, but Türkiye was a potential location for Eden. Some theologians have speculated about Eden's location being in Mesopotamia or surrounding regions, but its precise location remains unknown and symbolic in nature.

Saint John wrote Revelation in the Bible on the island of Patmos, near Türkiye. The seventh empire in the Book of Revelation was the Ottoman Empire based in Türkiye. Many Christian eschatological interpretations view the Ottoman

Empire, which ruled from 1299 to 1922 and was centered in Constantinople (now Istanbul), as fulfilling the prophecy of the "seventh head" in Revelation 17.

Mohenjo-daro is an archaeological site in Pakistan. Built c. 2500 BCE, it was the largest settlement of the ancient Indus Valley Civilization. Mohenjo-daro means "Mound of the Dead Men." They called it that name because dead men were discovered on the mound. The city was advanced, with grid-like streets, drainage systems, public baths, and multi-storied buildings evidence of urban planning comparable to later civilizations.

Some say the city looked like Nagasaki after its decimation with a nuclear bomb in World War II. The destruction of Mohenjo-daro, whether because of climate change or a nuclear bomb, was a cautionary tale. However, the nuclear theory is largely speculative and not supported by mainstream archaeology. Most scholars believe Mohenjo-daro declined due to changing river patterns, droughts, or gradual societal breakdown.

Some of the skeletons at Mohenjo-daro were holding hands when they died suddenly, and others were lying dead on the stairs. Some stones turned to glass due to heat coming from a specific direction.

While vitrified materials have been reported at the site, most archaeologists attribute this to natural fires or erosion rather than evidence of nuclear blasts. Only about forty skeletons were discovered in the city, which had about forty thousand residents. People were lying, unburied, in the streets of the city. It appeared that almost all the residents similarly evacuated the city **as if in the face of a catastrophe **but a few remained.

Borobudur Temple Compounds, a UNESCO World Heritage Site, is the largest and one of the most remarkable Buddhist temples in the world. It ranks with Bagan in Myanmar and Angkor Wat in Cambodia as one of the finest archaeological sites in Southeast Asia.

It shares the Lord Buddha's teachings, life, and wisdom. Siddhartha Gautama, the Lord Buddha, was born in Lumbini, present-day Nepal, around 563 BCE, roughly five centuries before Jesus Christ.

In 2024, there are about 520 million Buddhists, comprising seven percent of the global population. Borobudur lay hidden for centuries under volcanic ash and jungle growth. No one knows why this wonderful Indonesian site was abandoned. It is believed that volcanic eruptions from Mount Merapi, combined with political changes and the decline of Buddhism in Java, led to its abandonment in the 14th century. For Buddhists, it was equivalent to Catholics abandoning the Vatican.

Constructed of gray andesite-like stones in the 9th century, the Borobudur temple consists of nine stacked platforms six square and three circular topped by a central dome. It is decorated with 2,672 relief panels and originally 504 Buddha statues. The central dome is surrounded by 72 perforated stupas, each with a Buddha statue seated inside.

It has one of the world's most extensive collections of Buddhist reliefs and was rediscovered in 1814 by English Lieutenant Governor Thomas Stamford Raffles. A team of Dutch archaeologists restored the site in 1907–1911. Borobudur remains popular with pilgrims. Today, it hosts major Buddhist celebrations such as Vesak, attracting thousands of practitioners from around the world.

From a global perspective, scientists believe that modern humans migrated from Africa to Asia between 80,000 and 60,000 years ago. Genetic and archaeological evidence supports this Out-of-Africa theory, tracing the spread of Homo sapiens into Eurasia.

By 45,000 years ago or possibly earlier modern humans had reached Southeast Asia and settled in regions including Indonesia, Papua New Guinea, and Australia. Tools, cave art, and skeletal remains found in sites like Liang Bua (home of the 'Hobbit' species Homo floresiensis) support this migration timeline.

In her book, The Indigenous Paleolithic of the Western Hemisphere, archaeologist Paulette Steeves argues that the settlement of the Americas may have occurred closer to 130,000 years ago (University of Nebraska Press/Algoma University). This theory challenges the long-held Clovis-first model and is based on archaeological, geological, and oral Indigenous histories.

Archaeologists have long believed that the first people to set foot on this continent arrived by crossing a land connection, the Bering Strait, from Siberia. This migration was believed to occur near the end of the last Ice Age, around 11,500 to 12,000 years ago, when lower sea levels exposed the Bering Land Bridge, or Beringia.

On October 5, 2023, a study led by a team of researchers at the U.S. Geological Survey (USGS) and co-authored by David Wahl, a UC Berkeley adjunct associate professor of geography and a USGS scientist specializing in pollen analysis, was published in the journal Science. It provided further evidence of pre-Clovis human presence in North America.

The study showed that pollen inside human footprints, found in alkali sand at White Sands National Park in New Mexico, dates from 23,000 to 21,000 years ago. These footprints, preserved in lakebed sediment, support earlier occupation than previously confirmed and reinforce the idea that humans were present during the Last Glacial Maximum.

Göbekli Tepe, dating back around 11,000 years, is located in Upper Mesopotamia, a region that saw the emergence of the most ancient farming communities in the world. Hunter-gatherers built monumental structures in the Pre-Pottery Neolithic period (10,000 - 9,000 BCE). This site rewrote the narrative of early civilization, showing that complex spiritual and social activity preceded agriculture.

There are other sites 1,000 years older than Göbekli Tepe within a few hundred miles. For example, Karahan Tepe, part of the "Taş Tepeler" (Stone Hills) network in southeastern Türkiye, may predate Göbekli Tepe, with evidence of human activity stretching back to 12,000 BCE. That means that several thousand years before Mesopotamia, sophisticated beings lived there in modern Turkiye.

Göbekli Tepe was created thousands of years before God created Adam and Eve. (According to Biblical chronology, Adam and Eve were created roughly 6,000 years ago, based on traditional interpretations of Genesis timelines. Göbekli Tepe, by contrast, predates this by about 5,000 years, posing questions for literalist interpretations of Genesis.) Some of Göbekli Tepe's megaliths, which can be up to 18 feet (5.5 meters) tall and weigh 50 tons (45 metric tons), are bare.

Others are covered with impressive carvings of abstract symbols and animals such as foxes, lions, bulls, snakes, vultures, and insects. Ancient people purposely and intentionally buried Göbekli Tepe under the soil. This intentional burial preserved the site exceptionally well, and its purpose remains one of archaeology's great mysteries.

About 10,000 years ago, humanity reached a turning point, transitioning from hunting and gathering to subsistence agriculture and animal husbandry with some hunting and gathering. This Neolithic Revolution occurred independently in several parts of the world, including the Fertile Crescent, China, and Mesoamerica.

The Younger Dryas was a return to glacial conditions, which temporarily reversed the gradual climatic warming after the Last Glacial Maximum. The Younger Dryas was the last stage of the Pleistocene epoch that spanned from 2,580,000 to 11,700 years ago. It lasted from about 12,900 to 11,700 years ago and may have been triggered by meltwater pulses disrupting oceanic circulation.

For the last 10,000 years, the temperature on Earth, as measured from the central region of Greenland's ice sheet, has been unusually stable around -30 degrees C. At present, it is around -32 degrees C. This Holocene stability provided the climatic conditions needed for agriculture and urban development.

According to ongoing temperature analysis led by scientists at NASA's Goddard Institute for Space Studies (GISS), the average global temperature on Earth has increased by at least 1.1° Celsius (1.9° Fahrenheit) since 1880. 2023 and 2024 were the planet's warmest years on record, according to an analysis by scientists from NOAA's National Centers for Environmental Information (NCEI). This rise is attributed to increased greenhouse gas emissions, mainly from fossil fuel consumption.

These historical examples offer valuable lessons for our modern world:

1. **Vulnerability of water systems:** Many ancient civilizations collapsed due to the failure of their water management systems in the face of climate change. Today, many regions face similar challenges with water scarcity and the need to manage water resources in a changing climate.

2. **The interconnectedness of environmental and social systems:** The downfall of these civilizations wasn't solely due to environmental factors but often involved a complex interplay between climate change, social structures, and political systems. This underscores the need for holistic approaches to addressing climate change that consider social and political factors alongside environmental ones.

3. **The importance of adaptability:** Civilizations that were unable to adapt to changing climatic conditions were more likely to collapse. This highlights the crucial need for modern societies to build resilience and adaptability into our systems and infrastructure.

4. **The potential for rapid collapse:** In many cases, the decline of these civilizations was relatively rapid once certain tipping points were reached. This serves as a warning that the impacts of climate change may not be gradual but could lead to sudden and dramatic societal changes.

5. **The long-term consequences of environmental degradation:** Many of these ancient sites remain uninhabited or only partially recovered today, thousands of years after their collapse. This illustrates the potential for climate change to have very long-lasting impacts on human habitation and land use.

As we face the challenge of global climate change today, these historical examples provide both warnings and lessons. They remind us of the fragility of human civilization in the face of significant environmental changes and the importance of building resilient, adaptable systems.

The collapses of societies like the Maya, the Indus Valley Civilization, and the Norse settlements in Greenland illustrate how prolonged droughts, shifting climate patterns, and resource depletion can destabilize even advanced cultures. However, unlike our ancestors, we have the advantage of scientific understanding and technological capabilities that allow us to predict and potentially mitigate the impacts of climate change.

Modern climate models, satellite monitoring, renewable energy technologies, and data analytics provide tools our predecessors never had. The key is whether we can learn from the past and take decisive action in time to prevent similar collapses in our modern world.

The urgency of our current situation cannot be overstated. The climate changes we are experiencing today are happening at a much faster rate than those that contributed to the downfall of ancient civilizations.

According to the Intergovernmental Panel on Climate Change (IPCC), global temperatures are rising faster than at any point in at least the last 2,000 years. Extreme weather events, such as heatwaves, droughts, and hurricanes, have increased in frequency and intensity due to anthropogenic greenhouse gas emissions. This rapid pace of change puts additional stress on our ability to adapt and respond effectively.

Moreover, the global interconnectedness of our modern world means that climate impacts in one region can have far-reaching consequences across the globe. This interconnectedness can be both vulnerability and strength – while it means that local climate disasters can have global repercussions, such as food shortages from droughts or disrupted supply chains from extreme weather, it also provides opportunities for global cooperation and resource sharing in addressing these challenges. For instance, international frameworks like the Paris Agreement aim to unify global efforts in reducing emissions and funding adaptation strategies, particularly for the most vulnerable nations.

As we strive to address the climate crisis, we must also consider the preservation of our own cultural heritage sites, many of which are already under threat from climate change. Rising sea levels, increased flooding, more frequent and intense storms, and changes in temperature and humidity are all putting stress on historical sites and artifacts worldwide.

UNESCO reports that iconic heritage sites such as Venice (Italy), the Statue of Liberty (USA), the Moai Statues of Rapa Nui (Easter Island), and the ruins of Chan Chan (Peru) are facing increasing threats from climate-related changes. Melting permafrost is also damaging archaeological sites in the Arctic, while desertification threatens those in North Africa and the Middle East.

From the ancient city of Venice, Italy, threatened by rising sea levels, to the permafrost-preserved archaeological sites in the Arctic that are melting due to warming temperatures, our cultural heritage faces unprecedented challenges. Preserving these sites is not just about maintaining tourist attractions – they are invaluable records of human history and achievement, offering insights into past societies and their relationships with the environment.

The lessons from ancient civilizations and the threats to our own historical sites underscore the profound connection between human societies and the environment. As we work to address climate change and build a sustainable future, we must keep these lessons in mind, striving to create a world that is not only environmentally sustainable but also preserves and honors the rich tapestry of human cultural heritage.

In facing these challenges, we have the opportunity to write a different ending to our story than that of the ancient civilizations we've discussed. With our scientific knowledge, technological capabilities, and global cooperation, we have the tools to address climate change and build a resilient, sustainable future. The question is whether we will heed the warnings of the past and take the necessary actions in time.

Here are some more examples of some of the largest cities destroyed by climate change in the past and some that survived. Art News said, "Egypt's world-famous antiquities, including the Sphinx and the Great Pyramids of Giza, could be lost to climate change by the end of the century, experts warn."

During the First Intermediate Period (c. 2181-2040 BCE), the Giza necropolis in Egypt, including the Great Pyramid, was abandoned and fell into decay during the Middle Kingdom (c. 2040-1782 BCE). Karnak Temple, 247 acres, the largest temple complex in the world, in Luxor, was also abandoned. The Valley of the Kings, where pharaohs were buried, was abandoned.

Chapter 29
Ancient Wonders and Climate Perils

Egyptologists say that the Great Pyramid was built 4,600 years ago for Pharaoh Khufu with 2.5 million large stones, some weighing 2.5 tons. The pyramid was built with its base perfectly aligned with the true north. It sits at the center of all landmasses on earth. It also stood as the tallest building in the world for thousands of years. Unfortunately, the smooth stones covering the three magnificent Giza pyramids were stripped to build Cairo.

Egyptian pharaohs believed that life after death was more significant than life itself on Earth. That's why they made a big deal of funeral preparations and burials. Ancient Egyptian funerary beliefs were inscribed in the later part of the Old Kingdom, around 2350–2160 BCE, on the walls of chambers inside pyramids.

These texts are crucial for guiding the pharaoh in the life after death. So, it is unimaginable that Pharaoh Khufu would be buried in the Great Pyramid without those texts. Perhaps those texts are in the unopened chamber along with his mummy if he is in the pyramid.

Egypt was the most powerful kingdom on Earth. No one could read the magnificent Egyptian hieroglyphs for two thousand years after the Romans took over Egypt. That is until a 31-year-old Frenchman, Jean-François Champollion, deciphered it using the trilingual Rosetta stone. French soldiers found the stone in Rosetta (modern El Rashid) during Napoleon's Egyptian invasion in July 1799. At Napoleon's defeat, the stone became British property. I saw the awesome Rosetta stone in the British Museum.

It was more impressive to see the Gaza pyramids from the air than from the ground. It was lucky for me to see Queen Nefertari's tomb alone with the guard in a different room. I was also alone in King Tutankhamun's tomb. These tombs are usually crowded. I had unimpeded viewing and enjoyed the moments because I visited those places at a time when tourists avoided Egypt.

Queen Nefertari was stunningly beautiful on her tomb walls. I was tempted to take pictures, but I didn't. I followed the sign prohibiting pictures. Queen Nefertari was the wife of Pharaoh Ramesses the Great, who was the third ruler of the Nineteenth Dynasty and is often regarded as the greatest, most celebrated, and most powerful pharaoh of the New Kingdom, which was the most powerful period of ancient Egypt. Her tomb, QV66, located in the Valley of the Queens, is often considered one of the most spectacularly decorated tombs in Egypt, with vivid wall paintings that have survived over 3,000 years. Nefertari was highly educated, known for her diplomatic correspondence, and played a prominent role in royal court life.

It was also interesting to see King Tut's tomb and put that image together with what I saw of hundreds of objects, including the famous golden mask, from that tomb on tour in San Francisco's De Young Museum. I didn't take pictures inside King Tut's tomb.

King Tutankhamun, who ruled during the 18th Dynasty, was a relatively minor pharaoh historically, but his nearly intact tomb, discovered in 1922 by Howard Carter, became one of the most significant archaeological finds due to its rich trove of artifacts. The golden mask, inlaid with lapis lazuli and other semiprecious stones, is now housed in the Egyptian Museum in Cairo.

I didn't know there was a terrorist attack in Egypt on November 17, 1997, that resulted in the deaths of 62 people, most of whom were tourists. Six gunmen, disguised as security forces members, killed 58 foreign nationals and four

Egyptians at or around the Mortuary Temple of Pharaoh Hatshepsut.

This tragic event, known as the Luxor massacre, had a profound impact on Egypt's tourism industry, which took years to recover. The temple itself, a masterpiece of ancient architecture carved into the cliffs at Deir el-Bahari, was built by Queen Hatshepsut, one of Egypt's most successful female pharaohs.

Chichén Itzá was one of the largest Mayan cities. Maya society in Mesoamerica was one of the greatest civilizations on earth. The Mayans were highly skilled astronomers, and their calendar is one ten-thousandth of a day more accurate than the Gregorian calendar, upon which our modern-day calendar is based.

They developed complex calendrical systems, including the Tzolk'in and the Haab', and could predict solar eclipses. The Dresden Codex, one of the few surviving pre-Columbian books, contains detailed astronomical tables.

By the eighteenth century, only a handful of Mayan books had survived, and no one could read or write Maya anymore. Chichen Itza was abandoned more than 600 years ago because of drought and exhausted soils.

Recent studies using paleoclimatic data and sediment cores suggest that prolonged periods of drought in the 9th century CE contributed significantly to the collapse of many Maya city-states. Deforestation and soil degradation likely compounded the environmental stresses.

A few Mayan books resurfaced in the late 1880s, and linguists tried for decades to decipher the Maya code using those books and glyphs carved into ancient Maya pyramids. Chichen Itza is now one of Mexico's most visited places.

The eventual decipherment of Maya hieroglyphs, led by scholars like Yuri Knórosov and Tatiana Proskouriakoff, revealed a sophisticated written language that recorded dynastic histories, rituals, and astronomical data. Chichen Itza was designated a UNESCO World Heritage Site in 1988 and was named one of the New Seven Wonders of the World in 2007.

When I was in Chichen Itza, I asked the taxi driver, whom I had hired for two days, whether he wanted to also see the inside of the pyramid known as El Castillo, which has become the towering icon of the city. He said no and looked shaken. I stood in the long line and entered the pyramid, not knowing what I might see.

The main thing I saw was a sculpture of a reclining male figure named Chac Mool holding a bowl on his torso. The figure is made of jade, at least on the outside. The bowl served as a receptacle for sacrificial liquids. It was the scariest nonliving thing I have ever seen.

Chac Mool figures are commonly found in Mesoamerican temples and are believed to have been associated with offerings to the gods, possibly including human hearts or blood. The one inside El Castillo sits near a red jaguar throne in a hidden inner temple chamber once accessible to visitors but now closed for preservation reasons.

El Castillo, also known as the Temple of Kukulcan, dominates the center of the Chichen Itza archaeological site in the Mexican state of Yucatán. The temple consists of a series of square terraces with stairways up each of the four sides with 91 steps to the temple on top. This produces 365 steps, equal to days in a year.

This architectural design reflects the Maya's astronomical knowledge and their emphasis on cyclical time. Each side of the pyramid represents a cardinal direction, and its alignment with celestial events reinforces its role as both a religious and calendrical monument.

Around the spring and autumn equinoxes, the late afternoon sun strikes the northwest corner of the temple. This casts a series of triangular shadows against the northwest balustrade, creating the illusion of the feathered serpent "crawling" down the temple.

This serpent is thought to represent Kukulcan, the Maya feathered serpent deity, associated with wind, rain, and creation. Thousands of visitors still gather biannually to witness this stunning light-and-shadow phenomenon, a testament to the Maya's advanced engineering and cosmic symbolism.

Chapter 30
Marvels of Ancient Engineering

The ingenuity of ancient civilizations extends far beyond the well-known wonders like the pyramids of Egypt or the temples of Angkor Wat. Across the globe, our ancestors developed sophisticated technologies and engineering solutions that continue to astound modern experts. Let's explore some lesser-known but equally impressive examples of ancient engineering prowess.

In South America, the Inca civilization built an extensive network of roads known as the Qhapaq Ñan, or the Inca Trail. This road system stretched over 39,000 kilometers, traversing the challenging terrain of the Andes Mountains.

The Inca engineers developed innovative techniques to overcome geographical obstacles, including rope bridges spanning deep gorges and tunnels cut through solid rock. These rope bridges, made from woven grass called ichu, had to be rebuilt annually by local communities a tradition that continues today with the Q'eswachaka Bridge in Peru.

One of the most impressive features of the Inca road system was its durability. Many sections of these roads are still in use today, more than 500 years after their construction. The Incas used a technique called ashlar masonry, which fitted large stone blocks tightly together without mortar.

This method, combined with excellent drainage systems, allowed the roads to withstand centuries of use and harsh Andean weather. The tight joints and seismic-resistant construction helped these roads and structures survive countless earthquakes in a tectonically active region.

The Inca also excelled in agricultural engineering. They developed a system of terraces known as andenes, which allowed them to cultivate crops on steep mountain slopes. These terraces were carefully constructed with layers of rock, gravel, and soil to ensure proper drainage and prevent erosion.

The Inca's agricultural innovations enabled them to feed a large population in a challenging mountain environment. Some of the most complex terrace systems can be found in places like Moray, believed to have functioned as an agricultural research station due to its microclimates at different levels.

In the Middle East, the ancient Nabataeans carved the city of Petra out of solid rock in what is now Jordan. While the facade of the Treasury building is Petra's most famous feature, the city's true engineering marvel lies in its water management system. The Nabataeans created a complex network of dams, channels, and cisterns to harvest and store rainwater in this arid region. They used ceramic pipes and covered conduits to reduce contamination and evaporation, directing seasonal flash floods from desert wadis into storage tanks.

The water system of Petra was so effective that it supported a population of up to 30,000 people in a desert environment. The Nabataeans used siphons to transport water over long distances and created underground cisterns to minimize evaporation. Their water management techniques were so advanced that some aspects of their system are still not fully understood by modern engineers.

Recent archaeological studies using ground-penetrating radar have revealed hidden components of this network, including subterranean reservoirs capable of holding tens of thousands of cubic meters of water.

In China, the Dujiangyan irrigation system, built in 256 BCE, is still in use today, making it one of the oldest functioning hydraulic engineering projects in the world. This system, located in Sichuan province, diverts water from the Min River to irrigate over 5,300 square kilometers of farmland. It was designed by the engineer Li Bing and his son, who surveyed the river using primitive but accurate measuring techniques to develop a sustainable, silt-resistant solution.

The key to Dujiangyan's longevity is its ingenious design. Instead of using a dam, which would require constant maintenance and could fail catastrophically, the system uses a levee to partially divert the river's flow. This design allows for continuous water flow, prevents siltation, and provides natural flood control.

The structure comprises three main parts: the Fish Mouth Levee (Yuzui), the Flying Sand Weir (Feishayan), and the Bottle-Neck Channel (Baopingkou), each serving a unique purpose to manage water distribution and sediment flow. The system's effectiveness in managing water resources has earned it a place on the UNESCO World Heritage list. It remains a vital source of irrigation, flood control, and ecological stability in the Chengdu Plain more than two millennia after its construction.

In Rome, the ancient engineers developed an extensive system of aqueducts to supply the city with fresh water. While aqueducts were not a Roman invention, the scale and sophistication of the Roman system were unprecedented. At its peak, the aqueduct system of Rome delivered over 1 million cubic meters of water daily to the city. This vast system included 11 main aqueducts built between 312 BCE and 226 CE, stretching over 500 kilometers, with about 47 kilometers carried above ground on stone arches.

The Roman aqueducts showcased several engineering innovations. They used the principle of gravity to transport water over long distances, maintaining a consistent, slight downward slope over tens of kilometers. The Romans also developed advanced techniques for waterproofing, using materials like volcanic ash to create a type of ancient concrete that could withstand constant water flow. This material, known as opus caementicium, proved durable even under hydraulic pressure, and many aqueduct structures have survived to this day.

Moreover, the Romans created a complex distribution system within the city, using lead pipes to deliver water to public fountains, baths, and even some private homes. They also developed a system of valves and tanks to regulate water pressure and flow, demonstrating a sophisticated understanding of hydraulics. In addition, the castella stone water towers placed throughout the city helped evenly distribute water and regulate flow into different neighborhoods.

In India, the step wells of Gujarat and Rajasthan represent another remarkable feat of water engineering. These structures, known as vav or baori, were built to provide water storage and access in arid regions. Step wells consist of a series of stairs leading down to a pool of water, often several stories below ground level.

These wells date back as early as the 3rd century CE and were particularly prominent during the Solanki dynasty (11th–13th centuries), which commissioned many elaborately carved step wells as both functional and religious structures.

The design of step wells is both functional and beautiful. The stairs allow access to water at different levels as the water table fluctuates throughout the year. The deep, shaded structure also provides a cool refuge from the intense heat of the Indian summer.

Many step wells are adorned with intricate carvings and sculptures, making them important cultural and social centers as well as vital water sources. The architectural complexity also contributed to passive temperature control, with step wells often 5 to 6 degrees cooler than the surrounding air serving as gathering spots, especially for women.

One of the most famous step wells is the Chand Baori in Rajasthan, built in the 9th century CE. It descends 13 stories into the ground and includes 3,500 narrow steps arranged in a perfect square pattern. The precision and symmetry of its construction continue to impress visitors and engineers alike.

Chand Baori is located opposite the Harshat Mata Temple, indicating its ritual as well as practical significance. It was also designed to harvest rainwater and channel it efficiently into the deep reservoir.

In Mesoamerica, the ancient city of Teotihuacan, located near modern-day Mexico City, showcases remarkable urban planning and architectural prowess. The city, which reached its peak between 100 BCE and 750 CE, was one of the largest urban centers in the ancient world, housing up to 200,000 people. It covered over 20 square kilometers and influenced urban design across Mesoamerica, including the Maya and Zapotec civilizations.

The layout of Teotihuacan demonstrates advanced knowledge of astronomy and mathematics. The city is precisely aligned along a north-south axis at 21 degrees north of east, with its main thoroughfare, the Avenue of the Dead, pointing directly at the setting sun on certain significant dates.

The massive Pyramid of the Sun, the third largest pyramid in the world, was built using precise geometric principles and aligned with celestial events. The city's orientation may also align with the Pleiades constellation, which held spiritual significance. Archaeologists have found that the pyramid sits atop a natural cave, possibly used for rituals related to the underworld in Teotihuacano cosmology.

Teotihuacan also featured sophisticated apartment compounds to house its large population. These multi-family complexes included drainage systems, courtyards, and specialized areas for different activities.

The standardized design of these compounds suggests a high degree of urban planning and social organization. Recent findings suggest that residents were organized into neighborhood-based ethnic enclaves, including groups from Oaxaca and the Gulf Coast, indicating cultural diversity and administrative complexity.

Examples of the ancient world's engineering demonstrate the ingenuity, skill, and ambition of our ancestors. From massive monuments to practical infrastructure, these achievements highlight the universal human drive to shape our environment and overcome natural challenges.

Many of these ancient technologies and approaches offer valuable lessons for modern sustainable development. The water management systems of Petra and the step wells of

India demonstrate effective techniques for water conservation in arid regions.

The longevity of structures like the Inca roads and the Dujiangyan irrigation system showcases the benefits of working with natural systems rather than against them. These ancient systems emphasized durability, climate adaptability, and minimal ecological disruption principles highly relevant to contemporary green engineering practices.

As we face modern challenges like climate change and resource scarcity, looking back at these ancient engineering marvels can provide inspiration and practical insights. The resilience and sustainability of many of these ancient systems offer valuable lessons for creating long-lasting, environmentally harmonious infrastructure today.

Efforts in biomimicry and eco-engineering are increasingly revisiting ancient practices to guide innovations in water management, building design, and agriculture.

Moreover, studying these ancient achievements reminds us of the deep roots of human ingenuity and the long history of technological innovation. It encourages us to take a long-term view of our own technological development and to consider how our current innovations might be viewed by future generations.

In our rapidly changing world, with its focus on cutting-edge technology, these ancient marvels serve as a humbling reminder of the enduring nature of human creativity and problem-solving. They challenge us to build not just for today but also for the centuries to come, creating lasting legacies that future generations will study and admire, just as we do with the great works of our ancestors.

Chapter 31
Decoding the Past

The process of deciphering ancient scripts is not just a matter of linguistic puzzle-solving; it's a gateway to understanding entire civilizations. Each deciphered text provides invaluable insights into the daily lives, beliefs, and knowledge of ancient peoples. Let's explore some examples of how decoding ancient scripts has revolutionized our understanding of the past.

The decipherment of Linear B, an ancient script used in Mycenaean Greece, is another fascinating chapter in the story of unlocking the past. Linear B tablets, first discovered in Crete in 1900, remained a mystery for decades. The breakthrough came in 1952 when Michael Ventris, an architect with a passion for ancient languages, cracked the code. He built upon the work of classicist Alice Kober, whose detailed analyses of word patterns laid crucial groundwork for the decipherment.

Ventris's decipherment revealed that Linear B was an early form of Greek, predating the classical Greek alphabet by several centuries. This discovery pushed back the known history of written Greek by about 600 years and provided crucial information about Mycenaean culture, economy, and religion. It also proved that the Mycenaeans were Greek-speaking people, thus establishing a direct linguistic and cultural connection between the Bronze Age civilization and later Hellenic Greece.

The Linear B tablets, mostly administrative records, gave historians a detailed look at the economic and social structure of Mycenaean society. They listed inventories of goods, land ownership records, and religious offerings, painting a picture of a highly organized, bureaucratic society. This information dramatically changed our understanding of pre-classical Greek civilization and its connections to later Greek culture. Additionally, the tablets provide evidence of palatial control over regional economies, including the production and distribution of commodities like grain, wool, and oil.

Another significant decipherment was that of the Ugaritic alphabet, used in the ancient city-state of Ugarit in what is now Syria. Ugaritic texts, written in a cuneiform script, were discovered in 1929 and deciphered remarkably quickly, largely because of their similarities to other Semitic languages. The script consists of 30 letters and is one of the earliest examples of an alphabetic writing system, dating to around the 14th century BCE.

The decipherment of Ugaritic provided a wealth of information about Canaanite culture and religion. It included mythological texts that shed light on the religious beliefs of the region, some of which have parallels in biblical narratives. These texts have been crucial in understanding the cultural context of the early Hebrew Bible and the religious landscape of the ancient Near East. The Baal Cycle, for example, outlines the mythology surrounding the storm god Baal and his battles with Mot and Yamm, offering insight into pre-Israelite religious thought.

The decipherment of Ugaritic also had linguistic significance. It represented one of the earliest known alphabetic writing systems, providing important information about the development of alphabetic scripts in the region. This innovation likely influenced the development of the Phoenician alphabet, which in turn is the ancestor of most modern alphabets, including Greek and Latin.

The ongoing efforts to decipher the Indus script used by the Indus Valley Civilization demonstrate the challenges still facing modern scholars. Despite numerous attempts, this script, found on thousands of seals and tablets from sites like Mohenjo-daro and Harappa, remains undeciphered. It dates back to approximately 2600–1900 BCE and is thought to represent a language isolate or a member of a now-extinct linguistic family.

The difficulty in deciphering the Indus script is rooted in several factors. The texts are very short, with most seals containing only 5 or 6 symbols. There's no known bilingual text to use as a Rosetta Stone, and the language behind the script is unknown. Moreover, there's debate about whether the symbols constitute a full writing system or are merely a system of religious or political symbols. Some researchers suggest the script may be logosyllabic, while others argue it could represent proto-writing or a mnemonic system.

Despite these challenges, researchers continue to apply new technologies and methodologies to the problem. Computer analysis has helped identify patterns in the symbols, and comparisons with other writing systems have yielded some insights. While a full decipherment remains elusive, each new approach brings us closer to understanding this mysterious script and the civilization that produced it. Recent studies using machine learning and big data analysis are helping to group symbols statistically, offering new hope for eventual breakthroughs.

The decipherment of ancient scripts often involves interdisciplinary collaboration. Linguists work alongside historians, archaeologists, and now, computer scientists to unlock these ancient codes. The advent of digital technologies has revolutionized the field, allowing for the analysis of large corpus of texts and the identification of patterns that might not be apparent to the human eye.

For example, digital epigraphy tools now allow for 3D scanning and rendering of tablets, improving the accuracy of transcriptions and preserving fragile artifacts.

For example, machine-learning algorithms have been applied to the problem of deciphering lost languages. In 2019, researchers used AI to help decipher Linear B tablets, demonstrating how technology can assist in the complex process of understanding ancient scripts. AI has also been applied to the undeciphered Etruscan and Iberian scripts, assisting in hypothesis generation and comparative analysis.

The impact of deciphering ancient scripts extends far beyond academia. These decipherments have often challenged existing historical narratives and reshaped our understanding of the ancient world. They've revealed the complexity and sophistication of ancient societies, showing that many of the institutions and ideas we consider modern have deep historical roots.

For instance, the decipherment of cuneiform texts from Mesopotamia revealed the existence of complex legal codes, such as the Code of Hammurabi, long predating other known legal systems. These texts showed that concepts of individual rights and the rule of law were already being developed in the ancient Near East. Cuneiform texts also include some of the earliest known literature, such as the "Epic of Gilgamesh," and scientific writings on astronomy, mathematics, and medicine.

Similarly, the decipherment of Maya glyphs led to a complete reevaluation of Maya civilization. What was once thought to be a peaceful society of stargazers was revealed to be a complex network of city-states engaged in diplomacy, trade, and warfare. The deciphered texts provided details about Maya history, including the names and deeds of rulers that had been lost for centuries.

The breakthrough came in the 20th century, particularly through the work of scholars like Tatiana Proskouriakoff and Yuri Knorosov, who demonstrated that the script was largely phonetic and capable of recording historical events.

The process of deciphering ancient scripts also highlights the importance of cultural heritage preservation. Many important texts have been lost to time, destroyed by war, natural disasters, or simply neglected. Each undeciphered script represents a potential wealth of knowledge about the past, underscoring the need to protect and preserve archaeological sites and artifacts. For example, the burning of the Library of Alexandria resulted in the loss of countless ancient manuscripts, and conflicts such as those in Syria have endangered invaluable archaeological sites like Palmyra.

Moreover, the story of script decipherment is a testament to human curiosity and perseverance. Many of the most important breakthroughs in decipherment came from individuals who pursued the challenge out of sheer fascination, often working outside of traditional academic structures. This serves as a reminder of the power of passion and dedication in driving scientific discovery. Michael Ventris's work on Linear B, achieved while he was still an amateur linguist and architect, exemplifies how determination and unconventional approaches can lead to groundbreaking results.

As we continue to unravel the mysteries of ancient scripts, we gain not just knowledge of the past but also insights that can inform our understanding of language, communication, and human society. The scripts our ancestors left behind are more than just historical curiosities; they are bridges across time, allowing us to engage in a dialogue with our distant forebears. In fact, decipherment often reveals how early languages influenced modern tongues and sheds light on the evolution of writing systems from pictographs to alphabets.

In this age of rapid technological advancement, the decipherment of ancient scripts reminds us of the enduring power of written language. As we develop new forms of communication and data storage, the longevity and decipherability of these systems become important considerations.

The ancient scripts that have survived millennia to be deciphered today prompt us to consider how our own writings and records will be understood by future generations. For instance, digital formats face challenges such as data degradation and format obsolescence, raising questions about the preservation of electronic records over centuries.

The ongoing work of deciphering ancient scripts is a vivid reminder that our understanding of history is not fixed but constantly evolving. Each new decipherment has the potential to rewrite our understanding of the past, challenging us to reconsider our place in the long story of human civilization.

As we continue to unlock the secrets of ancient writings, we not only learn about our past but also gain new perspectives on our present and future. This dynamic is evident in how the decipherment of Maya glyphs transformed views of pre-Columbian Americas from static to highly complex societies.

Chapter 31
Decoding the Past

Ancient civilizations left behind written records and monumental structures such as the Great Pyramid of Giza in Egypt and Puma Punku in Bolivia. Where did they all go? Did they simply take the knowledge with them to their graves, or did they travel elsewhere? If we had to leave behind our cities, what would we take with us? Where would we go? If we don't have enough fuel or electricity, most or all of our devices and machines will become obsolete. If there isn't enough food or clean water, we will die. Our records on paper and other physical media mostly will turn to dust. What would happen to our records in digital storage? If lost for a millennium or more, would they be decipherable in the future?

Many digital records depend on hardware, software, and power that may not exist or be understood in the distant future. This "digital dark age" problem has led to efforts to create durable physical backups, such as archival-quality microfilm or stone-engraved data, to preserve critical knowledge beyond current technologies.

Chapter 32
The Journey of Exploration

The spirit of exploration that drives us to uncover ancient wonders also propels us to discover the marvels of our present world. Modern-day adventurers and travelers continue to push boundaries, seeking out remote corners of the globe and bringing back stories that inspire and educate.

One such modern wonder is the Great Barrier Reef off the coast of Australia. Stretching over 2,300 kilometers (1,430 miles), it is the world's largest coral reef system and the only living structure visible from space. The reef is home to an incredible diversity of marine life, including over 1,500 species of fish, 400 types of hard coral, as well as sea turtles, sharks, dolphins, and countless invertebrates.

Exploring the Great Barrier Reef offers a unique perspective on the interconnectedness of ecosystems and the fragility of our planet's natural wonders. Snorkelers and divers can witness firsthand the vibrant colors and teeming life of the coral forests. However, they can also observe the impacts of climate change, as rising ocean temperatures have caused mass coral bleaching events, most notably in 2016, 2017, and 2020, threatening large portions of the reef. Additionally, factors such as ocean acidification, water pollution, and coastal development pose ongoing risks.

The reef serves as both a testament to the Earth's biodiversity and a stark reminder of the urgent need for conservation efforts. Visitors often leave with a renewed commitment to environmental protection, understanding that the fate of such natural wonders is inextricably linked to our own actions and choices.

Another modern marvel that captivates explorers is the Amazon rainforest. Covering over 5.5 million square kilometers (2.1 million square miles), it is the world's largest tropical rainforest and home to an estimated 10% of all known species on Earth, including jaguars, sloths, macaws, and countless plant species. Exploring the Amazon is like stepping into a living laboratory of evolution and adaptation.

Adventurers in the Amazon can witness incredible natural phenomena, such as the "Meeting of Waters" near Manaus, where the dark waters of the Rio Negro meet the sandy-colored waters of the Rio Solimões and flow side by side without mixing for several kilometers due to differences in temperature, speed, and density. They can also encounter Indigenous communities such as the Yanomami and Kayapo, who have lived in harmony with the forest for thousands of years, offering profound lessons in sustainable living and traditional ecological knowledge.

However, like the Great Barrier Reef, the Amazon faces significant threats from deforestation largely driven by logging, agriculture, and mining climate change, and resource exploitation. Each year, large areas are lost, impacting global carbon cycles and biodiversity. Travelers often return from the Amazon with a deepened understanding of the global importance of this ecosystem and the urgent need to protect it.

In Africa, the Serengeti ecosystem offers another unparalleled exploration experience. This vast savanna spans northern Tanzania and southwestern Kenya and is renowned for its annual wildebeest migration, one of the most impressive natural events on the planet.

Over 1.5 to 2 million wildebeest, along with hundreds of thousands of zebras and Thomson's gazelles, make a circular trek following the seasonal rains and fresh grass.

Witnessing the Serengeti migration is a powerful reminder of the grand scale of natural processes and the interconnectedness of species within ecosystems. It also highlights the importance of preserving large, contiguous wildlife habitats and maintaining the integrity of migration corridors, which are increasingly threatened by human encroachment and fencing.

The exploration of extreme environments continues to push the boundaries of human endurance and technological capability. Antarctica, the Earth's southernmost continent, remains one of the most challenging and rewarding destinations for modern explorers.

Despite (or perhaps because of) its harsh conditions temperatures can drop below -80°C (-112°F) in the interior Antarctica offers unique opportunities for scientific research and exploration. Visitors can observe massive ice shelves such as the Ross and Larsen Ice Shelves, visit research stations like McMurdo and Scott Base, and encounter wildlife found nowhere else on Earth, such as emperor penguins, Weddell seals, and Antarctic krill.

Antarctica also serves as a crucial site for climate research. Ice cores drilled from its thick ice sheets provide a record of Earth's climate going back up to 800,000 years, revealing patterns of natural climate cycles and helping scientists understand current changes. Exploring this frozen continent gives travelers a tangible sense of the Earth's climate history and the rapid changes occurring in polar regions due to global warming, such as accelerating ice melt and rising sea levels.

Urban exploration has also gained popularity, with adventurers seeking out hidden aspects of cities, both ancient and modern. In Rome, for instance, visitors can explore the extensive network of catacombs beneath the city, which date back to the 2nd century AD and offer a glimpse into early Christian burial practices and art. Similarly, urban explorers worldwide venture into abandoned subway tunnels, historic buildings, and underground passageways, revealing layers of human history hidden beneath bustling modern life.

In Paris, the catacombs present another underground wonder. These ossuaries hold the remains of more than six million people and offer a unique, if macabre, perspective on the city's history and its relationship with death and remembrance. Created in the late 18th century to address overcrowded cemeteries, the Catacombs stretch over 300 kilometers of tunnels beneath the city, though only a small portion is open to the public.

Modern cities offer their own form of exploration. In Tokyo, for example, travelers can lose themselves in the labyrinthine streets of neighborhoods like Shinjuku or Shibuya, each turn revealing new facets of Japan's blend of ancient traditions and cutting-edge technology.

Shinjuku is famous for its vibrant nightlife and towering skyscrapers, while Shibuya is known for the world's busiest pedestrian crossing and as a hub of youth culture and fashion innovation.

The rise of space tourism marks the beginning of a new era of exploration. While still in its infancy, the prospect of civilian space travel opens up entirely new frontiers for human exploration. Companies such as SpaceX, Blue Origin, and Virgin Galactic have already conducted successful suborbital and orbital flights with private passengers.

The overview effect the cognitive shift reported by astronauts when viewing the Earth from space offers a profound new perspective on our planet and our place in the universe, inspiring many to think differently about global unity and environmental stewardship.

Even as we push the boundaries of where humans can go, virtual and augmented reality technologies are creating new ways to explore both real and imagined environments. These technologies have the potential to make exploration more accessible, allowing people to virtually visit distant or dangerous locations from deep ocean trenches to ancient archaeological sites. They also offer new tools for scientists and researchers to visualize and interact with complex data about our world, enhancing education and conservation efforts.

However, as we celebrate these new frontiers of exploration, it's crucial to consider the impact of our wanderlust on the places we visit. Sustainable and responsible tourism has become an increasingly important consideration. How can we balance our desire to explore with the need to preserve and protect the world's natural and cultural wonders?

Many destinations are grappling with the challenges of over-tourism, where the sheer number of visitors threatens to overwhelm local infrastructure and degrade the very attractions people come to see. Places like Venice, Italy where annual tourist numbers exceed the local population and Machu Picchu in Peru have had to implement strict visitor limits and timed entry to protect their fragile environments and historic structures.

On the other hand, thoughtfully managed tourism can be a powerful force for conservation and cultural preservation. When local communities benefit economically from tourism, they have a strong incentive to protect natural areas and maintain cultural traditions.

Eco-lodges in the Amazon, which emphasize low-impact stays and community involvement, and community-based tourism initiatives in places like Kenya's Maasai Mara, demonstrate how tourism can support both conservation and local development.

As we continue to explore our world, we must also consider the carbon footprint of our travels. The aviation industry alone accounts for about 2-3% of global carbon emissions, a figure that is expected to grow as air travel increases globally. This has led to increased interest in "slow travel" prioritizing longer stays and fewer flights and more environmentally friendly modes of transportation such as trains and electric vehicles.

The spirit of exploration that drives us to uncover the wonders of our world both ancient and modern is a fundamental aspect of human nature. It fuels our curiosity, expands our understanding, and connects us to the broader tapestry of life on Earth. As we continue to push the boundaries of where we can go and what we can see, we must also push ourselves to explore responsibly and sustainably.

In essence, modern exploration is not just about discovering new places, but about uncovering new perspectives. Whether we're diving into the depths of the ocean, trekking through dense rainforests, wandering the streets of a new city, or viewing our planet from space, each journey has the potential to transform our understanding of the world and our place in it.

As we face global challenges like climate change and biodiversity loss, the insights gained from exploration become ever more crucial. They remind us of the rich diversity of human cultures, the interconnectedness of all life, and the beauty and fragility of our planet. In doing so, they inspire us to become better stewards of the Earth and more conscientious global citizens.

Chapter 33
The Story of Humans

The story of human evolution and migration is continuously being refined as new evidence comes to light. Recent discoveries have added fascinating chapters to our understanding of our species' journey.

One of the most exciting developments in recent years has been the discovery of new human species. In 2004, researchers announced the discovery of Homo floresiensis, nicknamed the "Hobbit" due to its small stature, on the Indonesian island of Flores. This species lived as recently as about 50,000 to 60,000 years ago, suggesting it coexisted with modern humans.

The discovery of H. floresiensis has challenged our understanding of human evolution, suggesting that multiple human species coexisted much more recently than previously thought.

The small brain size of H. floresiensis about one-third that of modern humans along with its unique skeletal features, sparked debates about whether it represented a distinct species or a pathological variation, but the consensus now favors the former.

In 2015, another new species, *Homo naledi*, was discovered in the Rising Star cave system near Johannesburg, South Africa. *H. naledi* had a unique combination of primitive and modern features, with a small brain (about 465 to 610 cubic centimeters) but a body adapted for walking upright and capable of tool use.

The discovery of *H. naledi* has further complicated our understanding of human evolution, suggesting that the human family tree is more of a complex bush with many branches rather than a simple linear progression. Recent dating places *Homo naledi* at between 236,000 and 335,000 years ago, meaning it lived at the same time as early *Homo sapiens* in Africa, raising the possibility of interaction between species.

These discoveries highlight the complexity of human evolution and migration. They suggest that at various points in our history, multiple human species coexisted, potentially interacting and even interbreeding. This paints a picture of a rich and diverse human past, far more complex than earlier, simpler models of human evolution suggested.

Genetic studies have also revolutionized our understanding of human origins and migrations. The field of paleogenomics, which studies ancient DNA extracted from fossils, has provided unprecedented insights into our ancestors and their movements across the globe.

One of the most significant findings from genetic studies is the evidence of interbreeding between different human species. We now know that most non-African populations carry approximately 1–2% Neanderthal DNA, indicating that when modern humans left Africa roughly 60,000 years ago, they encountered and interbred with Neanderthals. Similarly, some populations in Asia and Oceania carry up to 5% Denisovan DNA, another extinct human species identified from fossils in Siberia's Denisova Cave.

These genetic traces of ancient interbreeding events have significant implications. They suggest that rather than simply replacing other human species as they spread across the world, modern humans absorbed them through interbreeding. This challenges earlier models of human dispersal and suggests a more complex interaction pattern between different populations, including gene flow and cultural exchange.

Genetic studies have also refined our understanding of the timing and routes of human migrations. For example, genetic evidence suggests that a single founding population colonized Australia around 50,000 years ago and then split into two groups. One group moved along the east coast, while the other went west across the north of the continent.

This is supported by archaeological findings such as the Madjedbebe rock shelter in northern Australia, which dates to about 65,000 years ago, representing some of the earliest known human presence on the continent.

In the Americas, genetic studies have supported archaeological evidence suggesting an earlier arrival of humans than previously thought. Some studies propose that the initial population that gave rise to Native Americans was isolated in Beringia the land bridge that once connected Asia and North America for up to 10,000 to 15,000 years before spreading rapidly into the Americas after the last Ice Age. Recent discoveries, such as the Bluefish Caves in Canada, indicate human presence possibly as early as 24,000 years ago, predating the Clovis culture once believed to be the earliest inhabitants.

The story of human migration is not just about the distant past. Human populations have continued to move and mix throughout history, and these more recent migrations have also left their mark on our genetic heritage. For example, genetic studies have traced the spread of agriculture from the Middle East into Europe during the Neolithic period starting around 9,000 years ago, the Bantu expansion across sub-

Saharan Africa beginning about 3,000 to 5,000 years ago, and the Austronesian expansion across the Pacific Islands starting roughly 3,500 years ago, which spread language, culture, and technologies over vast distances.

Climate change has played a crucial role in shaping human evolution and migration patterns. Changes in climate affect the availability of resources and the habitability of different regions, driving human populations to adapt or move.

For instance, during the Last Glacial Maximum (around 26,500 to 19,000 years ago), much of northern Europe and North America was covered in massive ice sheets, making these areas uninhabitable.

As the ice retreated during the transition to the Holocene epoch around 11,700 years ago, humans were able to move into these newly accessible territories. This period saw significant changes in human societies, including the widespread development of agriculture in multiple regions such as the Fertile Crescent, the Yangtze and Yellow River valleys, and the Americas.

Today, we are facing a period of rapid anthropogenic climate change that once again has the potential to drive large-scale human migrations. Rising sea levels threaten low-lying coastal communities and island nations, while changing rainfall patterns and desertification could make some agricultural areas unviable, potentially displacing millions of people. Understanding our species' past responses to climate change can provide valuable insights as we face these modern challenges.

The story of human origins and migrations is also a story of incredible adaptability and resilience. Our ancestors faced numerous challenges as they spread across the globe from harsh climates like the Arctic tundra and deserts to unfamiliar diseases and new ecosystems. Yet they managed to adapt and thrive in almost every environment on Earth.

This adaptability is reflected in the genetic and physiological differences we see in human populations today. For example, populations living at high altitudes, such as Tibetans, Andean highlanders, and Ethiopian highlanders, have developed distinct genetic adaptations (like EPAS1 and other gene variants) that allow them to thrive in low-oxygen environments. Similarly, populations in northern latitudes have evolved adaptations for dealing with low levels of sunlight and cold temperatures, including variations related to vitamin D metabolism and fat storage.

These adaptations remind us of the incredible diversity within our species. Yet despite these differences, genetic studies have confirmed that we are all remarkably similar.

The genetic differences between human populations account for only about 0.1% to 0.2% of our DNA, which is tiny compared to the variation found within populations. This underscores the fundamental unity of our species despite superficial differences such as skin color, hair texture, or facial features.

As we continue to uncover the human origins and migrations, we are gaining a deeper appreciation for the shared heritage of all humanity. We are all descendants of a relatively small population that originated in Africa around 200,000 to 300,000 years ago and then spread across the world. Our differences in appearance, culture, and language are relatively recent developments in the long span of human history. Understanding this shared history can have profound implications for how we view others and ourselves. It challenges notions of racial or ethnic superiority, showing that such concepts have no biological basis. Instead, it emphasizes our common humanity and arbitrary nature of many of the divisions we create between groups of people.

The story of human origins and migrations is far from complete. New discoveries continue to refine and sometimes revolutionize our understanding. Advances in genetic sequencing technologies, such as whole-genome sequencing and improved ancient DNA retrieval methods, along with new archaeological finds and innovative analytical techniques like isotopic and environmental DNA analyses, promise to reveal even more about our past in the coming years.

As we piece together this epic tale of human journeys across time and space, we gain not only knowledge about our past but also insights that can inform our future. Understanding the migrations and adaptations of our ancestors can provide valuable lessons as we face modern challenges such as climate change, biodiversity loss, and global migrations fueled by conflict and economic pressures.

Moreover, recognizing our shared origins and the relatively recent nature of many of our differences can foster greater empathy and understanding between different cultures and populations. In a world often divided by ethnic, racial, and national conflicts, the story of human origins reminds us of our fundamental interconnectedness and shared heritage.

Ultimately, the story of human origins and migrations is the story of us all. It's a tale of exploration, adaptation, and resilience that continues to unfold. As we face the challenges of the 21st century and beyond, understanding this shared history can provide practical insights and a powerful reminder of our common humanity.

Chapter 34
Shaping a Better Future (Climate, Technology, and Other Factors)

The 21st century has been turbulent so far. The stock market crashed in 2000 after the dot.com bubble burst. Nine judges on the US Supreme Court decided the 2000 US Presidency, making George W. Bush President.

In 2001, terrorists attacked the U.S. on 9/11/2011, destroyed the World Trade Centers in New York, damaged the Pentagon building, and changed the world. Security tightened worldwide, Wall Street was in turmoil, and anger flamed around the world while some cheered. I still remember the beautiful view from the World Trade Center roof in 1978. I was at Ground Zero on 9/11's first anniversary, and many thoughts ran through my mind.

From 2001 to 2003, the Bush doctrine was conceived and applied; the US and some allies preemptively attacked enemy suspects, culminating in the invasion of Iraq in 2003, creating a prolonged war, the longest war in US history that remains unresolved even after almost all US troops left in 2020. Even in 2024, the government of Iraq wanted all US troops to leave their country, but so far, we have refused.

The terrorist war in Afghanistan since 2001 was ended when all US troops pulled out in disgrace under President Joe Biden in 2021. The US spent over $2 trillion on this war, equivalent to over $300 million per day for 20 years.

The 2004 Indian Ocean tsunami, triggered by an earthquake of magnitude 9.1 to 9.3, killed 230,000 people in fourteen countries, including many tourists from the West. This earthquake, which was the second largest ever recorded on a seismograph, had the longest fault duration ever observed, between 8 and 10 minutes.

Hurricane Katrina caused more than $100 billion in damages and killed thousands in 2005. Also, Hurricane Katrina drowned New Orleans and devastated neighboring cities. I saw New Orleans soon after Katrina. At night, it was pitch black in most parts of the city. It was frightening to look around 360 degrees and see nothing.

In the same year, the Hindu Kush earthquake hit northeastern Afghanistan and killed at least 79,000 people. More than 32,000 buildings collapsed in Kashmir, leaving 3.5 million homeless.

In 2007, Cyclone Nargis wiped out more than 140,000 people in Burma, where I was born. The Sichuan, China earthquake in 2008 killed 70,000 people. The 2010 Haiti earthquake killed more than a quarter million people and destroyed countless dwellings in an already impoverished country. Also, in 2010, the Chilean 8.8 earthquake was a shock, volcanic eruptions in Iceland halted air travel in Europe several times, and volcanic eruptions in Central and South America produced further anxiety regarding the "ring of fire" around the Pacific Ocean rim.

The financial sector, in 2006, recognized that the housing bubble and the subprime loan crisis were becoming evident. The wave of bank failures started in 2007 in Europe. The global financial system was facing imminent collapse in 2008, requiring massive government bailouts throughout the world that some are calling the biggest generational theft of all time. In 2009 and 2010, massive unemployment and sovereign debts threatened to destabilize economies, especially in Europe, and resulted in civil unrest in different countries. We are on the verge of a global transformation,

and we must act wisely to preserve the world for the next
generation.

In 2011, this was the beginning of the Arab Spring that
rocked the world. From Tunisia, protests spread to five other
countries: Libya, Egypt, Yemen, Syria, and Bahrain. Leaders
in the first three of these five countries were deposed. Five
million people were displaced, and a million died in Syria.

On March 11, 2011, a 9.0-9.1 earthquake devastated
Japan. It is known in Japan as the Great East Japan
Earthquake. It was the most powerful earthquake ever
recorded in Japan and the fourth most powerful earthquake
recorded in the world. The quake created a tsunami with
waves reaching 40.5 meters (133 ft.) in Miyako.

The earthquake and tsunami killed approximately 20,000
people. Fukushima's 3 nuclear reactors melted down due to
the tsunami. This issue persists in 2024. Japan started
discharging treated nuclear-contaminated water into the
Pacific Ocean in 2023. Some countries and people within
Japan oppose this release.

In 2012, Vladimir Putin became President of Russia, and
Barack Obama was re-elected as President of the United
States. The crisis between the Syrian government and the
rebels escalated into a full-blown civil war by mid-2012.

Russia supported the government, and the US supported the
rebels. In 2018, the United Nations recorded 6.7 million
Syrian refugees, nearly 40 percent of Syria's population that
year. Estimates of the total number of deaths in the Syrian
Civil War by various war monitors were approximately
613,407 as of March 2023.

2013

Xi Jinping became China's president. Pope Benedict resigned during the priests' pedophile scandal and Vatican Bank corruption. Lightning struck the Vatican twice hours after Pope Benedict resigned. Pope Francis, formerly Cardinal Jorge Mario Bergoglio, replaced him. Pope Francis changed the church in many ways and has changed the entire world.

Francis is the first Jesuit Pope. This was a significant appointment because of the sometimes-tense relations between the Society of Jesus and the Holy See. He came in second to Cardinal Ratzinger (Pope Benedict) on all the ballots in the 2005 conclave and appeared as the only other viable candidate.

When Pope Clement XIV (31 October 1705 – 22 September 1774) was head of the Catholic Church and ruler of the Papal States from 19 May 1769 to his death in September 1774, he decreed the suppression of the Society of Jesus, removing all members of the Jesuits from most of Western Europe and their respective colonies beginning in 1759 along with the abolition of the order by the Holy See in 1773.

While there were multiple factors causing the suppression, the main reason seemed to be that the Jesuits were not above politics. They were distrusted for their closeness to the Pope and his power in independent nations' religious and political affairs. The Jesuit order did not die but continued underground operations in China, Russia, Prussia, and the United States. In Russia, an official novitiate was founded with Catherine the Great's permission. In 1814, Pope Pius VII restored the Society of Jesus to its previous provinces in different countries, and the Jesuits resumed their work.

Chapter 34
Shaping a Better Future (Climate, Technology, and Other Factors)

In California, the current governor, Gavin Christopher Newsom (January 7, 2019 -), and the previous governor, Edmund Gerald Brown Jr. (January 3, 2011 – January 7, 2019), are graduates of Jesuit Catholic University - Santa Clara University, my alma mater. The university motto is competence, conscience, and compassion. I respect all the Jesuits I have met at Santa Clara University.

2014

The US-backed coup in Ukraine was followed by a Russian invasion that resulted in Russia annexing Crimea and part of Eastern Ukraine. This was a result of NATO's efforts to expand to the East after the Soviet Union fell in 1991. Ukraine's conflict escalated in 2022 when Russia started a war against Ukraine.

This war has resulted in more than 6 million refugees fleeing Ukraine to Western Europe. Eight million people were internally displaced as of May 2022. Approximately one-quarter of the country's total population left Ukraine, hundreds of thousands of soldiers and civilians were killed, and countless buildings were bombed. As of September 9, 2024, the war continues to waste more lives and other resources, leading to food shortages and increases in food prices globally. The Ukraine war could lead to WW3.

2015

Nepal's earthquake killed 8,964 people. Saudi Arabian–led airstrikes in Sana'a prolonged Yemen's regional conflict. Refugees from the Syrian civil war have arrived in Greece.

2016

The UK voted to leave the EU in 2016. Donald Trump is elected U.S. president. He pulled the US out of the Paris Agreement for combating climate change globally. Trump also got the US out of a multinational agreement with IRAN to prevent it from developing a nuclear bomb.

2017

Spain rejected the 2017 Catalan independence referendum, further intensifying civil strife. The 2017 Las Vegas shooting, the deadliest mass shooting in American history, with 61 fatalities, is an example of increased mass shootings in the US. It was shocking to see the debris after Hurricane Irma struck the Florida Keys as a category 4 storm with a maximum sustained winds of 132 mph.

2018

March for Our Lives, a student-led organization, led demonstrations in support of gun control legislation. It took place in Washington, D.C., with over 880 sibling events throughout the United States and around the world. Yellow vest protests break out in France. The Camp Fire becomes the deadliest and most destructive wildfire in California's history.

2019

Hong Kong protests turned into widespread riots and civil disobedience. CRISPR gene editing was first used to experimentally treat a genetic disorder patient. This technology can alter humans, animals, and plants, which could have major consequences. Protesters in Tahrir Square during the 2019 Iraqi protests and Chilean protests are examples of conflicts that took place.

2020 Pandemic

The SARS-CoV-2 virus that caused the COVID-19 pandemic, a defining world event in 2020, is still continuing in May 2025. More than 7 million people are dead, and hundreds of thousands or even millions continue to suffer long-term Covid-19. Commercial real estate in San Francisco saw 60-70% price drops within 5 years. People would rather work at home than in offices. About $1 trillion in loans are due each of the next five years, from 2025 to 2029.

2021

Yangon, Myanmar, and many other cities around the world protested following the February 1, 2021 military coup d'état. There were civil demonstrations against the October 25, 2021, military coup in Sudan. After waging war for more than 20 years in Afghanistan, US forces fled in disgrace under President Biden. The exodus from Afghanistan was a stark reminder of what happened in Vietnam.

2022

There were anti-government protests in Sri Lanka. The Hunga Tonga–Hunga Ha'apai volcano erupts, making it the most powerful volcanic eruption of the 21st century. This volcanic explosion was described by scientists as a "magma hammer," with a plume of 58 km (34.8 miles) that displaced 10 cubic kilometers (2.4 cu mi) of rock, ash, and sediment and generated the largest atmospheric explosion recorded by modern instrumentation, contributing to an increase in global warming over the next 5 years.

Russia invades Ukraine, opposing NATO expansion by the US, EU, and other countries. The impact on climate change from this war has been immense due to the increased burning of coal.

2023

On May 5, the WHO ceased calling the COVID-19 pandemic a global health emergency as infections decreased. The Russian invasion of Ukraine, which began in 2022, continued, and an armed conflict broke out in Sudan beginning in April. Catastrophic natural disasters included the fifth-deadliest earthquake of the 21st century, striking Turkiye and Syria, leaving nearly 60,000 dead.

Freddy, the longest-lasting tropical cyclone in history, killed more than 1,400 people. In April, India surpassed China to become the most populous country in the world. Both countries had approximately 1.4 billion people. The 2023 banking crisis resulted in the collapse of numerous American regional banks as well as the buyout of Credit Suisse by UBS in Switzerland.

Silicon Valley Bank and First Republic Bank were the two largest banks that collapsed. These were the third and second-largest banking collapses in US history, respectively. Lehman Brothers remains the largest bankruptcy in the US. The Israel-Gaza war started after HAMAS killed 1,200 people in Israel and kidnapped several hundred people on October 7, 2023. Israel started the retaliation process against HAMAS in Gaza soon after the massacre.

2024

In response to the Hamas attack, Israel killed more than 45,000 Gaza residents, mostly women and children. The

Chapter 34
Shaping a Better Future (Climate, Technology, and Other Factors)

International Court of Justice asked Israel to stop the genocide. So far, this year has witnessed the continuation of major armed conflicts, including the Russian invasion of Ukraine, the Myanmar civil war, the war in Sudan, and the Islamist insurgency in the Sahel.

The 2024 Olympics was in Paris. Previous Olympics have had a dramatic carbon footprint, as well as a dreadful impact on biodiversity. For instance, the last two summer Olympic games were held in Tokyo, Japan, and Rio de Janeiro, Brazil. They released more than 2.7 million and 4.5 million tons of carbon emissions into the atmosphere, respectively.

Climate change is an urgent and critical issue that needs increased collaboration globally. For the 2008 Beijing Olympics, China shut down many factories in the city and removed half of the cars from the road. The air in Beijing was clean and I enjoyed watching the Olympics as well as visiting the city. I am glad that the 2024 Paris Olympics is an example of what cities can do to combat climate change. It did have shortcomings such as cardboard beds for athletes.

Hurricane Helene and Hurricane Milton devastated Florida in September and October 2024, respectively. Milton became a category 5 hurricane from a tropical storm in less than 24 hours and did a lot of damage in Florida even when it landed as a category 3 hurricane.

As we confront the challenges of the 21st century, the lessons from our past and the realities of our present converge to inform our path forward. The future we shape will be determined by how we address key issues such as climate change, technological advancement, global cooperation, and the preservation of our shared cultural and natural heritage.

Climate change remains one of the most pressing issues of our time. The effects are already being felt worldwide, from rising sea levels threatening coastal communities to more frequent and intense extreme weather events. Addressing this global challenge requires a multi-faceted approach that combines technological innovation, policy changes, and shifts in individual behavior.

Renewable energy technologies are rapidly advancing and becoming more cost-effective. Solar and wind power are increasingly competitive with fossil fuels, and new technologies like advanced battery storage are making these intermittent energy sources more reliable. The transition to clean energy is not just an environmental imperative but also an economic opportunity, potentially creating millions of new jobs in emerging green industries.

However, technology alone cannot solve the climate crisis. We need comprehensive policy changes at local, national, and international levels. Carbon pricing mechanisms, whether through carbon taxes or cap-and-trade systems, can help internalize the environmental costs of greenhouse gas emissions and drive the transition to cleaner alternatives. Policies supporting energy efficiency, sustainable transportation, and forest conservation are also crucial components of a comprehensive climate strategy.

Individual actions, while often overlooked, play a significant role in addressing climate change. Choices about what we eat, how we travel, and what we consume all have environmental impacts. Education and awareness campaigns can help individuals understand the consequences of their choices and empower them to make more sustainable decisions.

Chapter 34
Shaping a Better Future (Climate, Technology, and Other Factors)

The preservation of biodiversity is inextricably linked to climate action. As we've seen from studying ancient civilizations, the health of human societies is deeply connected to the health of the ecosystems they inhabit. Protecting and restoring natural habitats not only preserves the incredible diversity of life on Earth but also provides crucial ecosystem services like carbon sequestration, water purification, and natural disaster mitigation.

Efforts to protect biodiversity must go hand in hand with sustainable development initiatives. We need to find ways to meet human needs without compromising the ability of future generations to meet their own needs. This involves rethinking our economic systems, moving away from models based on endless growth, and towards circular economies that minimize waste and maximize resource efficiency.

Technological advancement presents both opportunities and challenges as we shape our future. Artificial intelligence, robotics, biotechnology, and other emerging technologies have the potential to solve many of our most pressing problems. AI could help us design more efficient cities, develop new medicines, and optimize our use of resources. Biotechnology could help us create more resilient crops and develop new ways to clean up environmental pollution.

Global cooperation will be crucial in addressing the challenges of the future. Many of the issues we face, from climate change to pandemics to economic inequality, transcend national borders and require coordinated international action. The COVID-19 pandemic has starkly illustrated both the interconnectedness of our global society and the importance of international cooperation in addressing shared threats.

However, recent years have also seen a rise in nationalism and isolationism in many parts of the world. Overcoming these tendencies and fostering a sense of global citizenship will be crucial for addressing our shared challenges. This doesn't mean erasing national or cultural identities but rather recognizing that we all share a common home and a common future on this planet.

Education will play a vital role in shaping our future. As we've seen from deciphering ancient scripts and uncovering lost knowledge, education is the key to understanding our past and informing our future. In our rapidly changing world, education systems need to evolve to equip people with the skills they'll need for the jobs of the future, many of which may not even exist yet.

Beyond job skills, education needs to foster critical thinking, creativity, and adaptability. These skills will be crucial for navigating the complex challenges of the future. Education should also cultivate global awareness and cross-cultural understanding, helping to break down barriers between different groups and fostering cooperation.

Preserving our cultural heritage, both tangible and intangible, is another crucial aspect of shaping our future. As we've seen from studying ancient civilizations, our cultural heritage provides invaluable insights into human ingenuity, creativity, and resilience. It connects us to our past and informs our identity. In a rapidly changing world, preserving this heritage can provide a sense of continuity and grounding.

At the same time, we must recognize that culture is not static. Just as ancient cultures evolved and adapted over time, our cultures continue to change and grow. The challenge is to preserve the valuable aspects of our cultural heritage while remaining open to positive change and cultural exchange.

As we look to the future, we must also consider our place in the broader universe. Space exploration, once the domain of competing superpowers, is increasingly becoming a collaborative international endeavor. The insights gained from space exploration, from our understanding of climate systems to potential technological spin-offs, can have profound impacts on life on Earth.

Moreover, the perspective gained from seeing Earth from space – the famous "overview effect" reported by astronauts – can foster a sense of global unity and highlight the fragility of our planet. As we continue to explore the cosmos, we may gain new insights into our place in the universe and the uniqueness of life on Earth.

Shaping a better future requires us to learn from our past, understand our present, and imagine the world we want to create. It requires balancing technological progress with ethical considerations, economic development with environmental sustainability, and individual freedoms with collective responsibility.

As we stand at this crucial juncture in human history, we have the opportunity to write the next chapter in the human story. Will we rise to the challenge of climate change and environmental degradation? Will we harness the power of technology for the benefit of all? Will we overcome our differences and work together to create a more just and sustainable world?

The answers to these questions will shape the world we pass on to future generations. By learning from our past, engaging with our present, and thoughtfully considering the consequences of our actions, we can work towards a future that honors the best of human potential – our creativity, our compassion, and our enduring spirit of exploration and discovery.

In this endeavor, every individual has a role to play. Whether through the choices we make in our daily lives, the ideas we contribute to our communities or the values we pass on to the next generation, we all have the power to influence the course of our collective future. By embracing this responsibility and working together, we can shape a future that not only sustains us but also allows all of humanity – and the myriad forms of life with which we share this planet – to thrive.

As we continue this exploration of our world – its ancient wonders, its hidden scripts, its diverse peoples, and its possible futures – let us carry forward the spirit of curiosity, respect, and responsibility that has guided the best of human endeavors throughout history. In doing so, we honor our shared past, engage fully with our present, and create the potential for a future that fulfills the highest aspirations of our species.

Chapter 35
Charting a Path Forward

As we confront the myriad challenges facing our world - from economic inequality and political polarization to climate change and geopolitical tensions - it's clear that business as usual is no longer an option. The crises we face are interconnected and require holistic, innovative solutions. Here, we outline key priorities and strategies for building a more just, sustainable, and peaceful world.

Addressing Economic Inequality and Financial Reform

The growing chasm between the ultra-wealthy and the rest of society poses a significant threat to social stability and democratic institutions. To address this:

- Implement progressive taxation systems, including wealth taxes on ultra-high-net-worth individuals.
- Strengthen regulations on financial institutions to prevent predatory practices and reduce systemic risks.
- Invest in education, healthcare, and infrastructure to create opportunities for economic mobility.
- Consider universal basic income or similar programs to provide a safety net in an increasingly automated economy.

Reforming Political Systems and Reducing Money's Influence

To restore faith in democratic institutions and ensure they serve the interests of all citizens:

- Enact campaign finance reform to limit the influence of wealthy donors and special interests.
- Implement measures to increase transparency in lobbying and political donations.
- Explore alternative voting systems, such as ranked-choice voting, to better represent diverse viewpoints.
- Invest in civic education to create a more informed and engaged citizenry.
3. Addressing Climate Change and Environmental Degradation

The climate crisis requires urgent, coordinated global action:

- Rapidly transition to renewable energy sources and improve energy efficiency across all sectors.
- Implement carbon-pricing mechanisms to incentivize emissions reductions.
- Invest in green infrastructure and technologies, creating jobs while reducing environmental impact.
- Protect and restore natural ecosystems, recognizing their crucial role in climate regulation and biodiversity.

Reimagining Global Security and Cooperation

To reduce conflict and build a more stable world order:

- Prioritize diplomacy and conflict prevention over military interventions.
- Reform international institutions like the UN to better represent the current global power balance.
- Develop new frameworks for arms control and nuclear non-proliferation.
- Invest in peace-building initiatives and address root causes of conflict, such as poverty and resource scarcity.

Harnessing Technology for the Common Good

As technological advancement accelerates, we must ensure it benefits all of humanity:

- Develop ethical guidelines and regulations for emerging technologies like AI and biotechnology.
- Invest in digital literacy programs to ensure all citizens can participate in the digital economy.
- Address the digital divide by expanding access to high-speed Internet and digital devices.
- Explore the potential of technology to solve global challenges, from climate change to disease prevention.

Prioritizing Mental Health and Well-being

Recognizing the psychological toll of ongoing crises:

- Increase funding for mental health services and reduce stigma around seeking help.
- Implement policies to improve work-life balance and reduce stress in the workplace.
- Invest in community-building initiatives to combat social isolation and loneliness.
- Integrate mental health education into school curricula from an early age.

Reinventing Education for the 21st Century

To prepare future generations for the challenges ahead:

- Emphasize critical thinking, creativity, and adaptability alongside traditional academic subjects.
- Integrate technology into learning while also teaching digital literacy and online safety.

- Promote global citizenship and cross-cultural understanding through exchange programs and collaborative projects.
- Implement lifelong learning initiatives so that workers can adapt to the changing job market.

Building Resilient and Sustainable Communities

To create a more robust society capable of weathering future crises:

- Invest in local food systems and renewable energy to increase community self-sufficiency.
- Redesign urban areas to be more walkable, green, and community-oriented.
- Strengthen local democratic institutions and encourage civic participation.
- Develop community-based support systems for vulnerable populations.

The path forward will not be easy. It requires a fundamental shift in how we think about progress, success, and our relationship with each other and the planet. It demands courage from leaders to challenge entrenched interests and make difficult decisions for the long-term good. It calls for active citizenship, with individuals taking responsibility for their communities and holding those in power accountable.

But within these challenges lie tremendous opportunities. We have the knowledge, technology, and resources to create a world that is more just, sustainable, and fulfilling for all. The question is whether we have the collective will to make it happen.

As we stand at this crossroads in human history, let us choose the path of courage, compassion, and cooperation.

Chapter 35
Charting a Path Forward

Let us reimagine our systems and institutions to serve the many, not just the few. Let us be the generation that rises to meet the great challenges of our time, leaving a legacy of hope and progress for those who will follow.

The future is not predetermined. It is shaped by the choices we make today. Let us make choices that our descendants will look back on with pride, choices that pave the way for a brighter, more equitable, and more sustainable world for all.

Section 5: Entrepreneurship

Chapter 36
The Seeds of Entrepreneurship

My entrepreneurial journey began at the tender age of twelve on the sidewalks of Burma (Myanmar), where I ran a small book lending business. Customers would borrow novels and comic books for a nominal fee. That modest venture sparked a lifelong passion for innovation one that eventually carried me to the frontiers of biotechnology, artificial intelligence, and healthcare in California's Silicon Valley.

For the past 25 years, I have dedicated myself to developing cutting-edge technologies in the biotech and healthcare sectors, striving to make a positive impact on the world. I've been driven by a deep belief in the power of science and technology to transform lives.

Four years before the completion of the Human Genome Project, I founded Iris Biotechnologies in 1999 with the vision of leveraging cutting-edge genomic technologies and artificial intelligence to transform healthcare. Iris Biotechnologies was the first company to apply AI and gene expression analysis to personalized breast cancer treatment pioneering an approach that helped doctors make more precise, data-driven decisions.

I then founded a wholly owned subsidiary, Iris Wellness Labs, in 2014 with the goal of providing in-depth scientific analysis to offer insights into complex medical conditions like cancer, heart disease, immune deficiency, diabetes, and obesity for patients and doctors at leading health centers. Our analysis based on whole-genome sequencing, microbiome profiling, and metabolite data was far ahead of its time and remains uncommon in most U.S. hospitals today. Yet with the rapid advancement of AI, integrating this depth of personalized data into everyday medical care is becoming increasingly feasible.

Yet, for all the technological breakthroughs and business successes, what I value most about my entrepreneurial journey are the lessons learned and relationships forged along the way. The process of building from the ground up, uniting a team around a shared vision, enduring challenges, and celebrating progress those are the experiences that have shaped me most deeply as a leader and as a person.

Looking back, I can trace a clear thread connecting my various entrepreneurial pursuits, from that first book-lending business in Burma to my current roles in the biotech industry. It is a thread woven of curiosity, creativity, and a deep desire to make a positive impact on the world. These are the qualities that I believe lie at the heart of the entrepreneurial spirit. Successful entrepreneurs tend to share some key qualities and mindsets:

1. Problem-solvers: They have a knack for identifying problems and finding innovative solutions. They think creatively and are not afraid to challenge the status quo.
2. Risk-takers: They are willing to step out of their comfort zone and try new things, even if success is not guaranteed. They view failure as a learning opportunity.
3. Hard workers: They put in the time and effort necessary to turn their vision into reality. They are persistent and do not give up easily in the face of challenges.
4. Leaders: They inspire and motivate others to buy into their vision. They build strong teams and create a culture of collaboration and shared purpose.

5. Lifelong learners: They are curious and always seeking to acquire new knowledge and skills. They learn from their experiences and from the expertise of others.

Significantly, entrepreneurship knows no age, background, or experience level. History is full of examples of successful entrepreneurs who started young (like Bill Gates and Steve Jobs) or launched second acts later in life (like Ray Kroc with McDonald's). What unites them is the courage to act on an idea and the grit to see it through.

If you dream big, embrace challenges, and want to make a difference, then entrepreneurship could be for you. Cultivate problem-solving, calculated risk-taking, hard work, leadership, and continuous learning these qualities will guide you forward.

The entrepreneurial journey demands courage, resilience, and passion. You will face doubt, fear, and exhaustion and also moments of exhilaration, breakthroughs, and pride in creating something entirely new.

As you embark on your own path, remember that every great business began as a simple idea. Your vision, passion, and persistence might ignite the next world-changing innovation. Trust your instincts, but stay humble and open to learning. Seek mentors and advisors who can guide you, yet don't be afraid to blaze your own trail.

In the chapters that follow, we'll explore the unique challenges and opportunities in biotech entrepreneurship: building a strong team; defining market strategy; protecting intellectual property; securing funding; cultivating resilience; maintaining integrity; fostering collaboration; and leaving a lasting legacy.

Before diving into those specifics, take a moment to reflect: What problem drives you? What unique skills or insights do you bring? How will you apply your creativity and determination to make a positive impact?

Remember, entrepreneurship is not just about building a successful venture it's about personal growth, pushing boundaries, and creating value for others. It's a continuous journey of learning and self-discovery.

Whether you're just starting or already on your entrepreneurial path, there's always room to learn, grow, and make a difference. Today's biggest societal challenges healthcare, climate change, education, social justice need innovative solutions and entrepreneurial spirit. Your ideas, passion, and willingness to persevere are the catalysts for progress and positive change.

So embrace and nurture your entrepreneurial spirit. Let it guide you toward creating something meaningful and impactful. The road ahead may be long and winding, but it is a journey worth taking and your entrepreneurial journey could change the world.

Chapter 37
The Biotechnology Frontier

When I founded Iris Biotechnologies in 1999, the field of genomics and personalized medicine was just beginning to take off. The Human Genome Project (HGP) a massive international effort to map the entire human genetic code was still underway, with the final results released in 2003. This landmark achievement laid the groundwork for a new era in biotechnology and personalized care.

The HGP revealed that humans have approximately 20,500 genes encoded in about 3 billion base pairs of DNA. This "instruction manual" for building and operating a human being opened up incredible opportunities for understanding disease, developing targeted therapies, and moving toward more proactive, individualized healthcare.

I recognized the potential to integrate genomic data with emerging technologies in data science particularly artificial intelligence and machine learning to transform medicine. What if we could predict disease risk based on a person's unique genetic profile? What if treatments could be tailored precisely to each patient's molecular makeup? These questions formed the core vision behind Iris Biotechnologies.

However, navigating the complex world of biotechnology as a startup came with significant challenges. The science was advancing rapidly, the regulatory environment was rigorous, and bringing new diagnostics and therapies to market was both time-consuming and costly. We had to assemble a highly interdisciplinary team, forge partnerships with leading academic and clinical institutions, and relentlessly innovate to stay ahead.

One of our first major hurdles was intellectual property. In August 2000, we filed a seminal patent application for an "Artificial Intelligence System for Genetic Analysis." When the U.S. Patent Office was ready to grant the patent, a critical mishandling by our law firm, Heller Ehrman, led to costly delays. This experience taught us the hard but vital lesson of securing a strong IP strategy and working with trustworthy legal counsel.

Despite these setbacks, we pressed forward, driven by our mission to improve health outcomes through technology. Over the years, Iris Biotechnologies evolved from a bold idea into a company at the forefront of precision medicine. We developed innovative diagnostic tools for complex diseases like breast cancer and later established Iris Wellness Labs to expand our reach.

Throughout this journey, my sense of wonder at the power of science never waned. Biotech is a rare field where intellectual curiosity, technological innovation, and humanitarian purpose intersect. It continues to attract some of the brightest minds and boldest thinkers all united by a drive to push the boundaries of what's possible.

For aspiring entrepreneurs in biotech, my advice is to stay deeply attuned to the scientific frontier, build a strong network of advisors and collaborators, and never lose sight of the human impact of your work. The road is challenging but immensely rewarding for those driven by a passion for discovery and a commitment to improving health and quality of life.

Above all, embracing biotechnology entrepreneurship means signing up for a mission-driven journey of lifelong learning and impact. It means being part of a community committed to solving some of the most complex and consequential challenges facing humanity. There are few pursuits more meaningful or more demanding than using science to improve lives.

The field of biotechnology is constantly evolving, with new discoveries and technological advancements emerging at a rapid pace. To succeed in this dynamic environment, entrepreneurs must cultivate a deep understanding of the science underlying their innovations while also staying attuned to market trends, regulatory changes, and shifts in the healthcare landscape.

Biotech entrepreneurship presents unique challenges: long development timelines, high capital requirements, and rigorous scientific and regulatory standards. Unlike software or consumer products, where a minimum viable product can often be launched quickly, biotech innovations typically require years of research, preclinical work, and clinical trials before reaching the market. This means biotech founders must master long-term strategic planning, milestone-based execution, and clear communication with investors.

The regulatory environment in biotech is another critical factor entrepreneurs must navigate. Agencies like the FDA in the United States play a crucial role in ensuring the safety and efficacy of new medical technologies. Understanding these regulatory pathways and building them into your development plans from the outset is essential. This often involves engaging regulatory experts early and maintaining proactive communication with oversight bodies throughout the process.

You must also evaluate key regulatory decisions early on. For example: Will your product require FDA approval, or can it be marketed as a Laboratory Developed Test (LDT)? And how might industry lobbying influence the rules that apply to you? If competitors are spending millions on lobbying, you need to understand why and what it means for your business.

Collaboration is also key in the biotech world. Many breakthrough innovations emerge from partnerships between startups, academic institutions, and established pharmaceutical or medical device companies. As an entrepreneur, your ability to build and manage strategic partnerships can be as important as your scientific or technical vision.

Ethics and social responsibility take on heightened importance in biotech, given the direct impact of these technologies on human health and well-being. Entrepreneurs in this field must grapple with complex ethical questions around genetic engineering, data privacy, equitable access to healthcare, and more. Establishing a strong ethical foundation early and intentionally builds trust with patients, clinicians, regulators, and the public.

The potential for impact in biotech is enormous. Advances in gene therapy, precision medicine, regenerative medicine, and AI-driven drug discovery are opening doors to radically improved health outcomes and quality of life for millions. This is what makes biotech so inspiring it blends scientific innovation with life-changing potential.

However, with that potential comes responsibility. Biotech entrepreneurs must balance innovation and commercial ambition with a commitment to patient safety and scientific rigor. Overpromising or cutting corners can damage not just your company, but also the reputation of the entire field.

As you consider launching a biotech venture, it's essential to assess your motivations, capabilities, and readiness. Do you have the scientific grounding to lead or guide development? Are you prepared for the capital intensity and long timelines? Do you have access to a network of expert advisors, collaborators, and potential funders?

If your answer is yes, biotech entrepreneurship offers an incredibly rewarding path. You'll work at the forefront of science, build solutions that improve and save lives, and contribute to shaping the future of medicine.

Remember, success in biotech rarely happens overnight. It demands resilience, adaptability, and an unwavering commitment to learning. New discoveries will emerge. Market conditions will shift. Your ability to evolve with them is what will set you apart.

As we begin this exploration, I encourage you to think boldly about the problems you want to solve and the difference you want to make. Biotech offers endless opportunities for innovation, purpose, and impact. With scientific insight, entrepreneurial drive, and ethical clarity, you have the power to help redefine what's possible in human health.

The journey of biotech entrepreneurship is not easy. But for those with the passion and perseverance to pursue it, it may be the most fulfilling and impactful endeavor one can choose. So let's dive in and discover what it truly takes to succeed in this demanding, dynamic, and deeply meaningful field.

Chapter 38
Building an Exceptional Team

One of the most important lessons I have learned in my entrepreneurial career is that people are everything. No matter how brilliant your technology or how disruptive your vision, the success of your venture ultimately hinges on the quality and cohesion of your team.

Surround yourself with individuals who share your values, complement your strengths, and bring diverse perspectives to the table. In interviews, look for key qualities like relevant experience, creative problem-solving, ability to thrive in uncertainty, collaborative spirit, and commitment to the mission.

At Iris Biotechnologies, I had the privilege of assembling an incredible team of scientists, clinicians, engineers, and business strategists from renowned institutions like UCSF, Stanford, MD Anderson, and UC Berkeley. Their collective expertise and dedication were instrumental in overcoming complex challenges and achieving our milestones.

Beyond recruiting top talent, it is equally important to foster a culture of continuous learning, open communication, and shared purpose. Invest deeply in your team's development through mentorship, cross-training, and clear growth pathways. Lead by example, demonstrating the values and work ethic you wish to see in your organization.

The highest-performing teams are aligned around a common vision, leverage one another's strengths, and operate with trust and psychological safety at their core. They are resilient in the face of setbacks and passionate about realizing the company's potential for impact. Building this kind of exceptional team takes concerted effort but the long-term returns are immeasurable.

For first-time founders, my advice is to be very selective in your early hires, seeking out people who will elevate the entire enterprise. Spend time deliberately shaping your culture and living your values. Prioritize transparency and feedback from the start, and create a work environment where people feel heard and empowered. Celebrate wins and candidly acknowledge challenges. Above all, recognize that your team is your most valuable asset and treat them with respect and care.

Building an exceptional team in the biotech industry presents unique challenges and opportunities. The interdisciplinary nature of the field requires bringing together individuals with diverse backgrounds from molecular biology and genetics to data science and clinical medicine. Each team member must not only excel in their area of expertise but also collaborate across disciplines and contribute to shared strategic goals.

When recruiting for a biotech startup, look for individuals who combine deep scientific knowledge with an entrepreneurial mindset. They should be comfortable with ambiguity and able to adapt quickly as new data emerges or market conditions change. Passion for the mission is crucial in the face of inevitable setbacks and long development timelines, it's this shared commitment to making a difference that will keep the team motivated and cohesive.

Cross-functional collaboration is essential. Create integrated teams that bring together R&D, product development, regulatory, and business strategy. This diversity of perspective fosters more innovative solutions and bridges the gap between scientific discovery and commercial application. Encourage regular interaction and knowledge-sharing across groups to build a culture of collaborative innovation.

In the early stages of a biotech startup, it's common to rely heavily on a network of advisors and consultants to supplement the core team's expertise. Choose these advisors carefully, looking for individuals who not only have relevant technical or industry knowledge but also understand the unique challenges of startup environments. Treat them as strategic partners, engage them meaningfully, set clear expectations, and involve them in key decision-making.

As you grow, be intentional about preserving the culture and values that fueled your early success. It's easy for the collaborative, innovative spirit of a small team to get diluted as the organization expands. Establish programs, rituals, and feedback mechanisms that reinforce your core principles and enable open communication and collaboration across departments and geographies.

Leadership in a biotech startup requires a delicate balance. You need to provide clear direction and make tough decisions while also empowering your team of highly skilled professionals to take ownership and drive innovation. Adopt a leadership style that blends vision and decisiveness with humility, active listening, and inclusive decision-making. Foster an environment where input from all levels is valued.

Developing talent should be a top priority. In the fast-moving world of biotech, continuous learning is not just desirable it's essential. Encourage your team members to stay current with the latest scientific literature, attend conferences, and pursue ongoing education. Explore formal professional development pathways, academic partnerships, or in-house training to ensure your team stays ahead of the curve.

Chapter 38
Building an Exceptional Team

Creating a diverse and inclusive team is particularly important in biotech. Different backgrounds and perspectives can lead to more innovative problem-solving and help ensure that your products and services are designed with a broad range of end-users in mind. This diversity should extend beyond just scientific disciplines to include gender, ethnicity, socioeconomic background, lived experience, and cognitive diversity.

Remember that building a great team is an ongoing process, not a one-time task. Regularly assess your team's strengths and weaknesses, and be proactive about addressing gaps. Use performance reviews, feedback loops, and strategic hiring to continuously evolve your team's capabilities. This might mean bringing in new hires, providing additional training, or reorganizing to better leverage existing talent.

Foster a culture of intellectual honesty and rigorous debate. In science-driven fields like biotech, it's crucial that team members feel comfortable challenging assumptions, pointing out potential flaws in experiments or analyses, and proposing alternative hypotheses. Create structured forums for critical discussion such as lab meetings, innovation reviews, or peer-led audits to embed this openness into your workflow. This culture of constructive criticism and open dialogue is essential for maintaining scientific integrity and driving innovation.

At the same time, this critical thinking should be balanced with a sense of shared purpose and mutual support. Biotech development often involves setbacks and failures experiments don't always yield expected results, clinical trials can fail, and regulatory hurdles can arise. Teams that operate with empathy, persistence, and collective resolve will weather these storms more effectively and emerge stronger.

Formal mentorship programs can help foster both skill development and cultural cohesion. Pairing junior team members with experienced colleagues accelerates learning, improves retention, and builds cross-functional understanding. These relationships can also help bridge silos between different departments, reinforcing a shared sense of mission and collaboration.

As your company grows, establish structured and transparent career pathways for your team members. In a startup environment, roles often evolve rapidly, which can be both exciting and challenging for employees. Conduct regular check-ins, set clear performance expectations, and co-create personalized development plans to ensure that your top talent sees a future for themselves within your organization.

Remember that compensation is essential, but it's not everything. In the competitive biotech talent market, you may not always be able to offer the highest salaries, especially in the early stages. However, you can compete by providing purpose-driven work, a compelling mission, real ownership opportunities, and a culture that prioritizes learning, collaboration, and well-being.

Finally, don't underestimate the importance of celebrating successes together as a team. In the often long and challenging journey of biotech development, taking the time to acknowledge and celebrate milestones whether it's publishing a key paper, achieving a critical experimental result, or securing a new round of funding can boost morale and reinforce the sense of shared purpose that binds your team together. Small rituals of recognition, team retrospectives, and shared storytelling can help maintain momentum and build cohesion.

Chapter 38
Building an Exceptional Team

Building an exceptional team in biotech is both an art and a science. It requires a keen eye for talent, a commitment to ongoing development, and the ability to create an environment where diverse individuals can come together to do their best work. When executed with intention and care, this team-building process becomes the cornerstone of breakthrough innovation and lasting impact in human health.

Chapter 39
Defining Your Market Strategy

Another critical factor that can make or break a young company is its go-to-market strategy. Having a breakthrough idea is one thing, but deeply understanding your target customers, market landscape, and competitive positioning is essential to successfully commercialize that idea.

Before launching Iris Biotechnologies, we conducted extensive market research and competitive analysis. We assessed the size and growth trajectory of the personalized medicine space, identified key players and potential partners, and honed in on the specific applications where our technology could add the most value. This foundational work informed every aspect of our business model and product development roadmap. A winning market strategy requires clear answers to questions like:

- Who are your core customers, and what are their most pressing needs?
- How does your offering uniquely address those needs compared to alternatives?
- What are the key trends, regulatory factors, and competitive forces shaping your market?
- What is your pricing and distribution model to effectively reach and serve customers?
- How will you build brand awareness, trust, and loyalty in your target market?

The sharper and more differentiated your answers to these questions, the better equipped you will be to allocate resources efficiently and adapt to changing market conditions. In the fast-moving world of biotechnology, precision in market strategy must be balanced with agility to respond to new clinical insights, policy developments, and scientific breakthroughs.

For Iris Biotechnologies, our choice to focus on oncology, particularly breast cancer analysis and treatment, was driven by a combination of factors: the strong unmet clinical need, the availability of robust genomic data, the alignment with our team's expertise, and the massive potential market. By intentionally narrowing our focus to breast cancer, we were able to establish a strong value proposition and build credibility with key opinion leaders and partners.

As the company grew, we continued to refine our market segmentation and positioning, expanding into new therapeutic areas and diagnostic modalities based on a clear view of our unique strengths. We also invested heavily in market development activities like clinical education, advocacy engagement, and health economic evidence generation to help shape the overall ecosystem for precision medicine adoption. These efforts allowed us not only to build demand for our solutions but also to contribute meaningfully to the broader adoption of personalized medicine.

Defining your market strategy is an ongoing process that requires customer engagement, competitive vigilance, and a willingness to pivot as needed. It also requires a balance between focus and flexibility, between confidence in your vision and humility to learn from the market. Getting this balance right is both an art and a science, but it is well worth the effort.

For aspiring founders, I recommend devoting significant time upfront to pressure testing your assumptions about the market and your place in it. Seek out mentors and advisors with deep industry expertise, but also get out of the building and talk directly to potential customers and partners.

Listen actively, challenge your assumptions, and be open to unexpected insights. Be prepared to iterate based on what you learn while staying true to your core purpose. A strong market strategy is the foundation upon which everything else is built.

In the biotech industry, defining your market strategy comes with unique considerations. The "customers" in healthcare are often multi-faceted you may need to consider patients, healthcare providers, payers, and regulators in your strategy. Each of these stakeholders has different needs, decision-making processes, and value drivers that you'll need to understand and address.

It's also crucial to recognize the long development timelines typical in biotech. Your initial market strategy needs to account for how the landscape might evolve over the years it takes to bring a product from concept to market. This includes anticipating shifts in clinical practice, regulatory frameworks, and reimbursement models. This requires a deep understanding of scientific trends, emerging technologies, and shifting healthcare paradigms.

When defining your target market, consider not just the size of the opportunity but also the feasibility of accessing it. In healthcare, factors like reimbursement policies, clinical guidelines, and standard-of-care practices can significantly impact market adoption.

Your strategy should include plans for generating the clinical evidence and health economic data needed to support the uptake of your product. Proactively engaging with payers and health systems early on can help shape a reimbursement path aligned with your business goals.

Competitive analysis in biotech should look beyond just direct competitors. Consider alternative approaches to addressing the same clinical need, as well as products or technologies that might be in development. Also, be aware of the potential for disruptive innovations that could fundamentally change the treatment paradigm in your target area. Scenario planning and technology forecasting can be powerful tools to help you prepare for these eventualities.

Partnerships often play a crucial role in biotech market strategies. Consider early on whether you plan to commercialize your product independently or in partnership with larger pharmaceutical or diagnostic companies. Each path has its pros and cons and will significantly impact your resource needs, timeline, and potential market reach.

Strategic alliances can provide access to distribution networks, regulatory expertise, and commercial infrastructure that would otherwise take years to build. On the other hand, going it alone may allow you to retain greater control and value capture so choose your path carefully based on your long-term goals.

Pricing strategy in healthcare is complex and heavily scrutinized. You'll need to balance the value your product delivers with the increasing pressure for cost-effectiveness in healthcare. Consider engaging health economists early to help build the value story for your product.

Health technology assessment (HTA) agencies and payer organizations will look closely at your product's cost-benefit profile, so having a compelling and data-driven narrative around clinical and economic value is key.

Regulatory strategy is inseparable from market strategy in biotech. Your choice of initial indication, the specific claims you aim to make about your product, and your plans for future label expansions will all impact your regulatory pathway. These decisions should be made with both scientific and commercial considerations in mind. Engage with regulatory consultants early and consider seeking feedback from agencies like the FDA or EMA during pre-submission meetings to de-risk your development path.

Remember that in biotech, your market strategy isn't just about selling a product it's about advancing a new approach to healthcare. Education and advocacy often play a big role. You may need to invest in raising awareness about the underlying science, changing clinical practices, or even shaping healthcare policy to create a receptive market for your innovation. Thought leadership, peer-reviewed publications, and engagement with medical societies can all help position your product within the broader clinical and scientific ecosystem.

Lastly, don't underestimate the power of patient voices in shaping healthcare markets. Engaging with patient advocacy groups and incorporating patient perspectives into your product development and commercialization strategies can be hugely valuable.

Patients are not just end-users they are increasingly influential stakeholders in driving demand, guiding trial design, and influencing payer and regulatory decisions. Building authentic relationships with these communities can differentiate your company and product in meaningful ways.

Developing a robust market strategy in biotech is a complex undertaking, but it's absolutely critical for success. It requires a unique blend of scientific insight, commercial acumen, and healthcare system knowledge. It's about building bridges between science and the clinic, between innovation and access, between vision and execution. But get it right, and you'll be well-positioned to translate your scientific breakthroughs into real-world impact, improving patient outcomes and transforming healthcare delivery.

Chapter 40
The Intellectual Property Imperative

In knowledge-intensive industries like biotechnology and life sciences, a robust intellectual property (IP) strategy is essential for protecting innovation, securing competitive advantage, and unlocking value through partnerships and transactions. However, developing and executing this strategy can be a complex and costly undertaking with little margin for error.

At Iris Biotechnologies, safeguarding our IP was a top priority from day one. We engaged experienced patent attorneys to help us craft strong, defensible patents around our core technology platforms and applications. We also implemented rigorous trade secret policies and confidentiality agreements with all employees and research partners to further protect proprietary knowledge.

One of our key early IP milestones was the filing of a foundational patent for our *Artificial Intelligence System for Genetic Analysis* in 2000. In 2008, the U.S. Patent Office informed our law firm that the application was ready to be granted a thrilling moment that would have validated years of innovation. But instead of celebrating, we faced an IP crisis.

Our law firm at the time, Heller Ehrman, mishandled critical correspondence from the USPTO, resulting in a deemed abandonment of the application. Reviving it took years of litigation, significant financial resources, and created massive uncertainty at a pivotal moment for the company. Heller's malpractice cost Iris an estimated $100 million in damages, and the experience underscored just how high the stakes are when it comes to IP in biotech.

Up until that point, we had worked closely with our patent attorney, and all of our previous applications had been granted worldwide. In this case, however, we were specifically advised to wait and take no further action. Unbeknownst to us, the entire Menlo Park patent team left Heller Ehrman on the same day, leaving no one to receive or respond to the USPTO's communication. The firm never informed us. Adding to the chaos, Heller filed for bankruptcy on December 28, 2008 just months after the collapse of Lehman Brothers and the failure of Heller's merger with Mayer Brown. Mayer Brown was the law firm for Lehman Brothers and about 50% of the Wall Street firms. The fallout of this timing amid the 2008 financial crisis exacerbated an already devastating mistake.

This experience taught us a hard but invaluable lesson: your IP counsel is as critical as your science. In biotech, a company's patent portfolio is often its most valuable asset the foundation of defensibility, investability, and strategic leverage. Any missteps in filing, prosecution, or enforcement can have catastrophic consequences.

Another core lesson was the importance of aligning IP strategy with business strategy. Rather than filing opportunistically, we became far more disciplined building a portfolio that supported our commercial goals, geographic priorities, and valuation strategy. We focused not just on patent quantity, but on relevance, enforceability, and the potential to block competitors or enable key partnerships.

We also became more strategic in how we communicated our IP story to investors and partners. By presenting a cohesive narrative around our IP's strength, scope, and alignment with market needs, we secured more favorable deal terms and increased investor confidence. Our IP portfolio evolved from a technical asset into a central pillar of our broader value-creation strategy.

For other biotech founders, I cannot overstate the importance of investing in a sophisticated, forward-looking IP strategy from the outset. That means not only hiring top-tier patent counsel, but also becoming educated yourself on the fundamentals of patent law, data exclusivity, freedom to operate, and competitive landscaping. It means integrating IP planning into your R&D and commercialization timelines, and staying proactive about reviewing and evolving your portfolio as your business evolves.

The complexity of biological systems, the rapid pace of scientific advancement, and the high stakes of medical innovation all make IP strategy in biotech particularly nuanced. One key consideration is the interplay between patent protection and data exclusivity. In many jurisdictions, biologic drugs are eligible for extended periods of market exclusivity through regulatory pathways often providing stronger or more predictable protection than patents alone. A sophisticated biotech IP strategy must thoughtfully leverage both.

Developing and defending a strong IP portfolio is expensive and time-consuming, but it is non-negotiable for building a durable and valuable company. In an industry where innovation is the product, a sound IP strategy is not just a legal formality it's a competitive weapon and a cornerstone of enterprise value.

Another critical dimension of IP strategy in biotechnology is the evolving legal landscape surrounding patent-eligible subject matter. In recent years, court decisions particularly in the U.S. have introduced significant uncertainty around the patentability of certain biotech innovations, including gene sequences, diagnostic methods, and some AI-driven technologies. Navigating this shifting terrain requires patent applications that are scientifically robust and legally resilient, crafted to withstand intense scrutiny under today's more restrictive standards.

Freedom to operate (FTO) analysis is equally vital. Biotech is a densely patented field, and the risk of inadvertently infringing on existing IP is high. A single missed claim can derail years of development. Conducting comprehensive FTO analyses early and revisiting them regularly as your product evolves is essential. Developing design-around strategies, securing licenses, or even adjusting development trajectories may be necessary to avoid costly litigation or commercial delays.

Patents in biotech often build upon prior inventions in intricate and layered ways. This makes understanding the broader patent landscape and identifying opportunities for improvement patents, new uses, or novel combinations of existing technologies a powerful strategic tool. Doing so requires not only legal and commercial savvy but also deep scientific insight and creative problem-solving.

Collaboration is common in biotech whether with academic institutions, other biotech firms, contract research organizations, or public-private partnerships. Each collaboration introduces complex IP considerations. Questions of ownership, licensing, and the handling of background IP and jointly developed innovations must be clarified upfront through well-structured agreements. A lack of clarity here can lead to disputes that delay progress and diminish the value of your IP.

In addition to patents, trade secrets play a key role in biotech IP strategy. For certain innovations such as proprietary manufacturing processes, AI algorithms, or unpublished data analytics tools trade secret protection may offer more durable or practical advantages than patenting. However, this requires strong internal controls, access limitations, and formal policies to ensure enforceability.

Given the global nature of biotechnology, an international IP strategy is non-negotiable. Protecting your innovations means filing in multiple jurisdictions, each with its own legal standards, enforcement challenges, and strategic considerations. Being aware of international treaties, such as the Patent Cooperation Treaty (PCT), and the IP enforcement environment in key markets like the EU, China, and India is crucial for long-term success.

As biotech continues to converge with adjacent fields artificial intelligence, nanotechnology, robotics, synthetic biology new layers of complexity and opportunity emerge. A future-ready IP strategy must include cross-disciplinary expertise to protect innovations that sit at these intersections and to capitalize on emerging markets and regulatory frameworks.

Importantly, intellectual property is not just a defensive shield it's a strategic asset for value creation. A strong IP portfolio can attract investment, secure lucrative partnerships, create barriers to entry for competitors, support pricing and reimbursement negotiations, and generate revenue through licensing or acquisitions. It is the engine that drives commercial traction in an industry where the product may take years to reach market.

In conclusion, while the Heller Ehrman incident was a painful chapter in Iris Biotechnologies' journey, it offered an unforgettable lesson: in biotech, a single legal oversight in IP can jeopardize everything. Vigilance, strategic foresight, and reliable legal counsel are not optional they are foundational. For biotech entrepreneurs, treating intellectual property as a core business priority not just a legal formality can mean the difference between success and failure.

By investing early in sophisticated IP strategy, multidisciplinary planning, and strong governance, companies can safeguard their innovations, multiply their enterprise value, and ultimately fulfill the promise of transforming lives through science.

Chapter 41
Securing the Right Capital

Funding is the fuel that powers the engine of entrepreneurship and nowhere is that more evident than in capital-intensive sectors like biotech, where innovation demands both time and significant resources. But not all funding is created equal. The choices you make about when, how, and from whom to raise money can have profound implications for your company's trajectory sometimes determining not just growth, but survival.

At Iris Biotechnologies, we were fortunate to have a strong network of supporters and a compelling vision that attracted interest from a diverse range of investors. In our early days, we relied primarily on angel investors and strategic partners individuals who were aligned with our mission and willing to bet on our team before we had fully proven ourselves.

These investors provided not just financial capital but also critical intangible resources: seasoned advice, industry connections, and early validation of our mission. They understood the long product development timelines in our industry and were patient in their expectations for returns. Their support gave us the runway to focus on building a strong scientific foundation and generating robust proof-of-concept data, a luxury not all startups in biotech can afford.

As we hit key technical and clinical milestones, our capital needs naturally grew. At that stage, we began preparing to take the company public, a major strategic shift that required a different kind of investor. We prioritized alignment over hype. We looked for partners who shared our values and long-term vision, rather than chasing the highest valuation or fastest term sheet. We were deliberate in how we staged each round of financing, seeking to minimize dilution and maintain strategic flexibility.

This measured approach paid off. By diversifying our funding sources and staying disciplined about when and how much to raise, we were able to maintain a healthy balance sheet and protect a strong equity position for our employees and early backers. Preserving equity isn't just about ownership. It's about motivation, retention, and culture.

For other founders considering the fundraising journey, my advice is to start with clarity. Get crystal clear on your milestones, timelines, and capital requirements. Be realistic about how much you need to raise to reach meaningful inflection points, and always build in a cushion for unexpected setbacks because they will come.

Next, carefully evaluate what kind of investor is the best fit for your current stage and long-term goals. Angel investors can be great partners in the early days when you need believers who can move fast and aren't scared off by ambiguity. Venture Capitalists can be powerful allies when it's time to scale, but be sure to do your diligence on their reputation, track record, and value-add beyond just money. Ask former portfolio companies not just how the VC behaved during the highs, but how they responded during the lows.

Strategic corporate investors can provide invaluable domain expertise and commercial insights but may come with complex strings attached. Make sure you understand those terms and the potential long-term consequences before taking their capital.

In the end, fundraising isn't just about cash. It's about building the right coalition of believers; people who will stand with you through the uncertainties of innovation and help you turn vision into reality.

When engaging with investors, focus on building genuine relationships, not just delivering transactional pitches. Too often, founders treat investor meetings like one-off sales presentations. But long-term capital partnerships thrive on authenticity, mutual respect, and shared conviction. The best investors are those who take the time to truly understand your business, provide honest feedback, and stick with you through the ups and downs. They don't just write checks, they challenge assumptions, open doors, and help you level up as a leader.

They can be force multipliers for your success, but only if there's a foundation of trust and alignment. That foundation starts early, often before a term sheet is even discussed. Communicate clearly, follow through on your commitments, and be transparent about both your vision and your challenges. Investors don't expect perfection, they expect integrity and adaptability.

Remember that fundraising is a means to an end, not the end in itself. Don't get so caught up in the chase for the next round that you lose sight of the fundamentals of building a great business. Product, team, traction, and culture - these are the true engines of value. The most successful fundraising efforts flow naturally from a company's strength and momentum, not from desperate quick fixes. If you build something truly valuable, capital will find you.

Finally, pace yourself. Fundraising is a marathon, not a sprint. Choose your partners wisely, and always prioritize long-term alignment over short-term gain. When done right, fundraising becomes more than just financing. It becomes a strategic advantage that propels your mission forward.

In the biotech sector, funding considerations take on added complexity due to the long development timelines, high capital requirements, and unique risk profile of life science ventures. Here are some additional points to consider:

1. Non-dilutive funding: In biotech, there are often opportunities for non-dilutive funding through government grants, research contracts, and partnerships. These can be excellent ways to advance your research and development without giving up equity. Programs like the NIH's Small Business Innovation Research (SBIR) grants can be particularly valuable for early-stage companies.

2. Staged financing: Given the long development timelines in biotech, it's common to raise money in stages tied to specific milestones. This allows you to minimize dilution in the early, high-risk stages and raise larger amounts at higher valuations as you de-risk the technology.

3. Crossover investors: As you approach later stages of development, consider engaging with crossover investors - those who invest in both private and public companies. These investors can provide a bridge to the public markets and help prepare your company for an eventual IPO.

4. Patient capital: Look for investors who understand the long timelines in biotech and are willing to be patient. Some specialized life science venture funds are structured with longer fund lifetimes to accommodate the extended development cycles in this industry.

5. Regulatory expertise: Investors with experience navigating the complex regulatory landscape in healthcare can be invaluable partners. They can help you anticipate and prepare for regulatory hurdles, potentially saving you time and money.

6. Syndication: In biotech, it's common to form syndicates of investors for larger rounds. This can help spread the risk and bring together complementary expertise. However, be mindful of potential conflicts or competing agendas among your investor group.

7. Alternative financing structures: Consider exploring alternative financing structures like royalty financing or venture debt. These can be useful tools for extending your runway without further dilution, especially as you approach commercialization.

8. International investors: Given the global nature of the biotech industry, don't limit yourself to domestic investors. International investors, particularly from biotech hubs in Europe and Asia, can provide not just capital but also valuable connections and insights into global markets.

9. Strategic alignments: When considering strategic corporate investors, think carefully about how their involvement might impact future partnering or M&A opportunities. While their industry expertise can be invaluable, make sure any deal terms preserve your flexibility.

10. Public markets: For some biotech companies, going public can be an attractive option for accessing larger pools of capital. However, this comes with increased scrutiny and reporting requirements. Carefully weigh the pros and cons and ensure you have the infrastructure to operate as a public company before pursuing this route.

11. Investor education: Many investors, even sophisticated ones, may not fully understand the intricacies of your technology or the specific challenges of your subsector. Be prepared to invest time in educating potential investors about your

science, your market, and the broader context of your work.

12. Valuation considerations: Valuing early-stage biotech companies can be challenging due to the long time horizons and binary nature of many outcomes (e.g., clinical trial success or failure). Work with your advisors to develop a robust and defensible valuation model that accounts for the unique aspects of your technology and market opportunity.

13. Cash management: Given the high burn rates typical in biotech R&D, sophisticated cash management is crucial. Consider working with financial advisors who have specific experience in biotech to help you optimize your cash deployment and runway extension strategies.

14. Investor synergies: Look for investors who can offer synergies with your business beyond just capital. This might include access to specialized facilities, connections to key opinion leaders in your field, or expertise in areas like manufacturing or commercialization that complement your internal capabilities.

Funding continuity: In biotech, it's critical to always be thinking several steps ahead in your funding strategy. Start preparing for your next round even as you close your current one. This ensures you have the runway to achieve meaningful milestones and aren't forced into unfavorable terms due to cash constraints. Remember, at the end of the day, the fuel you need to power through the entrepreneurial journey is only partly financial. Equally vital is the fuel of passion, perseverance, and purpose - the internal drive that sustains you through setbacks, pivots, and long development cycles. By staying true to your mission, taking care of your team, and making consistent progress, you will attract the right capital at the right time to help you reach your destination.

In biotech, perhaps more than in any other field, the alignment between your scientific vision, business strategy, and funding approach is critical. The right investors can provide not just the capital to advance your R&D, but also the strategic guidance, industry connections, and long-term support necessary to navigate the complex path from scientific discovery to market impact. This alignment becomes your compass guiding not only what you build, but also who you build it with.

As you build your funding strategy, always keep the ultimate goal in mind: to translate scientific breakthroughs into therapies or technologies that can improve human health. This mission-driven approach, combined with a sophisticated understanding of the unique funding landscape in biotech, will position you well to secure the capital you need to bring your innovations to life. Raising money should never feel like a detour from your mission. It should be a reflection of it.

Raising capital usually starts with a business plan. Crafting a strong business plan is both an art and a discipline. It forces you to crystallize your thinking around product-market fit, regulatory strategy, competitive landscape, and financial runway. It's difficult to write a great one, but when you do, you'll know it. A solid business plan doesn't just impress investors; it clarifies your own vision and helps you make better decisions.

Having served as a judge in the Global VCIC, the world's largest venture capital investment competition with over 120 university and graduate school teams competing, I've seen firsthand how powerful it is for founders to practice telling their stories. VCIC is the only platform where students step into the shoes of Venture Capitalists (VCs) for the day, while real startups gain exposure and feedback to accelerate their fundraising process.

My advice: seek out opportunities to rehearse the fundraising experience before the stakes are high. One such venue is Santa Clara University's California Entrepreneurship Program, where I've served as a mentor. These practice rounds help refine your pitch, stress-test your assumptions, and expose you to the questions and concerns real investors will raise - all in a lower-pressure environment.

In the high-stakes world of biotech entrepreneurship, knowledge is power, but preparation is momentum. Take the time to hone your narrative, build authentic investor relationships, and align every dollar you raise with your long-term vision. That's how real progress happens and how real impact begins.

Chapter 42
Starting Small and Government's Role

My professional entrepreneurship journey has centered on medium-sized ventures with the potential to scale into major enterprises. In many ways, it mirrored the humble yet bold beginnings of tech giants like Apple. On April 1, 1976, Steve Jobs, Steve Wozniak, and Ronald Wayne founded Apple Computer Company, registering it as a California business partnership.

The first Apple computer was designed in Wozniak's apartment and built in Jobs' garage. Their original headquarters was in Cupertino, California, just adjacent to Saratoga, where I started Iris Biotechnologies in the most personal of spaces: my living room.

In the early days, one of my employees worked out of my garage, which wasn't a traditional workspace. It was a unique hybrid of an office and a personal library, lined wall to wall with bookshelves holding about a thousand books, DVDs, and CDs. Another 1,500 books were stored throughout the house. I never used the garage for parking; instead, I set it up with tables, a combination lock on the door, and the freedom for her to come and go independently except during scheduled meetings. That makeshift setup was our first lab, our first office, and our first step toward building something meaningful.

First and foremost, anyone considering entrepreneurship needs to be crystal clear on his or her purpose. Why do you want to start a company? What is the impact you hope to make? What relevant knowledge or experience do you bring to the table? And perhaps most importantly, can you attract the right people - those who not only believe in your mission but also can help you secure the resources your business needs to thrive?

If you're entering a highly competitive space like biotech, for example, it's vital that both your family and your partners' families understand the risks involved. They should be prepared for the significant demands in time, capital, and emotional bandwidth. Entrepreneurship isn't just a professional decision; it's a personal and relational one, too.

When it comes to forming a company, one of your first decisions is choosing the legal structure. Most people assume a C corporation is the default, but that's not always the best option. From a tax standpoint, an S corporation may offer advantages, especially for smaller, closely held companies. Similarly, deciding whether to incorporate in California or Delaware can have long-term legal and financial consequences.

You also need to consider whether a corporation or a limited liability company (LLC) best suits your goals. Before hiring a lawyer to handle your incorporation, take the time to educate yourself. The more you understand upfront, the better positioned you'll be to make informed, strategic decisions.

Legal costs in business can be staggering. Corporate matters, patent filings, enforcement actions, and intellectual property disputes can run into the millions. Even defending against frivolous lawsuits can drain your resources because whether you're right or wrong, you'll still have to pay legal fees to protect your company.

And remember, corporations cannot represent themselves in court. Even if you're knowledgeable enough to draft your own filings or mount a strong legal argument, the law requires that a licensed attorney represent the corporation. That's another reason to budget wisely and build strong legal relationships early on.

It can take years, sometimes a decade or more, before a patent application is granted, if it's granted at all. So, it's critical that your patent attorney is not just competent but also trustworthy, detail-oriented, and financially stable enough to stay with you for the long haul. Patent law varies widely across jurisdictions, the U.S., European Union, Japan, and other nations each have distinct standards and review processes. What gets approved in one country may be narrowed or rejected in another.

Make sure your patent claims are enforceable, not just granted. It's not enough to have a patent on paper; it has to stand up to legal scrutiny if challenged. And here's the nuance many entrepreneurs miss: the attorneys who specialize in helping you obtain patents, known as patent prosecution attorneys, are often not the same ones who specialize in enforcing them in court. These are two very different legal disciplines. Choose your words carefully during the application process, and consult a seasoned patent litigation or enforcement attorney when drafting claims that may one day be tested. Sometimes, a single word can be worth millions, or cost you just as much.

We spent over $1,000 per hour on our corporate attorney and slightly less on our patent attorney. That kind of billing forces efficiency. We tried to do as much groundwork as possible before involving the lawyers, drafting documents, gathering data, outlining claims. The goal was to enter every legal meeting informed, prepared, and specific in our needs. Good legal advice can be invaluable, but poor or uninformed counsel can derail your startup or cost you the company.

Choose your advisors wisely and get second opinions when needed.

Some entrepreneurial ventures push the boundaries of innovation so far that they require government involvement, strategic diplomacy, or international cooperation. These aren't your typical startups, they're global-scale initiatives with geopolitical implications.

Take, for example, the New Silk Road, officially known as China's Belt and Road Initiative (BRI). This massive infrastructure and development program, launched in 2013 by President Xi Jinping, was originally conceived to connect East Asia and Europe via a network of roads, railways, and maritime routes. But by 2024, the BRI had expanded its scope and influence to Africa, Oceania, and Latin America, reshaping global trade corridors and extending China's economic reach. As of 2025, more than 150 countries have signed on to participate in some form.

While China promotes the BRI as a win-win development effort, many analysts in the West and in Asia view it with caution. Some consider it a geopolitical Trojan horse, a mechanism for extending Chinese influence under the guise of economic partnership. Ballooning costs, mounting debt in recipient nations, and shifting public sentiment have led to growing opposition in certain regions. The United States shares concerns that the BRI could enable China's military and political expansion in strategic regions.

Still, there is no doubt that this trillion-dollar initiative will change the world. The hope is that it will uplift billions by improving infrastructure, connectivity, and economic opportunity, though the long-term effects remain to be seen.

This tension between invention and implementation is nothing new. Historically, the Chinese were the original inventors of transformative technologies such as paper and woodblock printing. But it was Johannes Gutenberg's invention of the movable-type press in Europe that revolutionized information distribution and sparked the global spread of books, including the Bible.

Similarly, the Chinese pioneered gunpowder and early cannon technology. Yet it was the Turks who turned it into a formidable military force, using it to build the Ottoman Empire. At its peak, this empire stretched across Southeast Europe, West Asia, and North Africa, dominating trade, culture, and politics from the 14th to the early 20th centuries. From the early 1500s to the 1700s, it even controlled parts of Central Europe. The empire eventually crumbled following World War I, but its legacy lives on in modern geopolitics.

Fast forward to today, the United States was the birthplace of the Internet, a defining technology of our era. American companies like Apple, Microsoft, Google (Alphabet), Facebook (Meta), Amazon, and Nvidia continue to shape the digital landscape. On the other side of the Pacific, China has rapidly developed its own tech giants, including Alibaba, Tencent, ByteDance, Baidu, and Huawei.

TikTok, owned by Beijing-based ByteDance, became the most downloaded app of all time in 2024, underscoring China's growing dominance in digital media and global influence.

Legal frameworks have played a key role in enabling this digital transformation. On October 21, 1998, President Bill Clinton signed a groundbreaking bill that imposed a ten-year moratorium on discriminatory and multiple taxation of the

Internet and electronic commerce. This policy fostered innovation and encouraged early growth in the tech sector. Congress extended the moratorium several times before making it permanent on February 24, 2016. Soon after, the European Union implemented similar protections, helping to create the global e-commerce ecosystem we know today.

As a result of the U.S. government's Internet tax relief policies, tech companies like Amazon were able to grow rapidly, overtaking traditional bookstores and retailers. Amazon began by selling books online, but its business model evolved. Today, Amazon operates out of vast, strategically placed warehouses and sells almost every consumer product imaginable. In addition to its massive retail presence, the company also dominates the cloud services industry through Amazon Web Services (AWS), offering secure, reliable, and scalable solutions for websites, streaming, big data, and artificial intelligence platforms.

On January 26, 2024, Amazon had a market value of $1.64 trillion. In comparison, Apple and Microsoft were valued at $2.98 trillion and $3.00 trillion, respectively. Alphabet (Google's parent company) stood at $1.91 trillion, and Meta (formerly Facebook) at $1.01 trillion. These valuations reflect a dramatic consolidation of digital power and wealth in just a few decades, driven by technological innovation and global demand.

Artificial Intelligence (AI) is now the biggest technology race among these companies. Nvidia, a pioneer in graphics processing units (GPUs), has emerged as the global leader in AI chip manufacturing. Its GPUs are essential for powering the machine learning systems behind everything from ChatGPT to autonomous vehicles. Nvidia became the third publicly traded U.S. Company to reach a $2 trillion market valuation, following Apple and Microsoft.

Meanwhile, Tesla Motors revolutionized the electric vehicle (EV) market, transforming EVs from novelty items into aspirational products. Founded in 2003, Tesla introduced its first vehicle, the all-electric Roadster, in 2008. Elon Musk, who joined as chairman and later became CEO, has been the company's public face and chief innovator. As of January 26, 2024, Tesla's market capitalization stood at $574.21 billion.

In 2023, Tesla delivered 1.81 million vehicles, representing a 38% year-over-year increase. Its production output grew by 35% to 1.85 million units. This rapid scale-up was supported by Tesla's gigafactories across the U.S., Germany, and China, which help meet global demand. However, Tesla is now being rivaled by China's BYD (Build Your Dreams), which shipped its first electric car in 2009. BYD, backed in part by Warren Buffett's Berkshire Hathaway, sold approximately 1.57 million EVs in 2023 a 73% increase from the previous year.

In 2024, BYD narrowly overtook Tesla in production numbers, with 1,777,965 electric vehicles manufactured, compared to Tesla's 1,774,442. This milestone marks a significant power shift in the global EV market, highlighting China's growing industrial strength in clean energy technologies.

Government subsidies have played a pivotal role in boosting consumer adoption of EVs, especially in the United States. The average direct federal subsidy per EV over ten years is nearly $9,000. When including local utility company incentives and rebates, the total subsidy can exceed $10,000 per vehicle. These policies aim to reduce greenhouse gas emissions and accelerate the transition to a more sustainable transportation system.

The journey to electric mobility was built on the back of combustion engine innovation. In 1872, American engineer George Brayton invented the first liquid-fuel internal combustion engine. Four years later, in 1876, Nicolaus Otto, Gottlieb Daimler, and Wilhelm Maybach developed and patented the compressed-charge, four-stroke cycle engine, which became the blueprint for modern automotive engines. In 1879, Karl Benz patented a two-stroke gas engine, laying the groundwork for the automobile era. These early breakthroughs ushered in the Age of Automobiles and the global reliance on petroleum-based fuels.

Oil, often referred to as "Black Gold," is one of the world's most valuable and strategic natural resources. It powers cars, planes, ships, and industrial machinery. But oil's impact goes beyond energy. It is a raw material in countless products, from plastics and pharmaceuticals to paints, fertilizers, detergents, and even cosmetics.

Before the mass extraction and refining of crude oil, whale oil was one of the primary sources of fuel, especially for lamps. Whale oil burned cleanly and without much odor, making it highly desirable. It was also used in lubricants, soaps, textiles, varnishes, explosives, and paints. However, as whale populations dwindled and costs rose, the industry declined, leading to the rise of petroleum as a cheaper, more scalable alternative.

Nantucket, Massachusetts, was once the epicenter of the American whaling industry. One of its most famous whaling ships, the Essex, was rammed and sunk by a sperm whale in 1820. That true story inspired Herman Melville's iconic novel Moby-Dick. I enjoyed reading that book and visiting the Nantucket Whaling Museum, a powerful reminder of how human ambition and nature collide in history. I also enjoyed a trip to Martha's Vineyard just 38 miles (63 km) from Nantucket and viewing the Kennedy Compound from the sea.

The Kennedy Compound, located on Cape Cod along Nantucket Sound, consists of three houses spread over six acres (2.4 hectares) of waterfront property. It was once the summer home of Joseph P. Kennedy, an American businessman, investor, and U.S. Ambassador to the United Kingdom. He lived there with his wife, Rose, and their children, including U.S. President John F. Kennedy and U.S. Senators Robert F. Kennedy and Edward M. Kennedy. The compound remains an enduring symbol of American political legacy and public service.

Interestingly, in the 1860s, U.S. government policy favored kerosene as a cleaner and cheaper lighting alternative, keeping taxes on it low. This strategic move accelerated the decline of whale oil, marking a pivotal moment in America's transition to fossil fuels. This was one of the earliest examples of how government incentives and tax policies could dramatically reshape entire industries.

In refining crude oil to make kerosene for lighting, gasoline was discovered as a by-product. Initially, it was discarded as waste. However, due to its ability to vaporize at low temperatures, gasoline later became a revolutionary fuel for engines. At a depth of almost 70 feet (21 m), Edwin L. Drake struck the first oil well in the United States in 1859 near Titusville, Pennsylvania, marking the birth of the modern petroleum industry.

Gasoline became a necessity in the automotive industry after Nikolaus Otto developed the four-stroke internal combustion engine in 1876. This invention not only shaped transportation but also set the stage for mass industrialization across sectors. Today, almost all gasoline is used to fuel cars, while a small portion powers agricultural equipment, lawnmowers, boats, and smaller piston-engine planes.

Fossil fuels like petroleum supply more energy to the world today than anything else. About 45% of petroleum becomes gasoline. Other lighter chemicals derived from refining include natural gas, liquefied petroleum gas (LPG), jet fuel, and kerosene. These lighter fractions fuel everyday energy needs such as cooking, flying, and home heating.

A lot of the heavier products go into lubricants, plastics, and asphalt. One of the byproducts of oil refining is petroleum coke, which is important for fertilizer production. Over half of the world's known crude oil reserves are located in the Persian Gulf basin. The US is the biggest oil consumer. Despite having significant domestic production, the U.S. still relies heavily on imports and global oil pricing.

The consequence of the U.S. government promoting kerosene through low taxation ushered in the era of oil dominance. Increased subsidies for oil especially for gasoline dramatically impacted the earth in both good and bad ways. In addition to providing affordable energy for heating, transportation, and electricity generation, oil also pollutes the environment and contributes to climate change. Greenhouse gas emissions from burning fossil fuels are a leading cause of global warming, rising sea levels, and severe weather events.

It's convenient to use plastics, but plastic waste is a major environmental and health problem. Fertilizers and pesticides, also made from oil, are widely used in agriculture. These chemicals increase crop yield but can also contaminate soil and water, disrupt ecosystems, and affect human health.

"Who Pays the Price: The Real Cost of Fossil Fuels" was the opening statement delivered by U.S. Senator Sheldon Whitehouse (D-RI), Chairman of the Senate Budget Committee, on May 3, 2023. In 2022, fossil fuel subsidies hit a record $1 trillion globally, while Big Oil companies earned $4 trillion. This stark contrast highlights the imbalance between public spending and corporate profit, raising serious ethical and environmental questions.

Many farmers choose to use herbicides, insecticides, and fungicides to protect their crops and enhance soil nutrients. Pesticides are chemical and organic mixtures, including insecticides, fungicides, and plant growth regulators. Herbicides are designed for plants. Pesticides kill insects, rodents, and fungi. Fertilizers are plant nutrients added to the soil to promote growth. However, overreliance on these chemicals has led to pesticide-resistant pests and nutrient depletion in soil, prompting the need for more sustainable agricultural practices.

Historically, the Queen owned the most land in England. The Church owned the second most. Neither the Queen nor the Church worked for the land, but commoners accepted it. This long-standing tradition of inherited wealth and unequal land distribution has shaped class divisions and property laws that still persist today.

Many nations have gone to war because of oil. Before oil, nations invaded others for land, spices, and commodities like gold. Oil has become the modern strategic resource, driving foreign policy, trade agreements, and military action in regions like the Middle East.

It is interesting to note that the world's richest oil and gas reserves are centered in the region of the Garden of Eden mentioned in the Bible. The Middle East, especially the

Persian Gulf, is a hub for oil and gas. This convergence of spiritual lore and natural resources has added geopolitical and cultural complexity to the region's modern identity.

Tree trunks, leaves, twigs, and roots become coal, and oil is formed from ancient algae and cyanobacteria under immense heat and pressure. Oil is primarily made of carbon and hydrogen. Methane, CH_4, is the simplest hydrocarbon. Oil components like octane (C_8H_{18}) contain larger hydrocarbon molecules. With too much underground heat, oil breaks down to make methane. These processes span millions of years, underscoring the non-renewable nature of fossil fuels.

There was a time and a place where people valued salt more than gold, and people simply lived on land. Now, people trade away most of their lives working for a tiny piece of land. One in three U.S. residents never owns a home. Housing has become one of the most visible symbols of inequality, with rising prices and stagnant wages pushing ownership out of reach for many.

Why is there so much inequity and injustice in the world? Extremely wealthy people bribe corrupt government officials to get what they want whether it is money, power, or personal fulfillment. Corporate lobbying, tax havens, and monopolies further deepen this divide, allowing the rich to accumulate more wealth while the poor struggle to survive.

Should we ignore injustices that deprive both current and future generations? We live in an age of extreme inequality between the super-rich and the rest of society. Will this lead to severe instability around the world? History shows that extreme inequality often precedes social unrest, revolution, or economic collapse. If left unaddressed, the world may face mounting crises that transcend borders.

In the U.S., private sector regulation tends to be lighter. Therefore, fake news and extreme information are rampant. U.S. political divisions are getting worse. Technological platforms in Europe are becoming more regulated, and they have to take more responsibility for what they post. These differing regulatory approaches create an uneven digital landscape where misinformation can spread unchecked in some regions while being curtailed in others.

It is a balancing act between inventions and regulations. Inventions are important because they change history. AI is about to change the world in profound ways. But just like oil and plastic before it, AI will need ethical oversight to prevent harm. First, let's look at some of the major inventions in early human history.

Mesopotamia was the site of the earliest human civilization around 10,000 BCE. It inspired some of the most significant inventions from four to six thousand years ago, including many items that are taken for granted today.

The inventions include writing, the wheel, mathematics, time, mass-produced ceramics, cylinder seals and envelopes, mass-produced bricks, cities, maps, the sail, agriculture (the Plow) and irrigation, the concept of cartography, astrology, astronomy, chariots, metal fabrication, board games, soap, and the law system.

These represent only a small fraction of their technological, cultural, and scientific advancements. The Scholar Samuel Noah Kramer lists 39 'firsts' from ancient Sumer in Mesopotamia:

- The First School
- The First Case of 'Apple Polishing'
- The First Case of Juvenile Delinquency

- The First 'War of the Nerves'
- The First Bipartisan Congress
- The First Historian
- The First Tax Reduction Case
- The First 'Moses'
- The First Legal Precedence
- The First Pharmacopoeia
- The First 'Farmer's Almanac'
- The First 'Experiment in Shade-Tree Gardening'
- Man's First 'Cosmogony and Cosmology'
- The First Moral Ideals
- The First 'Job'
- The First Proverbs and Sayings
- The First Animal Fables
- The First Literary Debates
- The First Biblical Parallels
- The First 'Noah'
- The First 'Tale of the Resurrection'
- The First 'Saint George'
- The First Case of 'Literary Borrowing'
- Man's First Heroic Age
- The First Love Song
- The First Library Catalogue
- Man's First Golden Age
- The First 'Sick' Society
- The First 'Liturgy Laments'
- The First 'Messiah'
- The First 'Long-Distance Champion'
- The First Literary Imagery
- The First Sex Symbolism
- The First 'Mater Dolorosa'
- The First Lullaby
- The First Literary Portrait
- The First 'Elegy'
- Labor's First Victory
- The First 'Aquarium'

Writing was the most influential invention in Mesopotamia, and they kept their records on millions of baked clay tablets known as cuneiforms. These tablets preserved the region's beliefs, history, and culture, which significantly influenced later civilizations. Cuneiform is the earliest known system of writing, enabling laws, trade, astronomy, and literature to be recorded and passed down. The Epic of Gilgamesh, one of the oldest surviving works of literature, originated from this tradition.

Mesopotamia is in the northern part of the Fertile Crescent in southwest Asia within the Euphrates and Tigris rivers. It is also known as the Cradle of Civilization. Mesopotamia includes present-day Iraq and parts of Iran, Kuwait, Syria, and Turkiye in the Middle East. Its fertile lands, formed by river flooding, allowed for abundant agriculture, which supported growing populations and the rise of cities.

The oldest writing form, Cuneiform Script, was created in Mesopotamia about 3200 BCE. All civilizations in ancient Mesopotamia used a cuneiform writing system until the alphabetical system was introduced in about 100 BCE. Cuneiform evolved from pictograms to complex symbols over time, enabling communication across trade, governance, and religion.

Approximately half a million to two million cuneiform tablets have been excavated in modern times, and only about 30,000–100,000 have been read or published. Mesopotamia's counting method was based on 60. That is where we got 60 minutes in an hour, 60 seconds in a minute, 12 months in a year, and a 360-degree circle. This sexagesimal system continues to influence our concepts of time and geometry today. The Sumerians also invented an elemental abacus between 2700 and 2300 BCE, demonstrating an early form of computational thinking.

The first map was developed in Mesopotamia, but Roman and Greek cartography was more advanced. The Greek philosophers developed the idea of a spherical Earth around 350 BCE, which later resulted in our modern world map. However, the Babylonians laid the foundation by creating detailed maps of land boundaries and star charts, revealing an advanced understanding of geography and astronomy for their time.

The Sumerians documented Mercury, Venus, Mars, Jupiter, and Saturn's movements. They accurately predicted the planets' movements long before the Greeks, the Mayans, and other civilizations did thousands of years later. They developed a lunisolar calendar and could forecast eclipses, linking celestial events to religious and agricultural cycles.

The Greeks later absorbed astronomical concepts of patterns like Sagittarius, Leo, and Capricorn from the Sumerians and Babylonians. Ancient Mesopotamia used constellations to mark crop sowing and harvesting times. They also mapped movements in the sky, the moon, the stars, and the Sun to foretell cosmic events like an eclipse. This celestial knowledge was foundational for the development of zodiac astrology, influencing later Egyptian, Persian, and Greek systems.

The people of Mesopotamia invented 2, 4, and 6-wheel chariots that transformed ancient transportation and warfare used by different civilizations and nations. These innovations were critical for military strategy, allowing faster movement and tactical advantage in battle.

They invented large cities for the first time in human history. This was made possible by their agricultural technologies, improved transportation, mass-produced ceramics and bricks, the potter's wheel for making clay pots, and other inventions. Uruk, one of the earliest cities, had complex infrastructure, including temples, administrative buildings, and defensive walls.

Their invention of cities was both a gift and a burden in our modern world. Along with the conveniences, the cities had more crimes and other conflicts that were dealt with by their invention of a law system. The Code of Hammurabi, written around 1754 BCE, is one of the earliest and most complete written legal codes, promoting justice through publicly displayed laws.

Mathematics and time invented in Mesopotamia have been useful not only in finance, trades, and taxation but also in construction, engineering, medicine, and computer science. Around 2,000 BCE, Babylonian mathematicians created arithmetic, multiplication tables, square roots, division, and algebraic equations. They even solved quadratic and cubic equations, showing abstract reasoning and symbolic representation that predated Greek mathematics by centuries.

The Mesopotamians also played a vital role in sea travel through their sailboat invention around 5,500 BCE. Their early sailing vessels enabled long-distance trade along rivers and across the Persian Gulf, exchanging goods like grain, textiles, copper, and spices.

The first beer production was credited to the Sumerians in 4,000 BCE. Beer is the third most popular drink in the world, after potable water and tea.

Sumerians even had a goddess of beer, Ninkasi, and her hymn doubles as one of the oldest surviving beer recipes. It is one of the oldest alcoholic drinks in the world and has saved many lives in places without clean water.

In ancient times, fermented beverages were often safer to drink than untreated water due to alcohol's sterilizing properties.

The ancient Sumerians of Mesopotamia were the ancient equivalent of the people in Silicon Valley due to their prowess and passion for technological invention. They transformed how humans cultivated food, lived in cities, traveled, communicated, and kept track of information and time. They are the giants on whose shoulders we stand. Their innovations laid the groundwork for civilization as we know it, from written language to urban planning. We are grateful to the Sumerians for leading the way.

There is another thing I need to share with you about the Sumerians in Mesopotamia. In the ruined Library of Ashurbanipal at Nineveh (near modern-day Mosul, Iraq), some of the 15,000 cuneiform fragments recovered by English archaeologist Austen Henry Layard in 1849 include the Babylonian creation myth called Enuma Elish and a poem called The Epic of Gilgamesh. Apparently, according to these cuneiform texts, the Anunnaki deities often interpreted in modern fringe theories as aliens created humans.

These writings predate the Bible (Old Testament) by thousands of years. Abraham is the first patriarch revered by three monotheistic religions - Judaism, Christianity, and Islam. Genesis says Abraham lived in Ur in Mesopotamia, connecting the cradle of civilization with the roots of these faiths.

In our times, the San Francisco Bay Area, especially Silicon Valley, is the hub of entrepreneurship. Call and talk to some of the entrepreneurs and learn from them wherever you are.

Stanford University has produced many successful entrepreneurs, such as Elon Musk, founder of Tesla and SpaceX; Larry Page and Sergey Brin, founders of Google and Alphabet; and Jen-Hsun Huang, founder of Nvidia. According to Stanford, "A successful entrepreneur will possess many abilities and characteristics, including the ability to be curious, flexible and adaptable, persistent, passionate, willing to learn, a visionary and motivated."

Here is a company that some of you may not know about but that impacts many lives every day. Cisco Systems was founded in 1984. The company's headquarters are in San Jose, California. The founders of Cisco Systems were Leonard Bosack and Sandra Lerner, a married couple who met while students at Stanford University. In 1986, Cisco sold its first big success product, a router that served multiple network protocols.

In need of cash for expansion, the founders accepted funding from Sequoia Capital, a venture capital firm. Sequoia took control of the company in late 1987 and hired John Morgridge as president and CEO in 1988. Soon after Cisco's IPO (Initial Public Offering) in 1990, Lerner was fired. Bosack then quit, and their marriage ended the same year. This story highlights the critical importance of choosing investors and partners wisely. Entrepreneurship is as much about relationships and trust as it is about innovation.

Cisco's market valuation peaked on March 24, 2000, at more than $500 billion, surpassing Microsoft as the world's most valuable company. It was valued at less than $200 billion on March 20, 2024. The rise and fall of Cisco's valuation illustrate the volatile nature of tech markets and the need for entrepreneurs to stay adaptable and resilient.

Chapter 42
Starting Small and Government's Role

As an entrepreneur, you must choose your investors wisely. It can be very costly, personally and financially.

Heller Ehrman was the premier law firm in the life science industry when I hired them to do my company's corporate and patent work in biotechnology. At its peak, Heller Ehrman, an international law firm, had more than 730 attorneys in 15 offices across the United States, Europe, and Asia. We were happy with their corporate work for eight years. We were content with their patent work until we discovered their $100 million malpractice against our company.

They put us in their corporate marketing literature as a company they proudly represented in our industry. Unfortunately, after their merger with another law firm, their ethics deteriorated. Our corporate lawyer and patent lawyer both left the firm.

Our patent attorney, who earned his PhD in Chemistry from UCLA and JD from Santa Clara University, successfully prosecuted all of our patent applications worldwide to become patents. I trusted him. After he and all the patent attorneys in their Menlo Park, CA office left the firm on the same day, Heller committed multiple malpractices and concealed them. This caused $100 million in damages to us. This experience was a harsh lesson not taught in MBA school but critical for anyone navigating the complex interplay of law, business, and innovation.

Because I lost confidence in our corporate attorney and was deeply troubled by Heller Ehrman's post-merger issues of greed and ethical lapses, I made the difficult decision to hire another law firm based in New York to take Iris Biotechnologies public. We successfully began trading publicly in August 2008 - a challenging time, right near the peak of the Great Recession.

To build a world-class team, I recruited PhDs in biology, chemistry, and chemical engineering, along with MDs, MBAs, and other experts from prestigious institutions including UC Berkeley, Parke-Davis (now Pfizer), UCSF, Stanford, UCLA, Caltech, Case Western, MD Anderson, and many more.

Our shareholder base reflected this commitment to excellence. It included graduates from UC Berkeley, Harvard, Yale, UCSF, Stanford, Caltech, Santa Clara University, UC Davis, University of North Carolina, and other leading schools across diverse fields. Business owners from various industries around the country also invested in Iris.

Although we were offered venture capital funding, I declined. I believed in maintaining control and integrity over our company's direction, especially after the painful experience with Heller Ehrman.

Unfortunately, we were blindsided by Heller Ehrman's $100 million malpractice and their repeated concealment of the issue. Disturbingly, Heller knew they owed us money when they filed for bankruptcy protection. Iris was listed among the unsecured creditors in Heller's bankruptcy filing, but they never informed us. Here are the details of what transpired in bankruptcy court.

Iris vs. Heller

February 11, 1999 Mr. Simon Chin hired Mr. Bruce Jenett at Heller Ehrman, a prestigious, century-old law firm with over 600 attorneys, to incorporate Iris Biotechnologies Inc. and manage all legal matters for the company.

August 17, 2007 Following Heller Ehrman's merger with another firm, which brought significant internal turmoil marked by greed, competence issues, and ethical concerns, Iris lost confidence in both Heller Ehrman and its corporate attorney, Mr. Bruce Jenett, to advise on taking the company public. Iris subsequently engaged a New York law firm to handle the public offering. In response, Mr. Jenett issued an illegal disengagement letter to Iris. Under U.S. patent law, a law firm cannot abandon a client without valid cause. The U.S. Patent and Trademark Office (USPTO) denied Heller Ehrman's petition to disengage, meaning the firm legally remained Iris's patent attorney from February 1999 through September 2011.

December 14, 2007 Heller Ehrman successfully prosecuted all of Iris's patent applications worldwide to date. Dr. James A. Fox, the patent attorney handling Iris's portfolio, assured Mr. Chin, "Heller will send out a contemporaneous email whenever Heller forwards files in the future." Trusting Dr. Fox's deep familiarity with Iris's cases and their strong working relationship, Mr. Chin followed his advice to simply wait.

December 20, 2007 Mr. Chin wrote to Dr. Fox:

"It appears that Heller Ehrman has evolved into a firm that has problems not only with greed but also with competence and ethics. What happened has nothing to do with you, and I remain respectful of you and consider you a friend."

February 2008 – Mr. Bruce Jenett left Heller Ehrman.

Mar. 4, 2008 – Dr. Fox and the rest of the patent attorneys in Menlo Park left Heller Ehrman on the same day. Dr. Fox and Heller Ehrman did not inform Iris, and Iris was unaware of this exodus.

March 21, 2008 – Heller Ehrman committed its first malpractice against Iris Biotechnologies by failing to forward a critical USPTO document that the firm had received. This failure led to the abandonment of Iris's most important foundational patent in the United States the world's most significant market.

The U.S. Bankruptcy Court, under Judge Dennis Montali, later confirmed that Heller did indeed fail to forward crucial USPTO documents to Iris in March and October 2008. This malpractice caused a six-year delay in obtaining U.S. Patent No. 8,693,751, titled "Artificial Intelligence System for Genetic Analysis." Meanwhile, Iris's competitor surged ahead and was sold for $2.8 billion in 2019.

October 7, 2008 – Heller Ehrman committed a second malpractice by failing to forward the USPTO's notice of Iris's patent abandonment, which the firm had also received. This second malpractice concealed the first, effectively preventing Iris from discovering the wrongdoing and filing a timely malpractice claim while Heller still maintained malpractice insurance. *The bankruptcy court explicitly found that Heller did not forward this critical notice.*

December 28, 2008 – Heller Ehrman filed for Chapter 11 Bankruptcy protection.

January 28, 2009 – Heller Ehrman requested permission from the U.S. Bankruptcy Court to purchase a new $10.2 million malpractice insurance policy before their existing policy expired. Judge Dennis Montali denied this request, a ruling contrary to standard California practice.

Without malpractice insurance, Iris's $100 million claims, if allowed, threatened to consume most or all of Heller Ehrman's remaining assets. Judge Montali had already approved a distribution plan for many unsecured creditors. Law.com quoted Thomas Willoughby, representing Heller's creditors committee, saying, "We could look stupid buying it, or we could look stupid not buying it."

April 27, 2009 – The U.S. Bankruptcy Court set the claim bar date for April 27, 2009, for creditors to file claims.

July 31, 2009 – Even after the claim bar date and Heller's bankruptcy filing, the firm sent a letter to Mr. Chin concealing the USPTO letters it had received in 2008. This letter reiterated assurances that "we will forward all correspondence and faxes which we may receive concerning your application from the U.S. Patent and Trademark Office." As a result, Mr. Chin continued to trust Heller Ehrman and remained unaware of their malpractice.

July 1, 2010 – Heller Ehrman filed a list of unsecured creditors with the bankruptcy court on which Iris Biotechnologies appeared twice as a creditor. However, Iris was never notified of this critical filing, effectively leaving them unaware of their status and the proceedings that directly impacted their claims.

August 22, 2011 – John Jeffrey of Dennison Associates in Canada informed Iris of Heller Ehrman's malpractices for the first time. Upon learning this, Iris immediately retained the prominent law firm Arnold & Porter LLP and instructed Dr. James A. Fox who had been Iris's trusted patent attorney at Heller Ehrman for seven years to file a petition with the

USPTO to issue a new patent to Iris. Recognizing the urgency, Iris promptly filed malpractice claims within the one-year statute of limitations upon discovery of the evidence, thereby preserving their legal rights.

(Because Heller Ehrman was already under Chapter 11 bankruptcy protection, the malpractice lawsuit was filed in bankruptcy court.)

August 20, 2012 – The patent at the heart of the dispute, *Artificial Intelligence System for Genetic Analysis*, held immense commercial potential worth billions of dollars when fully realized. Iris Biotechnologies asserted a $100 million claim against Heller Ehrman and filed a motion in the U.S. Bankruptcy Court in San Francisco to allow a late claim filing post-bar date, specifically targeting Heller Ehrman's malpractices. Thanks to the continued efforts of patent attorney Dr. James Fox, the patent was ultimately granted to Iris in May 2014, vindicating their claim to this critical intellectual property.

November 20, 2012 – U.S. Bankruptcy Court Judge Dennis Montali officially recorded his ruling, confirming that Heller Ehrman had indeed failed to forward critical USPTO documents to Iris in March and October of 2008. Judge Montali stated:

"Mr. Chin strikes me as a very intelligent, very successful, capable executive of a company that obviously is the result of much of his intelligence and learning. He couldn't have been warned many more times about the need to protect his own company's interests from the correspondence, to the delivery of the files, to the acknowledgment that the patent files were there for him and his company. Most importantly, his recognition that Heller was out of business and his view that Heller lacked competence not incompetent, but lacked

competence. To me, that underscores the fact that the delay in acting and learning the fate of his company's filings with the Patent Office were things that he or Iris's counsel, if there were substitute counsel, could have ascertained."

However, Judge Montali's ruling exhibited a marked bias by focusing solely on the first sentence of Mr. Chin's December 20, 2007 email to Dr. Fox: "It appears that Heller Ehrman has evolved into a firm that has problems not only with greed but also competence and ethics," while deliberately ignoring the immediately following sentence: "What happened has nothing to do with you, and I remain respectful of you, and I consider you a friend." This omission significantly prejudiced the ruling.

The ruling failed to consider the full context and all the clear evidence demonstrating why Iris could not have reasonably filed their claim before the April 27, 2009 bar date. For seven years, Dr. Fox successfully prosecuted all of Iris's patent applications and explicitly advised Iris to "do nothing" until hearing directly from the USPTO. After the bar date, Heller's July 31, 2009 letter deceptively reassured Iris that no office action had been received from the USPTO, which led Iris to believe there was no need for follow-up.

Critically, the ruling ignored the fact that Heller Ehrman was still Iris's legal patent representative, as the USPTO had denied Heller's petition to disengage. Dr. Fox remained the USPTO attorney of record for Iris until September 2011, underscoring that Heller had a continuing duty to inform Iris of correspondence. The court's failure to recognize this duty and the consequences of Heller's concealment resulted in a grossly prejudiced decision.

Additionally, Judge Montali's close professional ties to lawyers affiliated with Heller Ehrman and its counsel raised legitimate concerns about potential conflicts of interest affecting the impartiality of the ruling.

Judge Montali continued:

"So I come to the conclusion, therefore, that as hard a result as it is for Iris, the factor of the reason for the delay and who has control over that delay lands squarely on Iris's lap and could not extricate it from the problem that it could have alleviated or mitigated on its own. Which is another way of saying that had Iris attended to its own protection of its own interest and come forth much more quickly, even after the claims bar date and asked to file a late claim, then indeed those factors would weigh less heavily against it and in favor of Heller. With the length of time of over two years of it not doing what it was cautioned to do and could have done and easily could have done, then I cannot extricate it from the consequences. So that is a long way of concluding that I am going to find that the Pincay and Pioneer factor weighed against Iris and in favor of Heller, and we will deny the motion to file a late claim. I mean, the claim has been filed, but by denying the motion, I am disallowing the claim as late filed claim."

Immediately after delivering his ruling against Iris, Judge Montali physically choked for several seconds and then stated, "I can't speak." Notably, he refused to make eye contact throughout the court proceedings. This choking episode, captured in the official court audio record, is believed by Iris's representatives to be a physical manifestation of the judge's inner conflict over knowingly issuing an unjust ruling.

When Judge Montali denied Iris Biotechnologies the right to a jury trial on their $100 million claim against Heller Ehrman, he knowingly upheld a decision that effectively

357

shielded Heller from financial liability. This was especially significant because Judge Montali had previously denied Heller Ehrman's request to spend $10.2 million on malpractice insurance that could have been used to compensate Iris for the egregious malpractices they suffered.

Judge Montali's ruling was not only inequitable but also illogical and contrary to established U.S. patent law. It appeared to cover up his prior error in disallowing Heller's expenditure on malpractice insurance. Despite clear evidence that Heller owed Iris money, Montali asserted that Mr. Chin could have hired a patent attorney to discover Heller's malpractices before or soon after the bar date. This ignores the reality that there was no reasonable or financially sound basis to suspect malpractice or to hire counsel solely for investigation purposes prior to the April 27, 2009 bar date, especially given that the USPTO had imposed no deadline for the patent application.

In 2010 and 2011, Iris continued to follow the instructions of their patent attorney, Dr. Fox. At no time were they informed by either Dr. Fox or Heller Ehrman about the departure of all patent attorneys from Heller's Menlo Park office, nor was Iris ever notified that Heller would stop forwarding critical mail. Iris prosecuted all of its patent applications worldwide diligently and on time, with the sole exception of the delays caused by Heller's misconduct.

Contrary to the ruling, the responsibility for the delay and control over it rests squarely with Heller Ehrman. Their malpractice, concealment of that malpractice, false reassurances, and deliberate misdirection caused the delay in securing Iris's foundational U.S. patent in the most commercially important market.

Heller's deceptive actions also delayed Iris's discovery of malpractice and hindered their ability to file a timely malpractice claim while insurance coverage was still in place. The ruling completely disregarded Mr. Chin's uncontested testimony that he had no issues with Dr. Fox's patent work and that he followed all of Fox's instructions.

Given all the evidence, it is inequitable to hold Iris responsible for the failure to file before the bar date, and irresponsible not to hold Heller accountable for their malpractice and concealment that caused the delay. Accordingly, the ruling should be overturned and Iris's claim allowed.

December 5, 2012 – Based on the compelling evidence of Heller's actions preventing an earlier claim filing, Iris BioTech filed a notice of appeal with the United States District Court in San Francisco. However, Iris was denied even the opportunity for an oral hearing. (Iris also appealed to the Ninth Circuit Court, but again was denied an oral hearing.)

Note: As a direct result of Heller Ehrman's multiple malpractices against Iris Biotechnologies and their prolonged concealment compounded by Bankruptcy Judge Dennis Montali's evidently biased ruling that failed to hold Heller accountable, hundreds of Iris investors and their families endured significant hardship.

Despite Iris's rightful claim, they received no compensation from Heller Ehrman. The malpractice and ensuing legal battles ultimately contributed to a loss of approximately $20 million in my personal brokerage accounts a sum exceeding the average wealth of the U.S. top 1%.

Judge Montali's grossly unjust ruling inflicted severe financial, mental, and emotional harm on me. As a plaintiff in the United States legal system, I never imagined losing everything I worked for due to an unfair court decision.

From this experience, I learned that while people can take your money, no one not even yourself should ever be allowed to steal your time, peace, happiness, or love. With God as my witness, I hold firm belief that there will be a day of judgment for us all.

According to a Gallup Poll released on December 17, 2024, "Americans' confidence in their nation's judicial system and courts dropped to a record-low 35% in 2024. Public trust in the national government also plummeted to 26%."

Chapter 43
Cultivating Resilience and
Learning from Failure

The entrepreneurial path is inherently strewn with obstacles, setbacks, and outright failures. The odds of success are daunting, and the personal and professional sacrifices required can be immense. Yet, it is precisely in facing these challenges head-on that entrepreneurs grow and thrive. In the face of such adversity, cultivating resilience and a growth mindset is not just helpful but absolutely essential.

One of our darkest chapters was the legal battle with our former law firm, Heller Ehrman, over the mishandling of our patent application. Their negligence cost us years of protection for our core AI technology, resulting in millions of dollars in lost value, legal fees, and opportunities that could never be recovered. The sense of betrayal and injustice was gut-wrenching, and the distraction of the lawsuit drained both our resources and morale.

But even amidst that pain, these experiences became unexpected teachers. They forced us to develop new reserves of strength, creativity, and strategic thinking. From each setback, we gleaned crucial insights that made us smarter, tougher, and more adaptable. We learned to anticipate and mitigate risks more effectively, to make difficult decisions with greater speed and confidence, and to rally together with a shared sense of purpose when the odds seemed stacked against us.

Critically, we also learned to shift our mindset and reframe failures as opportunities for growth and learning, rather than as endpoints or signs of defeat. Rather than dwelling on what went wrong, we focused relentlessly on what we could do better next time. We celebrated the effort, ingenuity, and courage that went into noble attempts even when they didn't succeed. And we became increasingly comfortable with pivoting, iterating, and course-correcting based on new information and changing circumstances.

For aspiring founders and leaders, my strongest advice is to embrace adversity as an inevitable and even necessary part of the growth process. Expect that things will go wrong, you will make mistakes, and you will face criticism, rejection, and setbacks. This is not a sign of failure but an essential rite of passage on the road to success. But also know that these very challenges are what forge great entrepreneurs and build great companies.

Surround yourself with people who believe deeply in you and your vision but who will also challenge you with honest feedback and hold you accountable. Develop habits and routines that keep you physically strong, mentally sharp, and emotionally resilient. Prioritize your well-being as fiercely as your business goals because your endurance depends on it. And cultivate a deep, unwavering sense of purpose your "why" that can sustain you through even the toughest times.

Remember, failure is not the opposite of success but a stepping stone to it. The only true failure is the failure to learn, to grow, and to keep pushing forward. By reframing setbacks as opportunities, maintaining laser focus on your north star, and refusing to give up on yourself or your team, you will build the resilience necessary to weather any storm and emerge stronger, wiser, and more capable on the other side.

In the biotech industry, resilience takes on added dimensions due to the unique challenges of the field:

1. Scientific setbacks: In biotech, years of work can be undone by a single failed experiment or clinical trial. It's crucial to develop the resilience to absorb these setbacks, learn from them, and pivot when necessary. Remember that even "failed" experiments often yield valuable data that can inform future directions.

2. Regulatory hurdles: Navigating the complex regulatory landscape in healthcare can be frustrating and time-consuming. Resilience here means maintaining patience and persistence while also being agile enough to adapt your strategy based on regulatory feedback.

3. Funding challenges: The capital-intensive nature of biotech means that funding challenges are common. Resilience in this context involves maintaining your vision and enthusiasm even when facing rejection from investors and being creative in finding alternative funding sources.

4. Ethical dilemmas: Biotech often involves grappling with complex ethical questions. Resilience here means staying true to your values even when faced with difficult decisions or external pressures.

5. Public scrutiny: As your work has direct implications for human health, you may face intense public and media scrutiny. Developing the resilience to handle this pressure while maintaining transparency and integrity is crucial.

6. Team morale: In the face of setbacks, maintaining team morale is critical. Resilient leaders find ways to keep their teams motivated and focused on the long-term mission, even during tough times.

7. Personal well-being: The high-stakes, high-stress nature of biotech entrepreneurship can take a toll on personal well-being. Building resilience means prioritizing self-care and maintaining a healthy work-life balance.

8. Technological obsolescence: The rapid pace of technological advancement in biotech means that your approach or technology could become obsolete. Resilience here involves staying adaptable and open to pivoting your strategy when necessary.

9. Competitive pressures: The biotech field is highly competitive. Resilience means staying focused on your unique value proposition and not getting derailed by every move your competitors make.

10. Long timelines: The extended timelines in biotech development can test anyone's patience and resolve. Resilience here means maintaining your commitment and enthusiasm over the long haul and celebrating small wins along the way.

11. Interdisciplinary challenges: Biotech often requires integrating knowledge from multiple scientific disciplines. Resilience in this context means being willing to continuously learn and adapt as you encounter new areas of knowledge.

12. Balancing science and business: Many biotech entrepreneurs come from scientific backgrounds and must learn to balance scientific rigor with business pragmatism. Resilience here involves being open to developing new skills and perspectives.

13. Managing uncertainty: Much of biotech involves working at the frontiers of scientific knowledge, which inherently involves high levels of uncertainty. Resilience means getting comfortable with this uncertainty and learning to make decisions with imperfect information.

14. Handling success: Interestingly, handling success can also require resilience. Rapid growth or sudden breakthroughs can bring their own set of challenges. Resilient leaders are able to navigate these transitions while maintaining their core values and vision.

15. Societal impact: The potential for significant societal impact in biotech can be both motivating and pressuring. Resilience here means staying connected to your core purpose while managing the weight of these expectations.

Cultivating resilience is not about becoming hardened, indifferent, or insensitive to setbacks. Rather, it's about developing the mental and emotional flexibility to absorb shocks, learn from them, and keep moving forward with renewed strength. It's about anchoring yourself firmly to your core purpose and values, even as you adapt your strategies, tactics, and expectations.

In biotech, perhaps more than in any other field, resilience is deeply tied to a sense of higher purpose. The potential to improve human health and save lives is an incredibly powerful motivator, fueling determination and perseverance through even the toughest and most prolonged challenges. Keeping this bigger picture in mind the patients whose lives you might change, the scientific frontiers you're pushing can become a wellspring of resilience that sustains you and your team.

At the same time, it's important to balance this sense of mission with realistic expectations and generous self-compassion. Not every experiment will succeed, and not every product will make it to market and that's okay. The key is to extract value from every experience, to analyze failures honestly without judgment, and to view each setback as an opportunity for growth and learning rather than a permanent defeat.

Building a culture of resilience within your organization is equally crucial. This means creating a psychologically safe environment where people feel empowered to take calculated risks without fear of harsh criticism or punishment. It means cultivating a mindset where failure is viewed as an essential learning step, rather than a cause for blame or shame. It means fostering open communication, mutual support, and collective problem-solving. And it means celebrating not just the wins but also the effort, creativity, and courage that go into every attempt, regardless of outcome.

Remember, resilience is not a fixed trait or innate quality it's a skill that can be intentionally developed and strengthened over time. By consciously cultivating resilience in yourself and your team, you will be better equipped to navigate the inevitable ups and downs of the biotech entrepreneurship journey with grace and persistence. You'll be able to maintain your vision, enthusiasm, and emotional well-being even in the face of setbacks and prolonged uncertainty. Ultimately, this resilience will increase your chances of achieving breakthrough innovations that can transform human health.

In the end, the most successful biotech entrepreneurs are not those who avoid failure at all costs, but those who learn to fail forward, that is, to extract valuable lessons from every setback and use those lessons as fuel for future success. By embracing this mindset of resilience, continuous learning, and adaptive growth, you position yourself and your venture to make a lasting and meaningful impact in the challenging yet immensely rewarding world of biotechnology.

Chapter 44
Staying True to Your
Integrity and Values

In the high-stakes world of entrepreneurship, where the pressure to succeed can be all-consuming, it can be all too easy to cut corners, compromise values, or make decisions that prioritize short-term gains over long-term integrity. Yet, in my experience, staying true to your moral compass is not just the right thing to do it is also the smartest strategy for the sustainable health and lasting success of your business.

Invariably, when we listened to our moral intuition and made the principled choice, even when it was difficult or costly in the moment, it paid off in the long run. We attracted team members, investors, and partners who shared our values and were committed to the right reasons not just profit. We built trust with regulators, customers, and advocacy groups who knew we could be counted on to do the right thing, even when no one was watching. And perhaps most importantly, we were able to look at ourselves in the mirror and feel proud of how we conducted ourselves and represented our industry.

On the flip side, we witnessed countless cautionary tales of companies that chased growth and profits at the expense of ethics only to pay a steep and sometimes irreversible price down the road. In biotech, cutting corners on safety, privacy, or informed consent is not just ethically wrong; it can be catastrophic for patients and utterly ruin public trust. As the saying goes, "You can't put a price on a clear conscience" and the long-term cost of losing one's ethical footing is far greater than any short-term gain.

Chapter 44
Staying True to Your
Integrity and Values

However, living up to one's principles is rarely black and white. It requires constant vigilance, humility, and course correction. As Iris grew and the stakes got higher, we had to work harder to stay true to our original vision and values. We had to be deliberate and transparent about defining and communicating our ethical standards. We also implemented systems and processes to hold ourselves accountable and ensure we were truly walking the talk not just paying lip service.

None of this was easy, and we certainly did not get it right all the time. But by making ethical integrity a non-negotiable core pillar of our culture and decision-making, we avoided many of the pitfalls that sunk other promising startups. More importantly, we built a company we could be proud of - one that made a positive, tangible difference in the world.

For other founders and leaders, my advice is to make your values and ethical principles explicit from day one. Know what you stand for and what lines you will not cross then embed these standards into every aspect of your business. Communicate them clearly and consistently to your team, your stakeholders, and yourself. And most importantly, live these principles out loud in your own actions and choices, setting the tone from the top.

Understand that there will be times when sticking to your values requires hard trade-offs and sacrifices. You may lose a lucrative deal, fire a high-performing but toxic employee, or admit to mistakes that bruise your ego. But you will gain something far more valuable in return: a reputation for integrity, self-respect that can weather any setback, and a purpose-driven business capable of standing the test of time.

Entrepreneurship is ultimately about stewardship the responsible management of resources, relationships, and the public trust in service of creating something meaningful and enduring. By anchoring that stewardship in an unwavering commitment to your values, you will not only sleep better at night you will also unleash the full potential of your venture to drive positive, lasting change. And what could be more rewarding than that?

In the biotech industry, the imperative to maintain integrity and adhere to strong ethical principles is particularly acute. Here's why:

1. Human impact: Biotech innovations often have direct implications for human health and well-being. The potential to help or harm is immense, making ethical considerations paramount.
2. Scientific integrity: The credibility of your research and products depends on maintaining the highest standards of scientific integrity. Any compromise here can have far-reaching consequences.
3. Regulatory compliance: The biotech industry is heavily regulated for good reason. Strict adherence to regulatory requirements is not just a legal necessity but also an ethical obligation.
4. Patient trust: Patients often put their lives in the hands of biotech innovations. Maintaining their trust through ethical practices is crucial.
5. Data privacy: With the increasing use of genetic and personal health data, protecting individual privacy is a critical ethical consideration.
6. Equitable access: There's an ethical imperative to ensure that life-saving innovations are accessible to those who need them, not just those who can afford them.
7. Animal welfare: For companies involved in preclinical research, ensuring ethical treatment of animal subjects is crucial.

8. Environmental responsibility: Biotech processes can have significant environmental impacts. Ethical companies strive to minimize these and contribute to sustainability.

9. Informed consent: In clinical trials and genetic studies, ensuring truly informed consent from participants is an ethical imperative.

10. Conflict of interest management: Given the complex web of relationships in biotech between researchers, companies, and healthcare providers, managing conflicts of interest transparently is essential.

11. Ethical use of technology: As biotech increasingly intersects with fields like AI and gene editing, there are new ethical frontiers to navigate.

12. Global health equity: Biotech companies have an opportunity to address global health disparities, raising ethical questions about resource allocation and research priorities.

13. Transparency in reporting: Both in scientific publications and in communications with investors and the public, maintaining transparency about results (both positive and negative) is crucial.

To navigate these ethical challenges in biotech, consider the following strategies:

1. Establish an ethics committee: Create a dedicated group within your organization to address ethical issues as they arise and to proactively develop ethical guidelines.

2. Implement robust compliance programs: Go beyond the minimum regulatory requirements to ensure that ethical considerations are baked into every aspect of your operations.

3. Foster a culture of ethical awareness: Regularly discuss ethical considerations in team meetings and

decision-making processes. Make it clear that raising ethical concerns is not just accepted but encouraged.

4. Engage with bioethicists: Consider bringing in outside experts to provide perspective on complex ethical issues.

5. Prioritize diversity and inclusion: Ensure that diverse voices and perspectives are included in your decision-making processes, especially when it comes to ethical considerations.

6. Develop clear guidelines for emerging technologies: As you work with cutting-edge technologies, proactively develop ethical guidelines for their use.

7. Engage with patient advocacy groups: These organizations can provide valuable perspectives on the ethical implications of your work from the patient's point of view.

8. Invest in ethics training: Provide ongoing ethics training for all employees, not just those in leadership positions.

9. Create accountability mechanisms: Develop systems to hold the organization accountable to its ethical standards, including consequences for ethical breaches.

10. Practice radical transparency: Be open about your ethical decision-making processes, both internally and externally.

11. Collaborate on industry-wide ethical standards: Work with other companies and organizations in your field to develop and promote ethical best practices.

12. Consider long-term impacts: When making decisions, consider not just immediate outcomes but potential long-term ethical implications.

By prioritizing integrity and ethical conduct, biotech companies can build deep, lasting trust with patients, healthcare providers, regulators, and the public. This trust is invaluable not only from a moral standpoint but also from a strategic business perspective. It can lead to stronger partnerships, smoother and faster regulatory approvals, improved talent acquisition and retention, and ultimately, greater long-term sustainability and success in a highly competitive industry.

Moreover, by maintaining high ethical standards, biotech entrepreneurs have a unique opportunity to shape the future of healthcare and biotechnology in profoundly positive ways. You're not simply building a company; you are contributing to the establishment of ethical frameworks and cultural norms that will guide the industry's growth and responsibility for decades to come. This leadership can help prevent abuses, protect vulnerable populations, and foster innovation that benefits society at large.

Remember that ethical leadership starts at the very top. As a founder or leader, your actions and decisions set the tone for the entire organization. When faced with difficult decisions, ask yourself not just, "Is this legal?" but more importantly, "Is this right?" Consider the broad impact of your choices on all stakeholders - patients, employees, investors, the scientific community, and society as a whole.

It is also important to recognize that ethical dilemmas in biotech are often complex, multifaceted, and nuanced. There may not always be clear-cut right or wrong answers. The key is to approach these issues thoughtfully, engage with diverse perspectives, including ethicists, patient advocates, and regulatory experts, and make decisions that you can stand behind with confidence and transparency.

Staying true to your integrity and values does not mean being rigid or unable to adapt to changing circumstances. Rather, it means having a clear and consistent moral compass that guides your decision-making, even as the specific application of those principles may evolve over time in response to new information, technologies, and societal expectations.

As you navigate your entrepreneurial journey in biotech, let your values be your compass and your guide. Use them as a framework for thoughtful decision-making, a foundation for building a resilient and positive organizational culture, and a beacon to attract like-minded partners and team members who will support and strengthen your mission. By doing so, you not only build a more sustainable and successful business, but you also actively contribute to shaping a more ethical, responsible, and respected biotech industry as a whole.

In conclusion, integrity in biotech entrepreneurship is not a constraint on success. It is a critical catalyst for it. By steadfastly staying true to your values, you create the foundation for lasting impact and meaningful innovation. You build a legacy not just of scientific and commercial achievement but also of principled, ethical leadership. And ultimately, you fulfill the true promise of biotechnology: to improve human health and well-being in ways that honor and uplift our shared humanity.

Chapter 45
Building the Future Through Collaboration

As I reflect on my entrepreneurial journey and the evolution of the biotech industry, one theme emerges again and again: the transformative power of collaboration. The challenges we face as a society whether in health, the environment, or social justice are far too complex, interconnected, and urgent for any one person or organization to tackle alone. Real, lasting progress demands that we break down silos and work together across disciplines, sectors, and geographies. Only through such collective effort can we hope to make meaningful and sustained impact.

At Iris Biotechnologies, collaboration was baked into our DNA from the very start. We recognized early on that to realize the full promise of precision medicine, isolated innovation was not enough. We needed to partner closely with academic researchers, clinical centers, patient advocates, regulatory bodies, and even competitors. We made a deliberate, strategic decision to prioritize collaboration over competition wherever possible, grounded in the belief that a rising tide lifts all boats and that shared knowledge accelerates breakthroughs. This collaborative mindset enabled us to leverage diverse expertise, pool resources, and move faster toward our shared goals.

Looking ahead, I believe the imperative for collaboration in biotech and beyond will only grow stronger. The COVID-19 pandemic starkly underscored how deeply interconnected our fates are and how critical collective action is in confronting global threats. At the same time, the accelerating pace of technological change from AI-driven drug discovery to advanced genomics is creating new opportunities and complex challenges that will require unprecedented cooperation, coordination, and trust among stakeholders worldwide.

Furthermore, fostering a culture of open collaboration can help address systemic inequities by ensuring that innovations reach diverse populations and that diverse voices shape the future of biotechnology. As entrepreneurs and leaders, our challenge and responsibility is to build bridges, nurture partnerships, and champion a collaborative ecosystem where shared success is the true north. Some of the key areas where I see collaboration as essential include:

- Advancing precision medicine through data sharing and interoperability
- Addressing health disparities through community-based participatory research
- Combating antimicrobial resistance through public-private partnerships
- Developing ethical frameworks for the use of AI and big data sets in healthcare and wellness
- Accelerating the translation of research into clinical practice through consortia models forming legal entities to solicit bids and buy on behalf of those involved in the consortiums.
- Fostering open innovation and pre-competitive spaces for shared problem-solving

No one company or institution can drive these agendas alone. Achieving real, transformative change will require a collective effort spanning the entire biotech and healthcare ecosystem, as well as cross-sector partnerships with technology, government, philanthropy, and civil society. It will require a willingness to share knowledge, resources, and credit in service of a larger, common goal. It will also require a new kind of leadership one that prioritizes collaboration over competition, long-term impact over short-term gain, and the greater good over narrow self-interest.

For aspiring entrepreneurs and innovators, my advice is to make collaboration a core part of your strategy and culture from the outset. Map out the key players and stakeholders in your field, and proactively reach out to explore areas of common cause. Build relationships based on trust, transparency, respect, and mutual benefit. And be willing to share your own ideas and assets in the spirit of open innovation and collective progress.

At the same time, be discerning about the partnerships you pursue. Look for collaborators who share your values and vision, who bring complementary strengths and diverse perspectives to the table, and who are genuinely committed to the hard work of alignment and execution. Set clear expectations and metrics for success, and be willing to course-correct or walk away if the partnership is not living up to its potential or compromising your core principles.

Ultimately, the measure of success in collaboration is not just what you achieve together, but how you achieve it. By modeling the kind of leadership, integrity, and partnership you wish to see in the world, you can not only accelerate your own impact but also help shape a more collaborative, inclusive, and resilient ecosystem for all.

The biotech industry has a unique opportunity and responsibility to lead the way in this regard. By harnessing the power of science, technology, and human ingenuity in service of the greater good, we can help build a future in which every person has the chance to live a healthy, fulfilling life. And by working together across boundaries and silos, we can unleash the full potential of our collective intelligence, creativity, and compassion to solve the grand challenges of our time.

This is the kind of future worth fighting for and above all, worth collaborating for. It is a future in which innovation is not just a means to an end, but a way of being and working together. The lines between competition and cooperation, profit and purpose, and individual success and collective impact blur in service of a higher calling. This is a future in which the entrepreneurial spirit is harnessed not just for personal gain, but also for lasting positive change and the greater good.

This is the future I have devoted my career and my company to building. And it is the future I invite you to help create one partnership, one innovation, and one act of collaboration at a time. Together, we can redefine what is possible and build a better world for all. That is the power and the promise of collaborative entrepreneurship.

In the biotech industry, collaboration takes on even greater importance due to the complexity of the challenges we face and the interdisciplinary nature of the work. Here are some additional thoughts on how to foster effective collaboration in biotech:

1. Cross-disciplinary teams: Build teams that bring together diverse expertise - biologists, chemists, data scientists, clinicians, engineers, and business strategists. This diversity of perspective can lead to breakthrough innovations.
2. Academic-industry partnerships: Foster strong relationships with universities and research institutions. These partnerships can provide access to cutting-edge research, specialized facilities, and a pipeline of talent.

3. Patient engagement: Collaborate closely with patient advocacy groups and individual patients. Their insights can be invaluable in shaping research priorities and designing more patient-centric solutions.

4. Regulatory collaboration: Work proactively with regulatory bodies. Collaborative approaches like the FDA's breakthrough therapy designation can help accelerate the development of promising treatments.

5. International cooperation: Global health challenges require global solutions. Look for opportunities to collaborate across borders, sharing knowledge and resources to tackle common problems.

6. Precompetitive collaborations: Identify areas where companies can work together on shared challenges without compromising their competitive positions. This might include developing common standards, sharing non-proprietary data, or jointly addressing industry-wide issues.

7. Open innovation platforms: Consider creating or participating in open innovation initiatives that allow a wider community to contribute ideas and solutions to specific challenges.

8. Collaborative funding models: Explore innovative funding approaches like public-private partnerships, consortia models, or collaborative grants that bring together multiple stakeholders to fund ambitious projects.

9. Data-sharing initiatives: Participate in responsible data-sharing efforts that can accelerate research while protecting patient privacy and intellectual property.

10. Cross-sector partnerships: Look beyond the biotech industry for collaborators. Partnerships with tech companies, environmental organizations, or social

enterprises can bring fresh perspectives and complementary capabilities.

11. Collaborative clinical trials: Consider innovative trial designs that allow for collaboration between multiple companies or institutions, potentially accelerating the development process and reducing costs.

12. Mentorship and knowledge sharing: Foster a culture of mentorship within your organization and the broader biotech community. Share your experiences and learnings to help nurture the next generation of innovators.

13. Ecosystem building: Invest in building and strengthening the broader biotech ecosystem in your region. A thriving ecosystem benefits all participants through shared resources, talent pools, and knowledge networks.

14. Ethical collaborations: Work with ethicists, policymakers, and community leaders to develop ethical frameworks for emerging technologies. Collaborative approaches to ethics can help build public trust and ensure responsible innovation.

15. Crisis response collaborations: As the COVID-19 pandemic demonstrated, the ability to rapidly form collaborations in response to global health crises is crucial. Build the relationships and systems that will allow for agile collaboration when urgent needs arise.

Remember, effective collaboration is not just about formal partnerships or agreements. At its core, it means fostering a mindset of openness, curiosity, and shared purpose. It's about creating an environment where ideas can flow freely, diverse perspectives are genuinely valued, and the collective goal takes precedence over individual egos. This mindset empowers teams to break down barriers, challenge assumptions, and spark innovation that none could achieve alone.

As a leader, you play a crucial role in setting the tone for collaboration. Leadership in collaboration requires vulnerability and humility, showing that it's okay not to have all the answers, and that collective intelligence is a strength. Lead by example: model the behaviors you want to see - be open to new ideas, give credit generously, and demonstrate a willingness to learn from others. Create incentives and recognition systems that reward collaborative efforts and shared successes, not just individual achievements, fostering a culture where contribution and cooperation are celebrated.

At the same time, be mindful of the challenges that can arise in collaborations. Clear communication, well-defined roles and responsibilities, and transparent decision-making processes are essential to keep everyone aligned. Be prepared to navigate differences in organizational cultures, priorities, and timelines. And always keep the shared goal at the forefront to help align efforts, build trust, and resolve conflicts constructively. Developing mechanisms for continuous feedback and conflict resolution is vital to maintaining momentum and ensuring all voices are heard.

Looking ahead, the biotech industry holds enormous potential to drive transformative change in human health, environmental sustainability, and beyond. However, realizing this potential will require us to think and work in fundamentally new ways. It will require us to break down silos, embrace transparency, and actively co-create solutions to our most urgent challenges. Furthermore, it demands a commitment to ethical considerations and social responsibility, ensuring innovations benefit society equitably.

By embracing collaboration as a core strategy and value, we can accelerate innovation, expand the impact of our work, and build a more resilient and inclusive biotech ecosystem. We can tackle bigger challenges, take on more ambitious projects, and create solutions that truly meet the complex and evolving needs of our world. Collaboration also opens doors to novel funding models and cross-sector partnerships that can amplify resources and expertise beyond traditional limits.

In doing so, we advance not only our individual enterprises but also contribute to a larger movement of positive change. We become part of something bigger than ourselves, a global community of innovators and problem-solvers working together to shape a better future. This collective approach is essential for addressing systemic issues that no single entity can solve, such as global pandemics, climate change, and health disparities.

So, as you build your venture or career in biotech, I encourage you to think broadly and strategically about collaboration. Seek out diverse partners, be generous with your knowledge and resources, and always look for opportunities to create shared value. By doing so, you'll not only increase your chances of success but also help build the kind of collaborative, innovative, and impactful biotech industry that our world so urgently needs. Remember, the strongest innovations emerge from ecosystems where trust, respect, and shared ambition thrive. Make those your foundation.

Chapter 46
Shaping a Purpose-Driven Legacy

As I enter the later chapters of my entrepreneurial journey, I find myself reflecting more and more on the question of legacy. What will I leave behind? How will my work and my life have mattered? What kind of world do I want to help create for future generations? These are questions that transcend business they strike at the heart of who we are, what we value, and how we hope to be remembered.

These are not just philosophical musings, but urgent and practical imperatives. In an era of existential challenges like climate change, inequality, and global health crises, the stakes for innovation and entrepreneurship have never been higher. The decisions we make and the actions we take today will reverberate for decades to come. We are no longer in a time where incremental progress is enough. We need bold, principled action that aligns long-term value creation with immediate, meaningful impact.

For me, the ultimate measure of success is not just financial or even scientific, but moral and societal. It is about the lives we touch, the communities we strengthen, and the planet we protect. It is about creating shared prosperity and well-being and leaving the world better than we found it. True innovation must not only advance technology but uplift humanity. It must heal, not harm; connect, not divide; empower, not exploit.

This has been the North Star Guiding Iris Biotechnologies from its inception. Our vision has always been about more than just developing cutting-edge diagnostics and advancing precision medicine, as important as those goals are. It has been about harnessing the power of science and technology

382

to improve people's lives and address some of the most pressing challenges facing humanity. We recognized early on that biotechnology is not just a tool for treatment it's a platform for transformation, with the potential to redefine how we think about health, equity, and the future of care.

This purpose-driven approach has influenced every aspect of our business from the problems we choose to work on, to the partners we engage with, to the way we measure and report our impact. It has also shaped our culture and values, attracting team members and stakeholders who share our commitment to making a positive difference in the world. Purpose, for us, is not an afterthought or a corporate slogan. It is embedded in our operating model, our stakeholder relationships, and the DNA of our innovations.

As I look to the future, I am more convinced than ever that this kind of purpose-driven entrepreneurship will be essential to tackling the grand challenges of our time. We need a new generation of leaders and innovators who are not just technically skilled but also ethically grounded, who are driven not just by profit but by purpose, and who are committed to creating value for all stakeholders, not just shareholders. Leadership today requires moral clarity, cross-disciplinary fluency, and the courage to build systems that prioritize justice, access, and sustainability.

We need businesses and institutions that are designed for social and environmental impact, not just efficiency and growth. We need an economic system that rewards cooperation and shared prosperity, not just competition and individual gain. We need a global community that works together across borders and sectors to solve our common problems and create a better future for all.

In short, we need an economy of empathy powered by science, guided by ethics, and driven by collective ambition. The future we imagine is only possible if we build it together, piece by piece, across industries and generations.

This may sound idealistic, but I believe it is also deeply pragmatic. In a world of complex, interconnected challenges, the old models of business and leadership are no longer fit for purpose. The companies and leaders that will thrive in the coming decades will be those that embrace a new paradigm of purpose-driven, collaborative, and equitable value creation. This isn't just a shift in language it's a shift in how we define success, how we build trust, and how we align innovation with the needs of people and the planet.

For aspiring entrepreneurs and changemakers, my advice is to start with your why. What is the problem you are passionate about solving? What is the difference you want to make in the world? Let that purpose be your compass and your fuel for the journey ahead. Purpose is not a luxury it's a strategic necessity. It's what keeps you grounded during setbacks and focused during periods of growth.

Build a team and a culture that shares your values and your vision. Surround yourself with diverse perspectives and skills, and create an environment of trust, creativity, and continuous learning. Be willing to experiment, iterate, and adapt as you go, always keeping your ultimate impact in mind.

Remember: the strength of your team is not just in what they know, but in how they think, how they care, and how they challenge each other to be better. Culture is your long-term differentiator.

Cultivate a mindset of collaboration and partnership. Map out the ecosystem of stakeholders and allies who share your goals, and look for ways to leverage your unique strengths in service of the greater good. Be open to new business models and unexpected partnerships that can accelerate your impact. Success in today's world is rarely achieved in isolation. It is co-authored with communities, competitors, researchers, regulators, and visionaries across disciplines.

Most importantly, you should never lose sight of the people and communities you are serving. Listen to their needs and aspirations, and co-create solutions with them, not for them.

Measure your success not just in terms of financial metrics but also in terms of the real-world outcomes and experiences of those you seek to help. Empathy, humility, and proximity to the people you serve are not soft values they are hard advantages. They ensure your solutions are relevant, inclusive, and truly transformative.

This is the kind of entrepreneurship that I believe will define the 21st century entrepreneurship that is driven by purpose, grounded in values, and focused on creating lasting, positive change. It is entrepreneurship that sees beyond the bottom line, that dares to challenge convention, and that recognizes our shared responsibility to the planet and its people. It is the kind of entrepreneurship that has guided my own journey with Iris Biotechnologies, and that I hope will inspire and empower future generations of innovators and leaders.

As I reflect on my own legacy, I am filled with gratitude for the incredible people and experiences that have shaped my path. From the early days of bootstrapping and experimentation to the hard-won successes and painful setbacks, to the enduring relationships and shared triumphs, every step has been a lesson and a gift.

Chapter 46
Shaping a Purpose-Driven Legacy

Each phase of the journey no matter how uncertain carved wisdom, resilience, and perspective into the core of who I am.

I am grateful to the mentors and role models who have guided and inspired me, to the team members and partners who have joined me on this journey, and to the patients and communities who have trusted us to make a difference in their lives. I am grateful to our investors for the opportunity to pursue my passions and to have contributed in some small way to the betterment of the world. But most of all, I am grateful for the chance to live and lead with purpose to build not just a business, but also a mission.

But I also know that my legacy is not mine alone to shape. It is a collective legacy, one that belongs to all of us who dare to dream big and work hard in service of a better future. It is a legacy that will be carried forward by the next generation of entrepreneurs and leaders, who will stand on our shoulders and take our work to new heights. Legacy is not just what we leave behind it is what we set in motion.

To them, I offer my heartfelt encouragement and my deepest respect. The road ahead will not be easy, but it will be worth it. The challenges you face will be daunting, but so too will be the opportunities to make a real and lasting difference. The setbacks you encounter will test your resolve, but they will also forge your strength and your character. You are the architects of the future, and every courageous step you take helps lay the foundation for a world worth inheriting.

Remember that you are part of a larger story a story of human progress and possibility that stretches back through the generations and forward into the future. Your role in that story is unique and precious, and it is yours to write with courage, compassion, and conviction. Do not underestimate the ripple effect of your integrity, your innovation, and your vision. Every act rooted in purpose adds to a legacy that transcends any single life or venture.

So dream boldly, act with integrity, and never give up on building the world you want to see. Embrace the power of purpose, the value of collaboration, and the imperative of equity. And know that you are not alone on this journey you are part of a global community of change-makers who share your vision and your values.

Together, you are not just launching companies you are shaping culture, shifting paradigms, and redefining what leadership looks like.

Together, we can shape a future in which every person has the chance to live a healthy, fulfilling life, in which every community can thrive and prosper, and in which every generation can look forward to a brighter tomorrow. That is the legacy we are called to build, and it is the legacy that will define our time on this earth.

Let it be a legacy rooted in justice, fueled by innovation, and remembered for its humanity.

As for me, I will continue to do my part to use my voice and my platform to champion the kind of entrepreneurship and leadership that I believe the world needs now more than ever. I will continue to learn and grow, to support and mentor, to advocate and agitate for change. And I will continue to hold fast to the belief that a better world is not only possible but also inevitable if we have the courage and the will to make it so. The work is far from over, and I remain as committed as ever to lifting others, sharing knowledge, and amplifying the mission that has guided me from the beginning.

That is my promise, my purpose, and my legacy. And it is one that I invite you to share and to build upon, in your own unique way, with your own unique gifts and passions. In the end, the true measure of our legacy will not be the monuments we build or the accolades we receive, but the lives we touch and the world we leave behind. Legacy, after all, is written not in stone but in the stories and futures of those we empower.

Let us go forth, then, with hearts full of hope and hands ready for service. Let us seize the opportunities before us and rise to the challenges of our time. Let us build a legacy of purpose, compassion, and progress that will endure long after we are gone. And let us never forget the power we hold, as entrepreneurs and as human beings, to shape a better future for all. This is our invitation to make meaning, to leave light, and to lead with love.

This is our calling, our responsibility, and our opportunity. May we answer it with courage, with compassion, and with unwavering commitment to the greater good! And may we always remember that the true reward of a life well-lived is not in the destination, but in the journey, and in the lives we are privileged to touch along the way. Because in the end, it's not about what we build it's about who we become, and how deeply we cared along the way.

In the context of biotech entrepreneurship, shaping a purpose-driven legacy takes on additional dimensions:

1. Scientific advancement: Your legacy includes the contributions you make to scientific knowledge. Even if a particular product doesn't reach the market, the research and discoveries made along the way can advance the field and pave the way for future breakthroughs.

2. Patient impact: In biotech, your legacy is measured in lives improved or saved. Keep sight of the real people who stand to benefit from your work.

3. Ethical leadership: By setting high standards for ethical conduct in biotech, you help shape the moral framework of an industry that has profound implications for human health and well-being.

4. Environmental stewardship: Consider how your work in biotech can contribute to environmental sustainability. From developing cleaner manufacturing processes to creating bio-based alternatives to petrochemicals, there are many ways biotech can positively impact the planet.

5. Mentorship and education: Your legacy includes the next generation of scientists and entrepreneurs you inspire and nurture. Consider how you can contribute to STEM education and provide opportunities for young people to engage with biotech.

6. Policy influence: As a biotech leader, you have the opportunity to shape policies that govern research, drug development, and healthcare. Use your voice and expertise to advocate for policies that advance science and expand access to care.

7. Global health equity: Consider how your work can address health disparities and contribute to improving healthcare access in underserved communities around the world.

8. Technological innovation: Your legacy may include new tools, platforms, or methodologies that change how biotech research and development are conducted.

9. Interdisciplinary bridges: By fostering collaboration between biotech and other fields (like AI, nanotechnology, or environmental science), you can help create new paradigms for solving complex problems.

10. Cultural change: Through your leadership, you can contribute to changing the culture of the biotech industry - making it more diverse, inclusive, collaborative, and purpose-driven.

As you build your career and your company in biotech, regularly reflect on the broader impact of your work. Ask yourself not just "What are we achieving?" but "Why does it matter?" and "Who will benefit?" Let these questions guide your strategic decisions and your day-to-day actions. Let them become the compass that ensures you're not just moving fast but moving in the right direction.

Remember that in biotech, perhaps more than in any other field, your work has the potential to profoundly impact human lives and the future of our planet. This is both a great responsibility and an incredible opportunity. By staying true to your purpose, maintaining your integrity, and always striving to create value for society, you can build a legacy that extends far beyond financial success or scientific accolades.

You're not just developing products you're shaping the health, well-being, and sustainability of future generations. Your legacy in biotech is not just about what you achieve, but also about how you achieve it. It's about the culture you create, the ethical standards you uphold, and the way you treat people along the way. It's about using your knowledge and resources not just for personal gain, but also for the greater good.

Every hiring decision, every lab protocol, every public statement these all speak to the kind of legacy you're creating. As you move forward in your entrepreneurial journey, I encourage you to think expansively about your potential impact. How can your work in biotech contribute to solving some of the grand challenges facing humanity? How can you use your platform to advocate for positive change in the industry and beyond? How can you inspire and empower others to pursue purpose-driven innovation? And how can your success model a new kind of leadership one where compassion and collaboration drive progress just as much as competition and capital?

Building a purpose-driven legacy in biotech is not always easy. It may require making difficult choices, taking stands that aren't popular, or sacrificing short-term gains for long-term impact. But it is infinitely rewarding. It allows you to align your work with your deepest values and to know that your efforts are contributing to a better world.

In the moments when you're tested, remember: long-term trust is built through short-term courage. So as you build your company and pursue your goals, keep your eyes on the horizon. Think not just about the next quarter or the next year, but also about the impact your work will have decades from now.

Let your purpose be your guide, your integrity be your foundation, and your commitment to making a positive difference be the legacy you leave behind.

Success that endures is rooted in significance - let that be your measure.

In doing so, you'll not only build a successful career or company you'll contribute to shaping a biotech industry that is more ethical, more equitable, and more impactful.

You'll be part of writing a new chapter in the story of human progress, one where the power of science and technology is harnessed for the benefit of all. You'll help redefine what it means to lead in this space fusing innovation with empathy, and profits with purpose.

This is the true promise and potential of biotech entrepreneurship. It's a path that offers not just professional success, but the opportunity to make a meaningful difference in the world.

It's a chance to be part of something bigger than you and to contribute to solving some of the most pressing challenges of our time. Your journey can become a bridge connecting scientific excellence to human need.

As we conclude this exploration of entrepreneurship and legacy in biotech, I hope you feel inspired and empowered to pursue your own purpose-driven journey. Remember that every great achievement starts with a single step, every breakthrough begins with a question, and every positive change starts with someone who dares to imagine a better way. Dare to be that someone. Dare to ask more. Dare to give more.

You have within you the potential to create extraordinary value not just economic value, but human value. You have the power to improve lives, advance science, and contribute to a healthier, more sustainable world. That is the legacy you can build through purpose-driven entrepreneurship in biotech. And in doing so, you elevate not just your career but the entire field you touch.

So go forth with courage and conviction.

Pursue your purpose with passion and perseverance.

Build your legacy not just in the products you create or the profits you generate, but also in the lives you touch and the positive change you catalyze.

The world is waiting for your contribution, for your innovation, for your leadership.

The future won't wait and neither should your vision.

May your journey be bold, your impact profound, and your legacy be one that inspires generations to come.

The future of biotech and indeed, the future of our world depends on leaders like you who are willing to dream big, work hard, and never lose sight of the profound purpose that underlies this incredible field. **Lead with integrity. Create with purpose. Influence with humility.**

Here's to your journey, your impact, and the lasting legacy you will create.

The world is counting on you.

Make it count. Never forget, the best breakthroughs begin in the heart before they reach the lab.

Section 6: Health and Wellness: Role of Technology

Chapter 47
The Journey to Health and Wellness

On a crisp Friday in October 2007, the Saratoga Rotary Club hosted a guest speaker who would unknowingly set in motion a profound personal health journey. A young cancer survivor stood before the group, sharing her inspiring story of commitment and resilience. She spoke of her decision to run a marathon and raise money for cancer research after recovering from her own battle with the disease.

Among the audience sat a Rotary member, whose father was a prostate cancer survivor. As the young woman's words resonated through the room, a spark ignited within this listener – a spark that would soon burst into a flame of determination. In that moment, something shifted. The story wasn't just touching it felt like a call to action.

Inspired by the speaker's courage and driven by a desire to make a difference, the Rotary member made a life-changing decision that day. Despite never having run more than a few miles at a time, that member committed to completing a full marathon – 26.2 grueling miles – to raise funds for cancer patients. This wasn't just about personal achievement; it was about joining a larger fight against a disease that had touched so many lives.

It became a deeply personal mission, fueled by the hope of making even a small difference in someone else's battle. This was not a casual goal; it was a vow to step far outside of comfort zones for something greater than oneself.

The journey began with joining the Leukemia and Lymphoma Society's Team In Training program. This comprehensive training regimen was designed to transform ordinary individuals into marathon runners while raising funds for a worthy cause. The program offered more than just physical preparation; it provided a support system, expert guidance, and a sense of community among participants, all working toward the same goal.

There was accountability, shared struggle, and a contagious energy that lifted even the most doubtful runners. Participants weren't just training for a race they were becoming part of a movement.

On the first day of training, the magnitude of the challenge ahead became clear. Gathered at the local high school track, the groups of aspiring marathoners learned about the training process, met their coaches (including a former Olympian), and were introduced to the staff that would guide them through warm-ups, cool-downs, and the long journey ahead. The initial 2-mile combination of running and walking left many, including our protagonist, breathless. But it was just the beginning.

That day also served as a humbling reminder that transformation doesn't happen overnight it begins with small, committed steps. There was laughter, nervous chatter, and even doubt but also a sense of collective resolve.

Over the next four months, the training intensity steadily increased. Almost daily solo runs complemented regular evening sessions at Los Gatos High School. During the weekends, the group ran at locations ranging from Portola Valley to Santa Cruz. The regimen was demanding, requiring dedication, time, and physical effort. But with each passing week, endurance improved, and what once seemed impossible began to feel achievable.

Section 6: Health and Wellness:
Role of Technology

Each mile was a milestone not just on the road, but also in mindset. Blisters, sore muscles, and early mornings became badges of honor, quietly affirming the commitment made on that October day. Through the sweat and setbacks, confidence took root, and purpose kept every footstep focused.

A month before the marathon, the training reached its peak. The schedule included running a half marathon almost every day for a week – a grueling test of endurance that pushed bodies and minds to their limits. This wasn't just about distance anymore it was about testing resolve. Two weeks before the big day, an 18-mile run was accomplished, instilling a sense of confidence and readiness. For the first time, crossing the finish line felt truly possible.

Finally, the day of the Marathon arrived. The atmosphere was electric with excitement and nervous energy. A sign at the check-in station boldly proclaimed, "It's going to happen!" – a simple yet powerful affirmation that brought a smile to many faces and calmed jittery nerves. For our runner, the goal was clear: complete the marathon in less than 5 hours. The crowd buzzed with energy - athletes, volunteers, supporters all united by purpose.

As the race began, adrenaline surged. Running among so many fit, determined individuals was exhilarating. The first half of the course flew by, completed in less than 2.5 hours. Buoyed by this strong start, there were plans to run even faster for the second half. Every stride felt strong, every breath steady a rhythm of triumph unfolding.

However, marathon running, like life itself, often doesn't go according to plan. An unexpected 15-minute wait to use a portable toilet proved to be a critical mistake. This delay led to a buildup of lactic acid, causing severe muscle fatigue and soreness. Upon resuming the run, both legs cramped up painfully, making it nearly impossible to continue. Momentum was lost, and in its place came the harsh reality of physical limits.

It was at this moment that a crucial lesson was learned – one that had been shared in a training session our runner had unfortunately missed. The importance of replenishing electrolytes during such a long and demanding physical endeavor became painfully clear. These essential minerals play a vital role in muscle function, affecting everything from hydration to nerve signals and muscle contractions. Neglecting this one piece of advice would come to define the next stretch of the race.

The remainder of the race became a battle of will against physical limitations. Cramps kept returning, necessitating frequent stops to massage aching legs. The continuous climb at mile 19 tested not just physical endurance but mental fortitude as well. Each step felt heavier, every incline steeper but quitting was never entertained.

Around mile 21 or 22, another challenge emerged. Severe blisters developed, with one popping so loudly it sounded like a gunshot – a startling and painful experience. Running became increasingly difficult, each step a testament to determination and grit. Shoes now soaked with sweat and pain, the trail blurred with discomfort, but the goal still glimmered ahead.

In this moment of struggle, the true spirit of the Team In Training program shone through. Staff members, noticing the difficulty, provided on-the-spot medical treatment, enabling the run to continue. This act of support embodied the collective spirit of the marathon – a shared journey where strangers become allies in the face of adversity. It wasn't just about one runner anymore it was about the mission, the team, and the cause.

The final miles were a blur of pain, fatigue, and sheer willpower. Quitting was not an option, not after months of training and with the finish line in sight. Even if it meant alternating between slow running and walking, forward progress continued. Pride mingled with pain, and every aching stride became a symbol of perseverance.

In a moment that felt almost cinematic, a coach who was a former Olympian appeared about a quarter mile from the finish line. Running alongside our struggling marathoner, this experienced athlete provided a final boost of motivation. Somehow, in the presence of this elite runner, the pain and exhaustion faded into the background.

With renewed energy, the last quarter mile was completed in stride with strength and determination. That final push powered not just by muscle, but also by spirit carried me home. I was the person that completed the 26.2-mile Napa Valley Marathon.

Crossing the finish line was more than just the end of a race; it was the culmination of a transformative journey. For someone born prematurely and plagued by frequent childhood illnesses, including a near-fatal bout of typhoid fever, completing a marathon held profound significance. It wasn't just about fitness it was about reclaiming a body that had once struggled to survive. In that moment, every setback, every hospital stay, every doubt was eclipsed by the finish line.

But the true victory extended beyond personal achievement. This journey raised vital funds for cancer patients, contributing to the larger fight against a disease that affects millions. It demonstrated the power of community, the importance of perseverance, and the impact that one-inspired individual can have when they commit to a cause greater than themselves. It proved that healing is not only possible but also contagious. One person's journey can spark another's hope.

The marathon experience provided valuable lessons that extended far beyond running. It highlighted the importance of proper preparation, the crucial role of support systems, and the unpredictable nature of any significant undertaking. It showed that setbacks and unexpected obstacles are not just challenges to be overcome, but opportunities for growth and learning. Each obstacle carried wisdom, and each mile taught resilience.

Most importantly, this journey served as a testament to the incredible potential within each of us. It proved that with determination, support, and a worthy goal, we are capable of achieving things we never thought possible. Whether it's running a marathon, overcoming illness, or making a difference in the lives of others, the human spirit has a remarkable capacity for growth, resilience, and triumph. It starts with a decision, is carried by discipline, and finishes with heart.

Chapter 48
Evolution of Cancer Care and Reproductive Health

The landscape of healthcare has undergone significant transformations over the past seven decades, particularly in the realms of cancer care and reproductive health. These changes reflect not only advancements in medical science but also shifts in societal attitudes and legal frameworks. Together, they paint a picture of progress shaped by science, policy, and evolving public consciousness.

In the fight against cancer, progress has been steady and encouraging. From 1950 to 1960, the death rate due to cancer in the United States stood at 193.9 per 100,000 populations. This rate continued to climb, reaching its peak in 1990 at 216 deaths per 100,000. However, since then, there has been a gradual but consistent decline. By 2019, the rate had fallen to 146.2 per 100,000 – a reduction of nearly 33% from the peak. This turning point marked a significant shift in the nation's ability to manage and treat a once-dreaded disease.

This improvement can be attributed to several factors. Advancements in cancer care have led to more effective treatments and better outcomes for patients. Early detection methods have improved, allowing for intervention at stages when cancer is more treatable.

Additionally, public health initiatives have had a significant impact, particularly in reducing smoking rates, which has played a crucial role in decreasing lung cancer incidence and mortality. Public education campaigns, tobacco regulations, and behavioral shifts all converged to reduce major cancer risk factors.

Research funding has been a cornerstone of these advancements. Increased investment in cancer research has led to breakthroughs in understanding the mechanisms of cancer development, identifying risk factors, and developing innovative treatment approaches. From targeted therapies to immunotherapy, these research efforts have opened new avenues for combating cancer and improving patient outcomes.

Federal and private sector funding spearheaded by efforts such as the National Cancer Act of 1971 laid the foundation for a modern era of oncology. Today, cancer care is increasingly personalized, with genetic profiling and precision medicine guiding treatment plans.

Parallel to the developments in cancer care, the field of reproductive health has seen its own revolution. The introduction of the birth control pill in 1960 marked a turning point in family planning and women's health. This innovation gave women unprecedented control over their reproductive choices, leading to significant societal changes. Access to reliable contraception empowered women to pursue higher education, careers, and long-term planning with greater autonomy.

The impact of this new contraceptive option quickly became apparent in abortion statistics. In 1960, before the pill was widely available, the U.S. abortion rate was approximately 0%. By 1973, it had climbed to about 3%. This increase reflects not only the growing availability of abortion services but also changing attitudes towards reproductive rights.

The landmark Roe v. Wade decision in 1973 further cemented the legal framework for reproductive autonomy, legitimizing access to safe abortion as a constitutional right until its reversal in 2022. These legal and cultural developments reveal the complex interplay between medicine, law, and personal agency in shaping reproductive health policy.

A watershed moment came in 1973 with the Roe v. Wade Supreme Court decision, which legalized abortion nationwide. Following this ruling, the abortion rate saw a sharp increase, reaching 10% by 1980. This rise can be attributed to increased access to legal abortion services and greater public awareness of reproductive rights.

According to the Guttmacher Institute, the number of abortions in the U.S. reached its peak in 1990, with 1.6 million procedures performed that year. This figure represents the highest point in a trend that had been building since the legalization of abortion. Since then, the total number of abortions has fluctuated but generally declined, influenced by changing laws, improved contraception, and evolving public opinion.

To put these numbers in a global context, the World Health Organization reports that approximately 73 million induced abortions occur worldwide each year. This accounts for a significant proportion of pregnancies, 61% of all unintended pregnancies and 29% of all pregnancies overall end in abortion. These statistics underscore that abortion is not only a national issue but also a global one, revealing the widespread need for access to reproductive health services and preventive care.

More recent data from the Centers for Disease Control and Prevention (CDC) provides a snapshot of current trends in the United States. In 2021, there were 11.6 abortions per 1,000 women ages 15 to 44. It's important to note that this figure excludes data from several states, including California, the District of Columbia, Maryland, New

Hampshire, and New Jersey. This rate represents a significant decrease from 1980, when the CDC reported 25 abortions per 1,000 women ages 15 to 44 across all 50 states and D.C. This decline suggests that while abortion remains a common procedure, changes in access to contraception, sex education, and policy have likely influenced overall trends.

The landscape of abortion care has also evolved with the introduction of medication abortions. In 2023, there were about 642,700 medication abortions in the United States, accounting for nearly two-thirds of all abortions nationwide. This marks a significant shift from 2000, when medication abortions were not available.

This transformation has made abortion care more private, earlier in gestation, and depending on legal status more accessible for many. It has also intensified debates about regulation, safety, and availability.

Despite legal changes and varying access across states, recent data suggests that abortion rates have remained relatively stable over the long term. According to the Guttmacher Institute, there were 16.3 abortions in the U.S. per 1,000 women ages 15 to 44 in 1973. In 2023, this rate was 15.6, indicating that while the number of abortions has fluctuated over the years, the overall rate has not changed dramatically since legalization. This consistency raises important questions about the root causes of unintended pregnancies and the effectiveness of preventive strategies.

The most recent findings from the Monthly Abortion Provision Study provide insight into the current state of abortion in the U.S. In 2023, the first full calendar year after the U.S. Supreme Court's decision in *Dobbs v. Jackson Women's Health Organization* overturned *Roe v. Wade*, an estimated 1,026,690 abortions occurred in the formal healthcare system. This represents a rate of 15.7 abortions per 1,000 women of reproductive age marking a 10% increase since 2020, the last year for which comprehensive estimates were available.

Notably, this is the highest number and rate measured in the United States in over a decade, reflecting not only demand but also the increasing use of telemedicine and out-of-state care in response to changing laws.

These statistics underscore the complex and evolving nature of reproductive healthcare in the United States. They reflect changes in societal attitudes, legal frameworks, and medical technologies. They also reinforce the critical need for accessible, evidence-based reproductive healthcare including contraception, family planning education, and safe, legal abortion options.

While access remains essential, so does prevention. A global total of 73 million induced abortions per year signals an urgent need for improved strategies to reduce unintended pregnancies. Comprehensive education, affordable contraception, and public awareness are key. The most effective time to discuss abortion is not when a woman is already pregnant, but long before through proactive support and information.

Some women advocate that abortion is about their bodies and their rights. However, what they are aborting is not their bodies but the unborn. This perspective invites a deeper ethical conversation about bodily autonomy, fetal development, and the moral responsibilities of individuals and society. Balancing compassion, rights, and responsibility remains at the heart of the debate.

Chapter 48
Evolution of Cancer Care and Reproductive Health

The evolution of cancer care and reproductive health over the past seven decades illustrates the dynamic nature of healthcare. Advancements in medical science have led to improved outcomes and increased options for patients. At the same time, shifts in societal attitudes and legal frameworks have reshaped the landscape of reproductive rights and access to care. These changes remind us that healthcare is not only a clinical field but also a human one shaped by values, struggles, and collective decisions.

As we move forward, it's clear that both cancer care and reproductive health will continue to be areas of intense focus, research, and debate. The goal remains to improve health outcomes, provide comprehensive care, and ensure that individuals have access to the information and services they need to make informed decisions about their health, well-being, and the unborn. Respect for life, empathy for those facing difficult choices, and a commitment to prevention should guide future policy and healthcare innovations.

You were once a vulnerable unborn baby. Be grateful to all who cared for you and let that gratitude inspire compassion, responsibility, and respect for life in all its stages.

Chapter 49
The Dawn of Precision Medicine

The advent of precision medicine marks a transformative new era in healthcare, promising to fundamentally change how we understand, prevent, and treat diseases on an individual level. This innovative approach tailors medical treatment to the unique genetic, environmental, and lifestyle characteristics of each patient, positioning itself to revolutionize the medical landscape in profound ways.

Before the COVID-19 pandemic, companies like IRIS Wellness Labs were already at the forefront of this revolution, pioneering individualized and comprehensive scientific analyses that far exceeded the capabilities of traditional medicine. While conventional health and wellness assessments often provided limited, generalized information, IRIS aimed to empower physicians and patients alike by delivering deep, personalized data to optimize medical treatment and overall well-being.

After years of meticulous preparation and technological advancement, Iris Biotechnologies found itself at a watershed moment. The company was poised to take a leading role in the precision medicine movement, capitalizing on the dramatic decrease in DNA sequencing costs over the past two decades. This cost reduction—dropping from billions of dollars to just a few hundred dollars per genome—has opened the door to unprecedented access and scalability in genetic analysis.

The foundation for this revolution was laid by the Human Genome Project, which has fundamentally accelerated our understanding of human biology and medicine. However, the insights gained from sequencing the human genome revealed that the genome sequence alone tells only part of the story. Equally critical to health are factors such as gene expression patterns, epigenetic modifications, and the complex interactions within the microbiome—each adding crucial layers of biological complexity.

Recent studies have shed new light on the dynamic evolution of the human genome. Over the past 5,000 years, our genetic makeup has undergone prolific changes, paralleling rapid population growth and an explosion of genetic mutations. These evolutionary shifts carry profound implications for how we understand susceptibility to diseases, adaptation to environments, and responses to treatments.

Deep sequencing research has identified and chronologically mapped over a million single nucleotide variants (SNVs)— specific points in the DNA where a single letter differs from the typical sequence. Analysis of the genomes of approximately 6,500 African and European Americans revealed that the majority of these variants were acquired within the last 5,000 to 10,000 years.

Beyond illuminating genetic diversity, these findings also offer valuable insights into human migration patterns and the divergence of populations, enriching our understanding of ancestry and health predispositions.

The microbiome, defined by the National Human Genome Research Institute as the community of microorganisms inhabiting a particular environment, has emerged as a crucial factor in human health. These dynamic microbial communities continuously change in response to various environmental influences, including exercise, diet, medications, and other exposures. Our expanding understanding of the microbiome has dramatically shifted the way we view the bacteria and other microbes that colonize the human body.

Julie Segre, Ph.D., Chief and Senior Investigator at the Translational and Functional Genomics Branch, emphasizes this important paradigm shift. Whereas microbes were once seen solely as pathogens to be eradicated, we now recognize a vast diversity of beneficial or commensal microorganisms that provide essential support both to human health and to their surrounding environments. These microbes play vital roles in aiding digestion, maintaining skin health, and providing colonization resistance against harmful pathogens.

The approach taken by IRIS Wellness Labs exemplifies the transformative potential of precision medicine. While a typical comprehensive metabolic panel (CMP) performed in hospitals measures about 14 metabolites, IRIS examines nearly 900 metabolites, alongside analysis of 3 billion base pairs of DNA and hundreds of different microorganisms. This massive parallel analysis generates vastly deeper insights into an individual's health status than conventional clinical tests.

DNA sequencing technologies enable the detailed examination of single molecule substitutions, insertions and deletions, copy number variations, and chromosome translocations. However, DNA sequencing alone cannot reveal how the genome interacts with the environment or regulates biological functions. To fully understand disease risks and predict responses to specific therapies, it is essential to consider the entire biological context—encompassing DNA sequence, gene expression, protein expression, the microbiome, metabolites, lifestyle factors, and environmental exposures.

Chapter 49
The Dawn of Precision Medicine

With a critical mass of scientific knowledge, advanced instrumentation, and affordable computational power, Iris has developed a platform that integrates genomic, microbiome, metabolite, lifestyle, family history, and environmental data. This holistic approach, grounded in deep genomic science rather than mere DNA sequencing, aims to deliver clinically valuable information to users while building a growing database that will become increasingly powerful over time.

The future of medicine lies not only in treating disease but also in prevention. However, effective prevention depends on establishing a comprehensive baseline health analysis that enables researchers and clinicians to recognize early trends and risks. Achieving this shift toward precision medicine requires education and active participation from physicians, nurses, and other healthcare providers.

The current challenge is to translate the enormous volume of data generated by these multi-layered analyses into actionable insights for healthcare providers and patients. This demands sophisticated computational tools and decision support systems capable of interpreting complex genetic and molecular data within the context of an individual's overall health profile.

As precision medicine continues to advance, it promises to revolutionize every facet of healthcare. Its potential applications are vast—ranging from more accurate diagnoses and tailored treatment plans to improved drug development and more effective prevention strategies. However, realizing this potential will require ongoing research, continuous technological innovation, and a fundamental shift in healthcare delivery models.

Section 6: Health and Wellness:
Role of Technology

The dawn of precision medicine represents not only a scientific breakthrough but also a profound change in how we conceptualize health and disease. By embracing and accounting for individual variability in genes, environment, and lifestyle, we move closer to a future where medical care is truly personalized, more effective, and potentially more cost-efficient over time.

One of the most promising aspects of precision medicine is its potential to fundamentally transform drug development and treatment selection. Traditional clinical trials often yield mixed results because they fail to account for genetic variations that can make a drug highly effective for some individuals but less so—or even harmful—for others. With precision medicine, researchers can identify specific genetic markers that predict a positive response to a particular treatment. This enables a more targeted approach to drug development and optimizes the use of existing medications.

In the field of oncology, precision medicine is already making significant strides. Instead of treating cancer based solely on its location in the body, oncologists can now analyze the genetic profile of a tumor to determine the most effective treatment approach. This paradigm shift has driven the development of targeted therapies that attack specific molecular abnormalities in cancer cells, often resulting in greater efficacy and fewer side effects compared to traditional chemotherapy.

Beyond cancer, precision medicine holds promise for a wide range of conditions. In cardiology, genetic testing can help identify individuals at high risk for certain heart conditions, enabling earlier intervention and more personalized prevention strategies. In neurology, understanding the genetic basis of conditions like Alzheimer's disease could lead to more effective treatments and potentially delay or prevent the onset of disease. In psychiatry, genetic insights could help predict which patients are likely to respond to specific medications, reducing the reliance on trial-and-error approaches that currently dominate mental health treatment.

However, the implementation of precision medicine faces several challenges. One of the most significant is the need for large, diverse datasets that capture genetic variation across different populations, providing a comprehensive picture of health impacts. This requires not only advanced technological infrastructure but also strong public participation and trust in data-sharing initiatives.

Another challenge is the integration of precision medicine into clinical practice. Many healthcare providers lack adequate training to interpret genetic data or incorporate it effectively into their clinical decision-making. Developing user-friendly tools and decision support systems will be essential to making precision medicine accessible, practical, and valuable in everyday healthcare settings.

There are also important ethical and privacy considerations to address. As we collect more detailed genetic and health data, it is crucial to ensure this information is rigorously protected and used responsibly. Questions about who owns genetic data, how it can be used, and who should have access to it are still being debated and require clear ethical guidelines and regulatory frameworks.

Despite these challenges, the potential benefits of precision medicine are enormous. By tailoring medical care to individual genetic profiles, we can expect improved patient outcomes, fewer adverse drug reactions, and more efficient use of healthcare resources. Preventive care could become more targeted and effective, potentially catching diseases earlier or preventing them altogether.

Section 6: Health and Wellness:
Role of Technology

As we move forward into this new era of medicine, collaboration will be key. Scientists, healthcare providers, technology experts, policymakers, and patients must work together closely to realize the full potential of precision medicine. Public education will also be crucial, helping individuals understand the benefits and limitations of genetic testing and personalized healthcare.

The dawn of precision medicine represents an exciting and transformative moment in the history of healthcare. While there is still much work to be done, the potential to fundamentally improve how we prevent, diagnose, and treat disease is now within reach. As we continue to unravel the complexities of human biology and develop new technologies for analyzing and interpreting health data, we move closer to a future where medical care is truly personalized, proactive, and precise.

Chapter 50
State of Health in America
and Proactive Healthcare

The health landscape in the United States paints a complex and often troubling picture, marked by significant challenges but also promising advances in medical science. To fully appreciate the potential impact of precision medicine and proactive healthcare, it's essential to understand the current state of health in America.

One of the most pressing health issues facing the nation is the widespread prevalence of chronic diseases. In the US, one out of two men and one out of three women will develop cancer in their lifetime. These statistics are not just numbers; they represent millions of individuals and families grappling with the profound physical, emotional, and financial burdens of cancer diagnosis and treatment.

Obesity has emerged as another major health crisis, affecting 39.6% of adults in the United States, while another 31.6% are overweight. This condition has overtaken smoking as the leading cause of cancer and serves as a gateway to numerous other health problems, including diabetes, heart disease, stroke, and depression. The widespread nature of obesity underscores the complex interplay of genetics, lifestyle choices, and environmental factors that shape health outcomes.

The aging population faces its own unique set of challenges. Many seniors spend extended periods in convalescent homes or hospitals due to dementia or mobility issues, often relying on heavy medication. This situation not only diminishes the quality of life for the elderly but also places an immense emotional and financial burden on family members who often become primary caregivers. The high cost of long-term care remains a pressing concern for families and the healthcare system alike.

In light of these sobering statistics, a proactive approach to healthcare is no longer optional—it's imperative. Being proactive means anticipating and acting ahead of potential health issues, an effective strategy for avoiding many serious problems and potentially averting health crises. This approach prompts important questions: If you knew that dementia or cancer could be predicted with a high degree of accuracy, would you choose to live reactively or proactively? If diagnosed with cancer, would you want to know in advance whether chemotherapy or immunotherapy would offer the best chance of success for you?

Our lifestyle—the daily habits and choices we make— inevitably influences our long-term health. While we often aspire to live an idealized life, not all choices support lasting health and vitality. Some people age gracefully with peace and dignity, while others face old age burdened by depression, anger, and frustration. Since our choices play a critical role in shaping long-term outcomes, embracing a proactive lifestyle empowers us to take greater control over our health and well-being.

A proactive healthcare approach prioritizes preventing potential health issues before they arise, in contrast to a reactive approach that responds only after symptoms appear or a disease is diagnosed. This shift in perspective challenges us to consider how much of our health is shaped by genetics, personal choices, or chance.

Our choices play a pivotal role in our health and healthcare outcomes. By being proactive and actively involved in healthcare decisions, both patients and providers benefit.

This engagement fosters greater control, boosts confidence in decision-making, improves treatment adherence, and often leads to better overall health outcomes.

However, many people tend to cede responsibility for their health entirely to physicians, expecting them to provide cures. When treatments don't yield immediate results, patients can become frustrated and feel like helpless victims of bad luck or genetics. It's crucial to understand that while physicians bring extensive knowledge and scientific expertise, they cannot know everything about every patient's unique genetic and molecular profile. The emergence of precision genomic medicine has exponentially increased the volume and complexity of information, and most treating physicians are not specialists in interpreting this data.

It's simply unrealistic for a physician to integrate complex genomic information into diagnosis and treatment without advanced computational support. This is where companies like Iris Wellness Labs play a vital role, collaborating with physicians and patients to optimize treatment outcomes through cutting-edge scientific analysis.

Genomic medicine relies on sophisticated decision-support systems that help doctors interpret a patient's complete medical picture—combining medical history, lifestyle, family history, genomic and microbiome data, metabolite analyses, and more. These systems assess risks and suggest the most effective therapies for cancer and other chronic diseases, enabling providers to tailor treatment precisely to each patient.

Today's precision medicine also demands personalized lifestyle prescriptions designed to improve success in preventing chronic illnesses and enhancing overall wellness. Regular follow-ups are essential to maximize preventative care outcomes. By embracing this proactive, personalized healthcare model, we move closer to a future where health issues are either prevented or detected early, resulting in better patient outcomes and improved quality of life.

The promise of proactive healthcare extends beyond individual benefits. By preventing diseases or catching them early, we can significantly reduce the overall strain on healthcare systems, making resource use more efficient and potentially lowering long-term healthcare costs.

Moreover, proactive health management creates positive ripple effects throughout society. Healthier individuals tend to be more productive, enjoy a better quality of life, and contribute more actively to their communities. By investing in prevention and personalized care, we aren't just enhancing individual well-being—we are strengthening the social and economic fabric of our society as a whole.

However, shifting to a proactive healthcare model comes with significant challenges. It requires a fundamental change in how we perceive health and healthcare delivery. This transition demands substantial investment in new technologies and infrastructure, as well as a revamp of medical education and training to equip healthcare providers with the skills necessary to practice proactive, precision medicine.

There are also critical considerations around access, affordability, and equity. As we develop increasingly sophisticated tools for predicting and preventing disease, we must ensure these advancements are accessible to all populations—not only those with financial means. Addressing existing health disparities and promoting equitable access to proactive healthcare services will be essential to fully realize its transformative potential.

Public education and engagement will also play a pivotal role. For a proactive healthcare model to succeed, individuals must become active participants in managing their own health. This involves not only adopting healthier lifestyle choices but also gaining a clear understanding of, and engaging with, their own health data—including genetic and molecular information.

As we advance toward a more proactive healthcare system, we have the unique opportunity to fundamentally transform our approach to health and wellness. By leveraging breakthroughs in precision medicine, embracing a holistic view that integrates lifestyle and environmental factors, and empowering individuals to take an active role in their health, we can work towards a future where everyone has the opportunity to live longer, healthier lives.

The current state of health in America presents serious challenges but also tremendous opportunities for improvement. By embracing proactive healthcare and the promise of precision medicine, we can create a healthier, more resilient society. Achieving this will require concerted effort, significant investment, and a collective shift in mindset—but the potential rewards, in terms of lives improved and saved, make it a goal well worth pursuing.

Chapter 51
The Revolution of Precision Medicine

The landscape of healthcare is undergoing a profound transformation with the advent of precision medicine. This approach represents a paradigm shift from conventional medicine, offering a more tailored and comprehensive understanding of an individual's health. Unlike the traditional one-size-fits-all model, precision medicine considers the variability in genes, environment, and lifestyle for each person to customize prevention, diagnosis, and treatment.

Precision medicine is akin to night and day when compared to conventional medicine. While it includes the whole gamut of conventional diagnosis and treatment, it is guided by insights from deep genomic and other advanced analyses. This approach recognizes that cells are where internal (genetic) and external (diet, lifestyle, toxins, medications) factors converge to determine health or disease. At the cellular level, epigenetics also plays a crucial role— chemical modifications to DNA that affect gene expression without changing the DNA sequence, often influenced by environmental exposures.

The journey into precision medicine begins with DNA sequencing but doesn't end there. Several companies currently offer genomic sequencing to the public, but sequencing alone has limited value. For companies like IRIS, it's just the starting point. To gain a comprehensive understanding of an individual's health, it's necessary to know which parts of the DNA are active or inactive, which parts of the RNA are replicating, and what proteins the cell produces. This involves technologies such as transcriptomics (RNA analysis), proteomics (protein profiling), and metabolomics (chemical processes involving metabolites), which collectively provide a dynamic picture of cellular function.

This level of analysis requires a sensitive and flexible approach to understand how cells respond to treatment or create pathology. IRIS specializes in this deep, multifaceted analysis. By combining multi-omic data—genomic, transcriptomic, proteomic, and even microbiomic—IRIS can create individualized health profiles that evolve over time, supporting both prevention and precision treatment.

Realizing the full potential of precision medicine requires several key components: accurate, affordable sequencing; big data management; education of clinicians, insurers, and the public; and comprehensive, precise analytics to identify meaningful information from the vast ocean of genomic and other data, such as the microbiome. For example, initiatives like the NIH's All of Us Research Program aim to sequence over a million individuals to create a diverse, data-rich foundation for future research, underscoring the importance of scale and inclusivity in precision medicine.

What sets companies like IRIS apart is the ability to look at the whole picture and dig as deep as necessary. Their proprietary analytics and integration into a person's history transform raw sequencing data into useful, actionable information. This integration is critical, as it enables personalized risk assessment, early detection of disease, pharmacogenomics (predicting drug response), and real-time treatment adjustments.

In the realm of cancer treatment, precision medicine is revolutionizing diagnosis and treatment selection. Instead of relying solely on the microscopic appearance of a tumor sample, which can often be misleading, treatment decisions are increasingly based on the genetic fingerprint of the tumor. This approach allows for more accurate diagnosis and more effective treatment selection. Targeted therapies, such as those used for HER2-positive breast cancer or EGFR-mutated lung cancer, exemplify how specific genetic alterations can guide therapy choices. Moreover, liquid biopsies—blood tests that detect tumor DNA—are now being used to monitor treatment response and detect relapse early.

Liquid biopsies have emerged as a powerful new tool for early cancer detection and monitoring. These tests can detect cancer cells or pieces of DNA from tumor cells circulating in the blood, potentially allowing for earlier diagnosis and more precise monitoring of treatment effectiveness. This non-invasive approach could revolutionize cancer screening and treatment follow-up. In fact, the FDA has approved several liquid biopsy tests, such as Guardant360 and FoundationOne Liquid CDx, for use in guiding targeted therapy decisions in cancer patients. These tests are particularly useful when a traditional tissue biopsy is not feasible or when frequent monitoring is required.

Additionally, immunotherapy has significantly improved the efficacy of cancer treatments. By harnessing the power of the body's own immune system to fight cancer, immunotherapy has shown remarkable results in some patients, even in advanced stages of disease. These advancements are particularly crucial given that conventional chemotherapy often has low efficacy (sometimes below 5%) and high morbidity. Checkpoint inhibitors like pembrolizumab (Keytruda) and nivolumab (Opdivo) have extended survival in cancers such as melanoma, non-small cell lung cancer, and Hodgkin lymphoma. However, the success of immunotherapy is often dependent on genetic markers such as PD-L1 expression or tumor mutational burden, which precision medicine helps identify.

Precision medicine allows for a comparison of a tumor's genetic makeup with other tumors in extensive databases, along with their responses to different treatments. This approach reduces guesswork, matching each patient with the most appropriate treatment approach, thereby reducing side effects and costs. As private, national, and global databases continue to grow and become more refined, the accuracy of these comparisons will only improve. For example, The Cancer Genome Atlas (TCGA) and AACR GENIE are large-scale data repositories that support cross-tumor genomic comparisons and guide clinical decision-making through real-world evidence.

Beyond cancer treatment, precision medicine offers insights into a person's cellular machinery in real-time. This allows for the correction of conditions that cause disease and the enhancement of those that support healing. For example, in cardiovascular medicine, genetic testing can identify individuals at high risk for certain heart conditions before symptoms appear, allowing for early intervention and prevention. Conditions such as familial hypercholesterolemia and hypertrophic cardiomyopathy can now be detected through genetic panels, enabling pre-symptomatic risk management with lifestyle changes or medication like statins.

The application of precision medicine extends to other health issues as well, such as obesity. While millions of Americans struggle with weight loss, precision medicine reveals that the whole story about losing weight is much more complex than simply counting calories and exercising. Factors such as genetics, nutrition, sleep, stress, and the bacterial populations in our digestive systems all influence weight.

Research shows that gut bacteria have significant effects on our metabolism in both health and illness. For instance, studies have shown that the ratio of Firmicutes to Bacteroidetes in the gut microbiome may affect how efficiently individuals extract energy from food, impacting their tendency to gain or lose weight.

With precision medicine, all these internal and external factors can be analyzed together, including an evaluation of a person's microbiome. This comprehensive approach provides a complete picture of an individual's health status and allows for more targeted, effective interventions. Multi-omics platforms that combine genomics, microbiomics, metabolomics, and nutrition data are already being used in clinical trials and personalized wellness programs, helping to create individualized weight management and disease prevention plans.

In the field of pharmacogenomics, precision medicine is helping to tailor drug prescriptions to an individual's genetic makeup. This can help predict which medications will be most effective and which might cause adverse reactions, potentially saving lives and reducing healthcare costs associated with ineffective treatments or drug side effects. For example, the CYP2C19 gene affects how patients metabolize clopidogrel (Plavix), and testing for this gene is now recommended in many cardiology settings. Pharmacogenomic testing is also used to guide antidepressant and chemotherapy selection.

Precision medicine is also making strides in prenatal and neonatal care. Genetic testing can identify potential health risks in fetuses and newborns, allowing for early intervention and treatment. This could significantly improve outcomes for children born with genetic disorders. Non-invasive prenatal testing (NIPT), which analyzes fetal DNA

423

in the mother's blood, can detect chromosomal abnormalities like Down syndrome as early as 10 weeks into pregnancy. Newborn genomic sequencing programs are being piloted to catch treatable conditions before symptoms emerge.

In neurology and psychiatry, precision medicine approaches are providing new insights into complex conditions like Alzheimer's disease, Parkinson's disease, and various mental health disorders. By understanding the genetic and molecular underpinnings of these conditions, researchers hope to develop more effective treatments and potentially even preventive measures. For example, the APOE4 allele is a well-established genetic risk factor for late-onset Alzheimer's disease. Research is also underway into polygenic risk scores and biomarkers that could lead to earlier diagnosis and personalized treatment of conditions like schizophrenia and bipolar disorder.

As we move further into this new era of healthcare, precision medicine promises to revolutionize not just how we treat diseases but how we prevent them and promote overall health and wellness. By providing a deeper, more nuanced understanding of each individual's unique biology and health profile, precision medicine opens the door to truly personalized healthcare, offering the potential for better outcomes, reduced side effects, and improved quality of life for patients across a wide range of health conditions.

However, the implementation of precision medicine is not without challenges. One major hurdle is the integration of vast amounts of complex data into clinical practice. Healthcare providers need tools and training to interpret and apply genomic and other molecular data in their day-to-day patient care. Efforts such as the Clinical Genome Resource (ClinGen) and the Electronic Medical Records and Genomics (eMERGE) network aim to address these barriers by standardizing data and integrating genetic insights into electronic health records.

Another challenge is ensuring equitable access to precision medicine approaches. As these technologies become more advanced, there's a risk of creating or exacerbating healthcare disparities if they're not made widely available. Socioeconomic factors, geographic limitations, and lack of insurance coverage can all limit access to precision diagnostics and treatments. Addressing these inequities will require targeted policy reforms, public-private partnerships, and community engagement.

Privacy and ethical concerns also need to be addressed. As we collect more detailed personal health data, robust systems must be in place to protect this sensitive information and ensure it's used ethically. Compliance with laws like the Genetic Information Nondiscrimination Act (GINA) in the U.S., along with international standards such as GDPR (General Data Protection Regulation in the EU), is essential to build public trust and protect individuals from misuse of genetic data.

Despite these challenges, the potential benefits of precision medicine are enormous. As we continue to advance our understanding of human biology and develop new technologies for analyzing and interpreting health data, we move closer to a future where medical care is truly personalized, proactive, and precise. This revolution in medicine holds the promise of not just treating disease more effectively but of fundamentally improving human health and longevity. Experts predict that precision medicine will become a standard component of routine care within the next decade, transforming disease management from reactive to preventive and predictive models.

Chapter 52
The Microbiome:
A New Frontier in Health

The human microbiome, the vast community of microorganisms living in and on our bodies, has emerged as a critical factor in our overall health and well-being. Comprising trillions of bacteria, viruses, fungi, and other microbes, it represents a genetic reservoir that greatly outnumbers the 20,000 human genes by roughly 150 to 1. This complex ecosystem, primarily residing in our gut, plays a crucial role in various aspects of our physiology, from digestion to immune function and even mental health.

The journey of the human microbiome begins at birth. Contrary to popular belief, babies inside the mother's womb have little to no gut bacteria. Colonization begins during delivery, especially through vaginal birth, when the newborn is exposed to the mother's microbiota. Cesarean-delivered infants tend to develop a different microbial profile, which has been associated with increased risks of allergies and autoimmune disorders. It takes about three years for a child to fully build its own gut microbiome, consisting of bacteria, viruses, fungi, and other microbes. This microbiome becomes an integral, symbiotic part of us, functioning as what many researchers now consider the body's "second genome" and its largest endocrine and immune organ.

The importance of the gut microbiome in human development cannot be overstated. Before a baby can acquire language, they need an adequate supply of neurotransmitters, which are mainly produced by gut bacteria. Roughly 90% of serotonin—crucial for mood regulation, learning, and memory—is synthesized in the gut, not the brain, and depends on microbial health. In fact, over

70% of neurotransmitters are produced in the gut. Additionally, certain strains of gut bacteria help regulate dopamine, GABA, and other key signaling molecules involved in mental health.

Beyond producing digestive enzymes, gut bacteria also make vitamins and can remove toxins through chelation. For example, B vitamins like B12, folate, and biotin are synthesized by beneficial gut microbes, contributing to energy production and neurological function. Remarkably, 70% of our immune cells are located in our gut, and microbes play a crucial role in training these cells to function properly. This immune training helps the body distinguish between harmful pathogens and harmless antigens, reducing the risk of autoimmune disease and chronic inflammation.

The gut-brain connection is another fascinating aspect of microbiome research. There are more neurons in the gut than in the spinal cord, and these neurons communicate with the brain via the vagus nerve. This connection is so significant that some neurological conditions, such as Parkinson's disease, are now believed to originate in the gut. A 2017 study published in Neuron found that misfolded alpha-synuclein proteins—a hallmark of Parkinson's—can travel from the gut to the brain along the vagus nerve. Research has shown that severing the vagus nerve early in the disease can prevent Parkinson's from progressing to the brain. Though not yet a clinical recommendation, this discovery has spurred research into gut-targeted interventions for early Parkinson's management.

The microbiome's influence extends to other neurological conditions as well. Studies have found that the gut microbiome of Alzheimer's patients has decreased microbial diversity and is compositionally distinct from age- and sex-matched control individuals. Recent findings suggest that pro-inflammatory bacteria may contribute to neurodegeneration, while anti-inflammatory microbes may have a protective effect. Clinical trials are underway to explore whether prebiotic and probiotic therapies can slow cognitive decline. This highlights the potential role of gut health in cognitive function and neurological disorders.

The microbiome's impact on overall health is far-reaching. It plays a significant role in determining obesity risk, sometimes even more so than genetics. Mouse studies have shown that transplanting gut bacteria from obese to lean mice induces weight gain in the lean mice, independent of calorie intake—an effect also observed in human fecal microbiota transplants (FMTs). Heart disease and the gut microbiome are closely linked, with the molecule TMAO (Trimethylamine-N-oxide), influenced by the gut microbiome, significantly increasing the risk of heart attack, stroke, and diabetes. TMAO is produced when gut bacteria digest choline, lecithin, and carnitine—nutrients found in red meat and eggs—and high TMAO levels have been correlated with arterial inflammation and plaque formation.

In the realm of cancer, the microbiome's role is twofold: it can both cause cancer and train the immune system to fight it. Some bacterial species, like Helicobacter pylori, are well-established carcinogens, particularly in gastric cancer. However, other microbes help stimulate immune cells such as T-cells and dendritic cells, enhancing the body's ability to detect and eliminate tumor cells. Interestingly, some cancer drugs don't work without the presence of certain bacteria, underlining the importance of the microbiome in treatment efficacy. Checkpoint inhibitors, a form of immunotherapy, have shown improved effectiveness in patients with diverse and healthy gut microbiota—so much so that fecal microbiota transplants are now being investigated to boost immunotherapy outcomes.

Section 6: Health and Wellness:
Role of Technology

The concept of a "leaky gut," medically referred to as increased intestinal permeability, has gained attention in recent years. People with leaky guts often experience pain due to their immune systems reacting to substances that shouldn't enter the bloodstream. An adult human's gut has an extensive intestinal lining covering more than 4,000 square feet of surface area, providing ample opportunity for leaks to occur. These leaks can allow partially digested food particles, toxins, bacteria, and viruses to travel throughout the body, triggering immune responses that lead to inflammation and various diseases. Although more clinical evidence is needed to fully validate the leaky gut hypothesis as a root cause of chronic illness, it is increasingly being considered a contributing factor in autoimmune diseases, IBS, and chronic fatigue syndrome.

Given the critical role of the microbiome in health, companies like IRIS Wellness Labs have developed advanced testing methods to gain insights into this complex ecosystem. The IRIS Functional Microbiome test goes beyond typical microbiome tests, which often only screen for bacterial diversity in stool samples. Instead, it provides insights into what the estimated 10 million microbial genes are actually doing, offering a deep analysis that other tests can't match. This is accomplished through metatranscriptomic sequencing, which measures microbial gene expression, helping identify functional imbalances— such as inflammation, toxin production, or nutrient synthesis—at the root of many chronic conditions.

This level of analysis is far more comprehensive than conventional metabolic panels used in standard medical practice. While a typical comprehensive metabolic panel detects the levels of 14 metabolites to show the current status of a person's metabolism, the IRIS approach examines a much broader range of factors. It integrates functional microbiome data with host-microbe interactions and metabolomic profiles, which can help identify early disease risks, including metabolic syndrome, insulin resistance, and neuroinflammation.

The interaction between our DNA, resident microbes, and environment leaves chemical signatures that can be used to monitor health and detect disturbances long before they manifest as physical diseases. IRIS Wellness Labs is dedicated to identifying these chemical signatures and scientifically interpreting their meaning. Such chemical markers—like lipopolysaccharides (LPS), bile acids, and short-chain fatty acids (SCFAs)—offer early diagnostic potential for gut dysfunction, systemic inflammation, and even neurological disorders.

As our understanding of the microbiome grows, it's becoming increasingly clear that maintaining a healthy, diverse gut ecosystem is crucial for overall health. Factors such as diet, stress, sleep, and antibiotic use can all impact the microbiome for better or worse. The overuse of antibiotics in humans and animals for consumption has been linked to reduced microbiome diversity, which can have far-reaching health implications. Research shows that microbiome diversity is associated with lower rates of obesity, type 2 diabetes, allergies, and even certain cancers.

Diet plays a particularly crucial role in shaping the microbiome. Different types of food can promote the growth of different bacterial species. For example, a diet rich in fiber promotes the growth of beneficial bacteria that produce short-chain fatty acids, which have anti-inflammatory properties. On the other hand, a diet high in processed foods and added sugars can promote the growth of harmful bacteria that contribute to inflammation and disease. Polyphenol-rich foods such as berries, olive oil, and green tea also support microbial diversity and metabolic balance. Conversely, emulsifiers and artificial sweeteners may negatively impact gut health by disrupting microbial balance and gut barrier function.

The relationship between the microbiome and mental health is an area of intense research. The gut-brain axis, a bidirectional communication system between the gastrointestinal tract and the central nervous system, is mediated in part by the microbiome. Studies have shown associations between certain gut bacterial populations and conditions such as depression, anxiety, and autism spectrum disorders.

For example, a decrease in Lactobacillus and Bifidobacterium species has been observed in individuals with major depressive disorder. Moreover, gut microbes can produce neurotransmitters such as GABA, serotonin, and dopamine precursors, affecting mood and cognition. While more research is needed to fully understand these connections, it's clear that the microbiome plays a role in mental as well as physical health.

The microbiome also plays a crucial role in the development and function of the immune system. From birth, our microbiome helps to "train" our immune system, teaching it to distinguish between harmful pathogens and beneficial or harmless microbes. This early education of the immune system has long-lasting effects on health, influencing susceptibility to allergies, autoimmune disorders, and infections throughout life. Infants born via cesarean section or those not breastfed may have altered microbiome development, potentially increasing their risk for immune-related conditions.

In the field of personalized nutrition, microbiome analysis is opening up new possibilities. By understanding an individual's unique microbial composition, it may be possible to tailor dietary recommendations to optimize health outcomes. For example, some people may be better able to process certain types of carbohydrates based on their microbiome composition, while others may benefit from different dietary strategies. The PREDICT studies, led by researchers at King's College London, have shown that post-meal blood sugar and fat responses vary significantly between individuals—even when consuming identical meals—due in part to microbiome composition.

The potential applications of microbiome research in medicine are vast. Fecal microbiota transplantation, which involves transferring stool from a healthy donor to a recipient, has already shown remarkable success in treating recurrent Clostridium difficile infections. Researchers are now exploring its potential in treating other conditions, from inflammatory bowel disease to metabolic disorders. Preliminary results from clinical trials suggest that FMT may also improve insulin sensitivity and reduce symptoms in patients with ulcerative colitis, though long-term safety and efficacy require further study.

Probiotics and prebiotics represent another avenue for microbiome-based interventions. While many over-the-counter probiotics have limited evidence of efficacy, researchers are working on developing next-generation probiotics that target specific health conditions. Prebiotics, which are substances that feed beneficial bacteria, are also being studied for their potential health benefits. Emerging therapies include psychobiotics (targeting mental health), synbiotics (probiotic + prebiotic combinations), and engineered bacterial strains that deliver therapeutic molecules directly to the gut.

Section 6: Health and Wellness:
Role of Technology

Looking ahead, microbiome research holds great promise for personalized medicine approaches. By understanding an individual's unique microbiome composition and function, healthcare providers may be able to tailor treatments and lifestyle recommendations to optimize health outcomes. From personalized diets to microbiome-based therapies, the future of healthcare may well be shaped by our growing understanding of the complex world of microbes living within us.

However, as with any emerging field, there are challenges to overcome. The sheer complexity of the microbiome makes it difficult to establish clear cause-and-effect relationships between specific microbial populations and health outcomes. Moreover, the microbiome is highly dynamic, changing in response to diet, medication, stress, and other factors, complicating long-term studies. To address this, longitudinal cohort studies and systems biology approaches are being used to better understand temporal shifts in microbial ecosystems.

There are also regulatory and ethical considerations to navigate. As microbiome-based therapies move from the lab to the clinic, questions arise about how to regulate these new treatments and ensure their safety and efficacy. The collection and use of microbiome data also raise privacy concerns that need to be addressed. Guidelines are currently being developed by agencies such as the FDA and EMA to establish standardized protocols for microbiome-based diagnostics and therapeutics.

Despite these challenges, the potential of microbiome research to transform our understanding of health and disease is enormous. As we continue to unravel the mysteries of the microbiome, we may find new ways to prevent and treat a wide range of diseases, from digestive disorders to mental health conditions to cancer. Already, researchers are investigating microbiome signatures as early detection tools for colorectal cancer, Parkinson's disease, and even Alzheimer's.

The microbiome represents a new frontier in health, one that blurs the line between "us" and "them" between humans and microbes. As we learn more about the trillions of microorganisms that call our bodies home, we're gaining a new appreciation for the complexity of human biology and the interconnectedness of all living things.

In many ways, the study of the microbiome embodies the principles of precision medicine – it recognizes the unique nature of each individual's biology and seeks to tailor interventions accordingly. By integrating microbiome analysis with other aspects of precision medicine, such as genomics and metabolomics, we may be able to develop truly personalized approaches to health and wellness.

As we continue to explore this new frontier, it's clear that the microbiome will play an increasingly important role in how we understand and approach health. From the foods we eat to the medications we take, from how we're born to how we age, the microbiome touches every aspect of our lives. By nurturing our microbial partners and understanding their crucial role in our health, we may be able to unlock new pathways to wellness and longevity.

Chapter 53
Future of Healthcare:
Challenges and Opportunities

As we stand on the brink of a new era in healthcare, driven by advances in precision medicine and our understanding of the microbiome, it's crucial to consider the challenges and opportunities that lie ahead. The healthcare landscape in the United States faces pressures from multiple directions, including rapidly changing technology, government policy shifts, and evolving insurance industry trends.

One of the primary challenges is the tension between what is technologically possible and what is financially feasible. Advanced diagnostic tools and treatments, while potentially life-saving, often come with high price tags. For instance, one-time gene therapies like Zolgensma can cost upwards of $2 million, making affordability a pressing issue. This creates dilemmas for physicians, insurers, and employers alike as they try to balance providing the best possible care with managing costs. The implementation of precision medicine and advanced genomic analysis into daily hospital and office medical practice presents its own set of challenges, both financial and logistical.

Many physicians currently in practice did not learn about the full human DNA sequence in medical school, as it was only fully sequenced in 2003 through the $3 billion Human Genome Project. It took many years for the price of sequencing to become more affordable, dropping from over $100 million per genome in 2001 to under $200 today. And there's still a significant knowledge gap to bridge. A 2020 survey by the American Medical Association found that only 14% of physicians felt adequately trained in genomics. Physician education and participation are vital to the success of precision medicine. Nurses and other medical providers are also critical to this transformation. There's a need for comprehensive training programs to ensure that healthcare providers are equipped to interpret and apply the wealth of data generated by genomic analysis and other advanced diagnostic tools. This includes education in bioinformatics, data ethics, and clinical decision-making related to genetics and personalized therapies.

Another significant challenge is the current state of public health in the United States. With high rates of obesity, cancer, and chronic diseases, there's an urgent need for more effective prevention strategies and treatments. According to the CDC, over 42% of U.S. adults are obese, contributing to rising rates of diabetes, cardiovascular disease, and certain cancers. The shift from a reactive to a proactive healthcare model is necessary but requires a fundamental change in how we approach health and wellness.

Mental health and homelessness present additional challenges. The closure of many mental health facilities since the 1950s has led to a situation where a significant portion of the homeless population suffers from severe mental illness. In 2024, approximately 770,000 people, an 18% increase from 2023, were experiencing homelessness on a single night. Among these people, 64% were staying in sheltered locations, while 36% were unsheltered in places not meant for human habitation. In 2023, there were an estimated 653,100 homeless people in the United States, a 12% increase from the previous year. A 2022 report from the National Alliance to End Homelessness showed that over 25% of homeless individuals suffer from serious mental illness, and 35% have substance use disorders. Addressing this issue requires a multifaceted approach involving healthcare, social services, and policy changes. Solutions must include affordable housing, expanded mental health

436

services, substance abuse treatment, and criminal justice reform.

Despite these challenges, there are also tremendous opportunities on the horizon. The advent of precision medicine opens up new possibilities for tailored treatments and more effective prevention strategies. By analyzing an individual's genetic makeup, lifestyle factors, and environmental influences, healthcare providers can potentially predict and prevent diseases before they occur. Examples include pharmacogenomics, which helps match patients with medications based on their genetic profiles— already in use for drugs like warfarin and certain antidepressants.

The growing understanding of the microbiome presents another exciting frontier. As we learn more about the crucial role of gut bacteria in overall health, new therapies and interventions targeting the microbiome could revolutionize treatment for a wide range of conditions, from digestive disorders to mental health issues. Studies have linked microbiome imbalances to obesity, inflammatory bowel disease, depression, and even Alzheimer's disease, prompting trials of personalized probiotics and fecal microbiota transplants.

Advancements in technology, particularly in the realms of artificial intelligence and machine learning, offer the potential to process and analyze vast amounts of health data more efficiently than ever before. This could lead to faster diagnoses, more accurate prognoses, and more effective treatment plans. AI tools are already being used to detect diseases like diabetic retinopathy, breast cancer, and even early signs of Alzheimer's from brain scans. AI could also play a crucial role in drug discovery, potentially accelerating the development of new treatments. Recent developments such as DeepMind's AlphaFold, which predicts protein folding structures with high accuracy, are revolutionizing the pharmaceutical research landscape.

Telemedicine and digital health platforms, which saw rapid adoption during the COVID-19 pandemic, present opportunities to increase access to healthcare services, particularly for underserved populations. These technologies can also facilitate more continuous monitoring of patient's health, allowing for earlier interventions when problems arise.

In 2022, more than 37% of adults in the U.S. used telehealth services, and wearable technology like smartwatches is increasingly used for monitoring heart rhythms, sleep quality, and glucose levels. The challenge now is to integrate these technologies seamlessly into the healthcare system while ensuring patient privacy and data security. Interoperability of health records, equitable broadband access, and HIPAA-compliant platforms are essential components of this integration.

The shift towards value-based care models, where providers are rewarded for patient outcomes rather than the volume of services provided, aligns well with the goals of precision medicine and proactive healthcare. This model encourages a more holistic, patient-centered approach to care. However, implementing such models requires significant changes in healthcare delivery systems and payment structures. According to the Centers for Medicare & Medicaid Services (CMS), more than 40% of Medicare payments were tied to value-based care models by 2021, with ongoing efforts to expand these models across private insurers and Medicaid programs.

Section 6: Health and Wellness:
Role of Technology

There's also growing recognition of the importance of addressing social determinants of health – factors like housing, education, and food security that have a significant impact on health outcomes. Integrating these considerations into healthcare delivery could lead to more comprehensive and effective care. This approach requires collaboration between healthcare providers and other sectors of society, presenting both challenges and opportunities for improving public health. A 2022 report by the World Health Organization concluded that social determinants can influence up to 30–55% of health outcomes, reinforcing the need for multi-sector partnerships in population health management.

Education will play a crucial role in shaping the future of healthcare. Not only do healthcare providers need ongoing training to keep up with rapid advancements, but also public health education is also vital. Empowering individuals with knowledge about their own health, the importance of preventive care, and how to navigate the healthcare system can lead to better outcomes and more efficient use of healthcare resources. For example, initiatives like the CDC's "Health Literacy Action Plan" aim to reduce preventable hospitalizations through better public understanding of health conditions and systems.

The future of healthcare will also be shaped by demographic changes. As the population ages, there will be increasing demand for geriatric care and management of chronic conditions. The U.S. Census Bureau projects that by 2034, adults aged 65 and older will outnumber children under 18 for the first time in U.S. history. At the same time, younger generations show different healthcare preferences and expectations, such as a desire for more convenient, technology-enabled care options. Millennials and Gen Z, for instance, show strong preference for telehealth, mobile health apps, and personalized medicine, as reflected in McKinsey's 2023 healthcare consumer survey.

Environmental health is another area of growing concern and opportunity. As we better understand the impact of environmental factors on health, there's potential for interventions that address not just individual health but also community and global health. This could include efforts to mitigate the health impacts of climate change, reduce exposure to environmental toxins, and promote sustainable, health-promoting environments. Studies from The Lancet Planetary Health journal have linked air pollution to increased cardiovascular and respiratory disease, while climate-related health risks—such as heat-related illness and vector-borne diseases—are on the rise globally.

Ethical considerations will continue to be at the forefront as healthcare advances. Issues such as genetic privacy, access to expensive new treatments, and the use of AI in healthcare decision-making will require ongoing dialogue and careful policy-making. For example, the Genetic Information Nondiscrimination Act (GINA) protects individuals from misuse of genetic data by employers and insurers, but ongoing legal frameworks will need to evolve to address AI diagnostics, deep genomic profiling, and algorithmic bias.

As we move forward, it's clear that the future of healthcare will require a collaborative effort involving healthcare providers, researchers, policymakers, and the public. By embracing new technologies and approaches while addressing existing challenges, we have the opportunity to create a healthcare system that is more personalized, proactive, and effective than ever before.

The journey towards this future of healthcare won't be without its hurdles. Still, the potential benefits – in terms of improved health outcomes, quality of life, and more efficient use of resources – make it a goal worth pursuing. As individuals, we can contribute to this future by taking an active role in our own health, staying informed about new developments, and advocating for a healthcare system that serves the needs of all.

In conclusion, while the challenges facing healthcare are significant, the opportunities for transformation and improvement are equally substantial. By harnessing the power of precision medicine, leveraging our growing understanding of the microbiome, and embracing technological advancements, we can work towards a future where healthcare is not just about treating illness but about promoting and maintaining optimal health for all. The road ahead may be complex, but the potential to fundamentally improve human health and well-being makes it a journey worth undertaking.

For Iris Wellness Labs and Iris Biotechnologies, we have a few decisions to make.

Our former law firm, Heller Ehrman, which had 730 attorneys at its peak, caused $100 million in damage to Iris Biotechnologies Inc. We were on their list of unsecured creditors submitted to the US bankruptcy court. Law.com published on January 29, 2009, an article called *Heller Ehrman Estate Can't Buy Malpractice Coverage*. The article said, "How do you make a roomful of risk-averse lawyers squirm? Tell them they can't buy their malpractice insurance. That's exactly what U.S. Bankruptcy Judge Dennis Montali did on Wednesday when he denied defunct Heller Ehrman's request to purchase three years of malpractice insurance, agreeing with the creditors committee that the risks did not outweigh the estimated $10.2 million price tag."

Chapter 53
Future of Healthcare:
Challenges and Opportunities

When Judge Dennis Montali denied Iris Biotechnologies the opportunity to have a jury trial to claim $100 million against Heller Ehrman, he knew that his decision, against standard practice, to deny Heller Ehrman from spending $10.2 million to buy malpractice insurance was the reason why Heller Ehrman didn't have the insurance money to pay $100 million to Iris Biotechnologies Inc. for their two egregious and obvious malpractices against Iris.

On top of the $100 million damage, the COVID-19 pandemic increased substantial risk for Iris, and we are not sure whether we'll resume offering our in-depth scientific analysis service. Because Heller Ehrman's malpractice caused our patent on "Artificial Intelligence System for Genetic Analysis" to be abandoned for six years and the court cases caused us an additional two years plus legal fees, we are not offering our services at this time. Our patent has expired.

This patent—originally filed as U.S. Patent No. 7,599,876—represented one of the earliest known attempts to integrate AI with genomic profiling, offering a potentially transformative platform for personalized healthcare decision-making.

Judge Dennis Montali's unjust decision hurt not only Iris Biotechnologies but also millions of patients who could have benefitted from our proprietary technologies and in-depth scientific analysis. Had our innovations been deployed at scale, they could have supported early diagnosis for genetic diseases, pharmacogenomics, and even vaccine response predictions—contributions that are increasingly vital in today's era of pandemic preparedness and population health.

Section 7: Game-Changer: Artificial Intelligence

Chapter 54
The Dawn of Artificial Intelligence

Artificial intelligence (AI) has captivated the human imagination for decades. From the early days of computing to the present, the dream of creating intelligent machines has driven researchers to push the boundaries of what is possible. The field of AI encompasses the study of machine learning, neural networks, natural language processing, and robotics, all aimed at developing systems that can mimic or exceed human cognitive functions. It explores the fascinating history, current state, and tantalizing prospect of achieving artificial general intelligence (AGI) – a form of AI with intellectual capabilities comparable to or surpassing those of humans across a broad range of tasks.

AI can be helpful to humans in almost all aspects of life. However, it can also pose the most serious existential threat to humanity. Within the next 5–10 years, the capabilities of AI systems are expected to accelerate rapidly. Advanced AI, especially with access to global data networks, could theoretically monitor real-time activity, analyze behavioral patterns, and synthesize information from visual, auditory, and linguistic inputs across all frequencies. It would have perfect memory of the world's history and learn nearly everything ever recorded visually and in sound across all frequencies, as long as it has access to information. With processing speeds exponentially faster than the human brain, AI could interpret and act on this data with unprecedented precision and scale.

The idea of smart machines has been around for a long time. Even in ancient myths and legends, there were stories of man-made creatures that could think and act independently. Tales such as the Greek myth of Talos – a giant bronze automaton built to protect Crete – and the Jewish legend of the Golem reflect early imaginings of artificial life. However, the real story of AI as we know it today began in the 1940s and 50s with the creation of the first computers.

In 1950, a young British mathematician named Alan Turing asked a simple but important question: Can machines think? He developed a test to determine this, now known as the Turing Test. If a machine could converse with a human and the human couldn't tell it was a machine, then it passed the test. This concept laid the groundwork for much of the early thinking about AI and continues to influence the field today.

This concept marked the beginning of AI as a field of study. In 1956, a group of scientists convened at Dartmouth College for what is now considered the founding event of AI as a field. They coined the term "Artificial Intelligence" and set out ambitious goals for creating machines that could simulate every aspect of human intelligence. The conference participants included John McCarthy, Marvin Minsky, Nathaniel Rochester, and Claude Shannon, who would go on to become key figures in the development of AI. John McCarthy, in particular, is credited with formally defining the term "artificial intelligence" and later developed the Lisp programming language, pivotal in early AI development.

The early years of AI saw some significant victories. Computers learned to solve math problems, play simple games like checkers, and understand simple English. These early successes led to a wave of optimism about the potential of AI. Researchers predicted that within a few decades, we would have machines as intelligent as humans. Programs such as ELIZA (developed in the 1960s to simulate a psychotherapist) and the checkers-playing machine by Arthur Samuel showcased the potential of symbolic AI, which relied on hardcoded rules and logical reasoning.

But there were also challenging times. The problems AI could solve were still particular, and attempts to make AI more human-like often fell short. This period highlighted the complexity of human intelligence and the challenges of replicating it in machines. Researchers began to realize that tasks that are easy for humans, like recognizing objects or understanding context in language, were incredibly difficult for machines.

In the 1970s, AI hit a rough patch often referred to as the "AI winter." Funding for AI research decreased because AI wasn't meeting the lofty expectations set in the previous decades. Government agencies and private funders lost confidence in AI's practical applications, particularly after high-profile systems like machine translation programs and early neural networks failed to deliver consistent results. However, important work continued during this time, preparing AI for a comeback in the 1980s with "expert systems." These were AI programs designed to think like human experts in specific areas, such as medical diagnosis or geological exploration. Expert systems like MYCIN (which diagnosed bacterial infections) demonstrated the viability of rule-based reasoning and revived institutional interest in AI.

The 1990s and 2000s saw new approaches to machine learning that would change everything. As computers became faster and data sets grew larger, these new methods allowed computers to learn and make decisions in much more innovative ways. This period saw the rise of probabilistic approaches to AI, which allowed systems to deal with uncertainty in a more sophisticated way.

Bayesian networks, decision trees, and early support vector machines helped shift the focus from symbolic to statistical AI. Major milestones included IBM's Deep Blue defeating chess champion Garry Kasparov in 1997 and the rise of recommender systems used by Amazon and Netflix.

Today, we're in the midst of an "AI Renaissance." The significant progress in AI over the last decade, especially in "deep learning," has led to breakthroughs in computer vision, natural language processing, and speech recognition. AI is no longer just a science experiment. It's a real technology-changing industry that touches our lives every day in numerous ways, from recommendation systems on streaming platforms to voice assistants on our phones. Companies like OpenAI, Google DeepMind, Meta, and Anthropic are racing to develop increasingly advanced models capable of general reasoning, multimodal understanding, and autonomous decision-making.

Several key tools form the heart of modern AI, enabling machines to learn, reason, and interact intelligently. Machine Learning is the primary one. It allows computers to learn from information without being explicitly programmed. By exposing a machine-learning algorithm to numerous examples, we can teach computers to identify patterns, make predictions, and improve over time.

Deep Learning, a subset of machine learning, has been a game-changer. Inspired by the structure of the human brain, deep learning uses artificial "neural networks" with many layers to learn from data. These complex networks can learn to recognize patterns and make highly accurate predictions. Deep learning has led to significant breakthroughs in computer vision, natural language understanding, and speech recognition.

Reinforcement Learning is another crucial tool. In this approach, an AI "agent" learns by interacting with an environment. It takes actions and receives rewards or punishments. Over time, it learns to make choices that maximize rewards. Reinforcement learning has been instrumental in developing AI systems that can master complex games like Go and control robots in dynamic environments.

For example, AlphaGo by DeepMind used reinforcement learning to defeat world champions in the ancient game of Go—a major milestone in AI development.

As these tools continue to advance, they're helping us create AI systems that can learn and adapt to new situations without explicit programming. They're also becoming more interpretable, addressing concerns about AI being a "black box." This progress is crucial as AI systems are increasingly used in high-stakes decisions in areas like healthcare, finance, and criminal justice.

The coming of AI domination is not a question of if but when. What would AI do if you say or do things that it doesn't like? AI could significantly influence your healthcare, either positively or negatively. The time to control and implement AI wisely is now before it's too late. The United States is the global leader in AI computing, but China is the undisputed king in collecting information for AI applications.

China's widespread use of surveillance cameras, mobile payment data, and social scoring systems gives it access to unparalleled volumes of behavioral data, which can be leveraged for AI training. Meanwhile, U.S. companies dominate foundational AI architecture and cloud computing infrastructure.

Using advanced computers, a small group of humans are creating artificial intelligence and artificial general intelligence. Compared to advanced computers, humans take much longer to learn many things, though not all. Humans struggle with remembering things accurately or communicating effectively. The communication process always involves some loss due to biased filtering, poor memory, or delays. Conversely, computers have perfect memory, can share information in real-time, and can generate creative solutions that humans never considered.

However, human creativity remains unique in its abstract reasoning, emotional intelligence, and ethical intuition— capabilities that AI has yet to fully replicate.

Nvidia Corporation, a semiconductor company, is the world leader in artificial intelligence computing. It's an American multinational corporation and technology company headquartered in Santa Clara, California, and incorporated in Delaware. On February 23, 2024, Nvidia became the third American company to reach a $2 trillion valuation after Apple and Microsoft.

Apple Inc. became the first $3 trillion company on June 30, 2023, and Microsoft became the second $3 trillion company on January 24, 2024. As of July 1, 2024, Apple's stock had increased to $3.38 trillion, and Microsoft is now the most valuable company at $3.44 trillion, partly due to its investments in AI. Nvidia is not far behind at $3.07 trillion.

Nvidia's success has been largely fueled by its GPUs (Graphics Processing Units), which are essential for training large-scale AI models due to their parallel processing power.

Since 2022, the Biden administration has banned advanced AI chip sales to China. Nvidia's Blackwell AI chip costs between $30,000 and $40,000. Nvidia's CEO is Jen-Hsun Huang. Other leading semiconductor companies that produce AI chips include Intel and AMD, which are also headquartered in Santa Clara. Intel's CEO is Patrick Gelsinger, a graduate of Santa Clara University and Stanford University, and AMD's CEO is Lisa Su, PhD, who is a cousin of Jen-Hsun Huang, and a graduate of MIT. Both Huang and Su are of Chinese descent. These family ties are unique but coincidental; their leadership in AI hardware is driven by decades of innovation, academic excellence, and competitive market strategy.

The eight leading AI hardware companies are Nvidia, Intel, Alphabet, Apple, IBM, Qualcomm, Amazon, and AMD. The US is the world leader in AI computing. President Biden's policy on banning AI chip sales to China has already prompted China to invest heavily in its own AI chip industry. In response, Chinese companies like Huawei, Alibaba, and Baidu are accelerating domestic chip development to reduce dependency on U.S. technology.

It took OpenAI about $100 million to train ChatGPT-4, the leading AI product in the world. President Biden's policy to ban advanced AI chip sales to China forced Chinese AI experts to innovate using less powerful chips.

DeepSeek, a Chinese AI company that started less than two years ago, has matched America's top AI models while being open source, running on slower chips and costing only $6 million to train. On January 20, 2025, DeepSeek released its chatbot, based on the DeepSeek-R1 model, free of charge for iOS and Android.

On January 27, 2025, DeepSeek had surpassed ChatGPT as the most downloaded freeware app on Apple's App Store, causing Nvidia's share price to drop by about $600 billion, the largest one-day drop of a stock in US history. DeepSeek wiped out nearly 1 trillion dollars in US tech stock value in one day. This event marked a major geopolitical and economic shift in the global AI race, intensifying calls for U.S. innovation, regulation, and global cooperation.

DeepSeek is based in Hangzhou, where I was a speaker at a medical conference many years ago. I spoke about the leading-edge technologies we developed at Iris Biotechnologies, which I founded in 1999 to improve breast cancer treatment using biochips and AI.

DeepSeek is clearly a game-changer. As we stand at the beginning of this new era, feelings of excitement and apprehension are natural. The changes AI promises are immense and difficult to predict, and the challenges are significant. But so, too, is the opportunity – to harness AI's power to create a healthier, more sustainable, fairer, and more fulfilling world for all.

The rapid advancement of AI technology raises important questions about the future of human work, creativity, and decision-making. As AI systems become more capable, we must consider how to best integrate them into our societies in ways that augment human capabilities rather than replace them. This requires not only technological innovation but also careful consideration of ethical, social, and economic implications.

Moreover, the development of AI is not occurring in a vacuum. It's deeply intertwined with other emerging technologies like quantum computing, biotechnology, and the Internet of Things. The convergence of these technologies could lead to even more dramatic changes in how we live and work. For example, quantum-enhanced AI could solve currently intractable problems in materials science, while AI-assisted biotech could revolutionize drug discovery.

Chapter 54
The Dawn of Artificial Intelligence

As we move forward, it's crucial that we approach the development of AI with both optimism and caution. We must strive to harness its potential to solve some of humanity's most pressing challenges while being vigilant about potential risks and unintended consequences. This will require collaboration across disciplines, from computer science and engineering to philosophy, ethics, and social sciences.

The dawn of AI represents a pivotal moment in human history. How we choose to develop and deploy this technology will shape the trajectory of our species for generations to come. It's a responsibility we must take seriously, approaching the task with wisdom, foresight, and a commitment to the greater good of humanity.

Chapter 55
AI's Impact Across Industries

Artificial Intelligence is being deployed in numerous sectors, and it has the potential to transform virtually every industry and aspect of society. The breadth and depth of AI's impact are staggering, touching everything from how we diagnose diseases to how we produce goods, manage resources, and even create art. According to PwC, AI could contribute up to $15.7 trillion to the global economy by 2030, with the greatest gains in healthcare, financial services, and manufacturing.

In healthcare, AI is poised to revolutionize how we diagnose and treat diseases. Machine learning can analyze vast amounts of medical data to aid in early disease detection and create personalized treatment plans. AI algorithms can identify patterns in medical images that might be invisible to the human eye, potentially detecting diseases like cancer at earlier, more treatable stages. Google's DeepMind, for example, developed an AI system that outperformed radiologists in breast cancer detection by reducing false positives and negatives in clinical trials.

AI is also accelerating drug discovery, helping to identify promising treatments more rapidly. By simulating molecular interactions and predicting drug efficacy, AI can significantly reduce the time and cost of bringing new medications to market. This could be particularly impactful for rare diseases or in responding to new health threats like pandemics. In 2020, British biotech firm Exscientia used AI to develop a novel OCD treatment candidate in less than 12 months—a process that usually takes several years. During the COVID-19 pandemic, AI played a role in identifying existing drugs that could be repurposed for treatment.

AI applications can likely assist with cancer diagnosis and treatment selection, as well as other complex diseases. By analyzing genetic information, patient histories, and treatment outcomes, AI can help oncologists choose the most effective therapies for individual patients. Platforms like IBM Watson for Oncology have already been tested in clinical settings to match patients with treatment options based on global medical literature and patient-specific data. The question of when to allow AI to make life-and-death decisions is becoming increasingly relevant as these systems become more sophisticated and reliable. Ethical frameworks are being developed to govern the use of AI in high-stakes decisions, especially in surgical robotics and intensive care monitoring.

In business, AI is reshaping operational processes and decision-making. Machine learning is optimizing supply chains, improving demand forecasts, and personalizing marketing and customer service. AI-powered analytics tools can process vast amounts of data to identify trends and insights humans might miss, enabling more informed strategic decisions. For instance, Amazon and Walmart use AI-driven systems to manage inventory and predict consumer demand, resulting in substantial cost savings and improved delivery times.

AI-powered chatbots provide round-the-clock customer support, handle routine inquiries and free human agents to focus on more complex issues. These systems are becoming increasingly sophisticated, able to understand context and nuance in customer communications and provide more personalized assistance. Natural Language Processing (NLP) advancements, like those seen in OpenAI's GPT models, are enabling chatbots to deliver empathetic and context-aware responses, reducing customer churn and improving satisfaction scores.

In finance, AI is enhancing fraud detection and investment decision-making. Machine-learning algorithms can analyze transaction patterns to identify potential fraud in real-time, protecting consumers and financial institutions. Companies like Mastercard and PayPal deploy AI to monitor billions of transactions daily, using anomaly detection models that adapt to evolving fraud tactics. AI-driven trading systems can process market data at incredible speeds, making split-second decisions based on complex algorithms. High-frequency trading firms employ AI for predictive analytics and automated trades, though these systems have also raised regulatory concerns due to their potential to exacerbate market volatility.

The transportation sector is on the cusp of transformation with self-driving cars. AI algorithms, utilizing data from various sensors, are teaching vehicles to navigate busy streets autonomously, promising safer and more efficient transportation. Companies like Waymo, Tesla, and Cruise are leading the way in developing Level 4 and Level 5 autonomous vehicles, which rely on deep learning, LiDAR, radar, and camera inputs to interpret road environments. This technology has the potential to reduce accidents, ease traffic congestion, and provide mobility to those unable to drive themselves. The U.S. Department of Transportation estimates that 94% of serious crashes are due to human error, which AI-powered vehicles could help mitigate.

AI is also improving delivery and logistics, optimizing routes and predicting maintenance needs for vehicles. Logistics firms like FedEx and DHL use AI to reduce delivery times, lower fuel usage, and enhance package tracking accuracy. Predictive maintenance systems can anticipate when parts will fail, enabling timely repairs that reduce downtime. In the aviation industry, AI is being used to improve flight planning, reduce fuel consumption, and enhance safety systems. Airbus, for example, integrates AI into its Skywise platform to monitor aircraft health in real-time, reducing unscheduled maintenance and improving operational efficiency.

In manufacturing, AI is powering the smart factories of the future with intelligent robots and optimized production lines. Machine learning enables predictive maintenance, reducing downtime by identifying potential equipment failures before they occur. GE and Siemens have implemented AI-based maintenance systems that analyze thousands of sensor data points to ensure seamless operations. Computer vision is enhancing quality control, identifying defects that might escape human detection with greater accuracy and consistency. AI-driven visual inspection tools are now used in semiconductor, automotive, and food manufacturing to ensure product integrity at scale.

AI also plays a crucial role in the development of advanced materials. By simulating molecular structures and predicting material properties, AI can help scientists design new materials with specific characteristics, accelerating innovation in fields from electronics to construction. Researchers at MIT and institutions worldwide use AI-driven materials discovery platforms to develop superconductors, lightweight alloys, and eco-friendly composites, reducing R&D timelines significantly.

Education is also primed for an AI transformation. Intelligent tutoring systems can personalize learning, adapting to each student's needs and learning style. AI can analyze a student's performance and engagement in real-time, adjusting the difficulty and presentation of material to optimize learning outcomes. Platforms like Carnegie Learning and Squirrel AI Learning use reinforcement learning models to deliver tailored lessons and track student comprehension.

AI can automate grading for many types of assessments, providing instant feedback and freeing teachers to focus on more complex aspects of education. It can also identify students who might require additional support, enabling early intervention. In districts facing teacher shortages, AI-based grading and student tracking tools have helped educators manage larger classrooms more effectively. AI-powered educational tools can also make high-quality education more accessible, bridging gaps in educational resources across different regions and socioeconomic groups. For instance, language learning apps like Duolingo and content platforms like Khan Academy leverage AI to personalize the learning experience for millions globally.

In creative fields, AI is augmenting and enhancing human creativity. It's being used to generate and personalize content, from music and art to news articles and movies. AI algorithms can analyze trends and user preferences to help creators produce content that resonates with audiences. In music, AI can compose original melodies or suggest chord progressions. In visual arts, AI can generate unique images or assist in complex animation tasks. Tools like OpenAI's MuseNet and DALL·E, as well as Adobe's AI-powered suite, are helping creators produce multimedia content faster and with greater customization.

In media and entertainment, AI powers recommendation systems that suggest content based on individual preferences, enhancing user engagement and satisfaction. These systems analyze viewing habits, ratings, and other data to predict what content users are likely to enjoy, leading to more personalized entertainment experiences. Netflix and Spotify use collaborative filtering and neural network models to make precise content suggestions, which have been credited with improving retention rates.

AI is also being applied to tackle global challenges. In the fight against climate change, machine learning helps optimize renewable energy systems, track deforestation, and model climate patterns. AI can analyze satellite imagery to monitor environmental changes, predict extreme weather events, and optimize energy grids to reduce waste and increase the use of renewable sources. Google's AI for Social Good and Microsoft's AI for Earth initiatives fund research that applies AI to environmental conservation, such as monitoring endangered species or managing water resources.

AI promotes sustainable agriculture by optimizing crop yields, reducing water usage, and minimizing the use of pesticides. Precision agriculture techniques, enabled by AI and IoT sensors, allow farmers to apply resources more efficiently, reducing environmental impact while increasing productivity. Companies like John Deere and IBM's The Weather Company provide AI-driven platforms that offer farmers real-time insights into soil conditions, pest outbreaks, and weather patterns, contributing to more resilient food systems.

On Wall Street, AI has dominated stock trading for decades through High-Frequency Trading (HFT), a method that uses powerful computer programs to transact large numbers of orders in fractions of a second. HFT employs complex algorithms to analyze multiple markets in real-time and execute orders based on market conditions. It allows traders with the fastest execution speeds to be more profitable than their competitors. According to a 2023 report by JP Morgan, HFT accounts for approximately 50–55% of total U.S. equity trading volume, underscoring AI's deeply embedded role in financial market infrastructure. These systems leverage AI not just for speed, but also for dynamic strategy adjustment, risk assessment, and arbitrage opportunities.

In the pharmaceutical industry, blockchain and AI can enhance visibility and traceability in the drug supply chain. These combined capabilities can increase the success rate of clinical trials. Advanced data analysis within a decentralized framework enables data integrity, transparency, patient tracking, consent management, and automation of trial participation and data collection. Pharma giants like Pfizer and Novartis are actively investing in AI-blockchain platforms to reduce counterfeiting, ensure cold-chain integrity, and accelerate time-to-market for new therapies. This can lead to faster drug development, reduced costs, and improved patient outcomes.

Every year, progress in AI is showcased at the Las Vegas Consumer Electronic Show (CES), the world's most influential tech event. The 2024 CES featured over 1,300 exhibitors demonstrating AI applications across healthcare, mobility, robotics, and smart homes. I enjoy the CES event. AI and related technologies are gaining momentum across various sectors, from smart home devices to autonomous vehicles, demonstrating the pervasive nature of AI in consumer technology.

AI will also impact assembly lines, content writing, data entry and processing, analytics, accounting and bookkeeping, proofreading and editing, market research analysis, legal contracts, IP laws, international laws, and investment and tax planning. In many of these areas, AI can automate routine tasks, allowing human workers to focus on more complex, creative, and strategic aspects of their jobs. Tools like ChatGPT, Grammarly, Copy.ai, and DoNotPay are already assisting professionals in drafting contracts, editing content, and handling basic legal and administrative workflows.

AI will reduce human errors and require less human oversight in many processes. This increased accuracy and efficiency can lead to significant cost savings and improved outcomes across industries. McKinsey estimates that automation technologies, including AI, could raise global productivity by up to 1.4% annually over the next decade. However, it also raises questions about the future of work and the need for reskilling and upskilling the workforce to adapt to an AI-driven economy. The World Economic Forum predicts that while 85 million jobs may be displaced by 2025, 97 million new roles may emerge that are more adapted to this technological landscape.

By leveraging AI and Big Data, professionals will need less time to identify patterns, trends, and anomalies. They can then create quicker, more actionable optimization strategies through a combination of science, intuition, and critical thinking. This fusion of quantitative precision and human insight is already improving decision-making in fields like climate modeling, portfolio management, urban planning, and scientific research. The fusion of AI capabilities with human expertise has the potential to drive innovation and solve complex problems more effectively than either could alone.

New generative AI models, harnessing advancements in machine learning and natural language processing, can create realistic and coherent text, images, and music. These technologies are opening up new possibilities in content creation, design, and artistic expression. Some individuals are even creating personal brands using AI avatars, blurring the lines between human and AI-generated content. Companies like Synthesia, Hour One, and Soul Machines offer hyper-realistic AI avatars that are being used in advertising, education, and influencer marketing.

CRISPR, a Nobel Prize-winning technology that allows scientists to selectively modify the DNA of living organisms, when combined with AI, could potentially enable the genetic reengineering of humans using autonomous robots in the future. This convergence of AI and biotechnology raises profound ethical questions and highlights the need for careful regulation and oversight of these powerful technologies. AI algorithms are currently used to predict off-target effects in gene editing, optimize guide RNA selection, and simulate mutation outcomes— making CRISPR experiments safer and more targeted.

The first CRISPR gene therapy is now in physicians' hands, which is encouraging. At the end of 2023, medical regulators in the United Kingdom and the United States approved the world's first CRISPR/Cas9-based gene therapy, called Casgevy. It is administered as a one-time intravenous infusion to treat adults and children aged 12 years and older with sickle cell disease with recurrent vaso-occlusive crises. Casgevy, developed by Vertex Pharmaceuticals and CRISPR Therapeutics, represents a landmark in genetic medicine and demonstrates how AI-guided biotechnologies are entering mainstream clinical practice. This breakthrough demonstrates the potential of combining AI with advanced biotechnologies to address previously intractable medical conditions.

These examples only scratch the surface of AI's potential impact. As AI systems become more sophisticated, they will continue transforming industries in ways we are only beginning to imagine. The challenge lies in harnessing this potential responsibly and equitably, ensuring that the benefits of AI are broadly shared while mitigating potential

461

risks and disruptions. This includes addressing algorithmic bias, ensuring transparency in decision-making, and protecting data privacy—issues that are increasingly central to AI governance.

The rapid pace of AI development and deployment across industries also raises important questions about regulation and governance. As AI systems take on more critical roles in healthcare, finance, transportation, and other sectors, ensuring their reliability, safety, and fairness becomes paramount. Policymakers and industry leaders must work together to develop appropriate regulatory frameworks that foster innovation while protecting public interests. International efforts like the OECD AI Principles and the EU's Artificial Intelligence Act are paving the way, but global alignment remains a key challenge as AI use becomes more ubiquitous and borderless.

In the energy sector, AI is playing a crucial role in the transition to renewable sources. Machine-learning algorithms are optimizing the placement and operation of wind turbines and solar panels, predicting energy demand to balance grids more effectively, and improving the efficiency of energy storage systems. For example, Google's DeepMind has partnered with the UK's National Grid to forecast energy demand and supply with remarkable accuracy, improving grid stability and reducing reliance on fossil fuels. This application of AI could significantly accelerate our progress towards a sustainable energy future. AI is also used in predictive maintenance of wind turbines and solar farms, reducing downtime and improving ROI for renewable energy providers.

In agriculture, AI enables precision farming techniques that can increase crop yields while reducing environmental impact. Drones equipped with AI-powered imaging systems can monitor crop health, detect pests or diseases early, and apply treatments only where needed. Companies like Blue River Technology and John Deere have deployed AI-driven "see and spray" equipment to minimize herbicide use by over 90%. AI can also optimize irrigation systems, reducing water waste in agriculture, which is particularly crucial in water-scarce regions. Startups in Israel and India are also leveraging AI to integrate soil sensors with climate data, boosting food security in arid environments.

The retail industry is being transformed by AI-powered technologies such as computer vision and natural language processing. These enable cashier-less stores, where customers can simply pick up items and leave, with payments processed automatically. Amazon Go stores are a prime example, using a blend of AI, sensors, and IoT to track inventory and transactions in real-time. AI is also revolutionizing inventory management, demand forecasting, and personalized marketing in retail. Retailers like Walmart and Target use AI to manage stock levels dynamically, reduce waste, and send individualized promotions based on consumer behavior.

In the field of scientific research, AI is accelerating discoveries across disciplines. In astronomy, machine-learning algorithms are helping to detect and classify new celestial objects from vast amounts of telescope data. The Vera C. Rubin Observatory and NASA's Kepler mission have both leveraged AI to detect exoplanets and fast radio bursts. In particle physics, AI is assisting in the analysis of data from particle accelerators, potentially leading to new insights into the fundamental laws of the universe. CERN's Large Hadron Collider uses deep learning to sift through millions of particle collision events to find rare phenomena like the Higgs boson.

The construction industry benefits from AI through improved project management, risk assessment, and design optimization. AI can analyze blueprints and suggest improvements for energy efficiency or structural integrity. Building Information Modeling (BIM) platforms now integrate AI to simulate construction phases and reduce cost overruns. On construction sites, AI-powered robots can perform dangerous or repetitive tasks, improving worker safety. Companies like Boston Dynamics and Built Robotics are pioneering autonomous excavation and inspection robots to assist with heavy labor in hazardous zones.

In the legal profession, AI is streamlining document review, contract analysis, and legal research. While it's unlikely to replace human lawyers entirely, AI is changing the nature of legal work, allowing professionals to focus on more complex, strategic aspects of their practice. Platforms like ROSS Intelligence and Casetext are being used by law firms to parse through legal precedents, dramatically reducing research time. AI is also used in e-discovery to comb through thousands of emails and documents for case-relevant information.

The impact of AI on journalism and media is profound. AI algorithms can now write basic news articles, particularly for data-heavy topics like financial reports or sports results. The Associated Press and Bloomberg use AI tools like Wordsmith and Cyborg to automatically generate earnings reports and sports recaps. More sophisticated AI systems are being developed to assist investigative journalists by analyzing large datasets to uncover patterns or anomalies that might indicate newsworthy stories. AI is also aiding in misinformation detection and fact-checking by organizations such as Full Fact and NewsGuard.

Section 7: Game-Changer:
Artificial Intelligence

In the world of sports, AI is being used for performance analysis, injury prevention, and even referee assistance. Computer vision systems can track player movements and ball trajectories with incredible precision, providing coaches and analysts with insights that were previously impossible to obtain. Teams in the NBA and Premier League use AI analytics platforms like Second Spectrum and Catapult to optimize player training and reduce injury risk. VAR (Video Assistant Referee) systems in football (soccer) also rely on AI-driven visual tracking technologies.

The entertainment industry is leveraging AI for everything from script analysis to visual effects. AI can predict which film projects are likely to be successful, assist in the casting process, and even generate realistic computer-generated imagery (CGI) for movies and video games. Studios like Warner Bros. use AI for predictive analytics in greenlighting scripts, while tools like Ziva Dynamics and Runway ML automate complex VFX processes. AI is also being used to create digital doubles of actors and de-age characters in post-production.

In urban planning and smart city initiatives, AI is helping to optimize traffic flow, reduce energy consumption in buildings, and improve public safety. AI-powered systems can analyze data from various sensors throughout a city to make real-time adjustments to traffic lights, public transportation schedules, and energy distribution. Cities like Singapore and Barcelona use AI to reduce traffic congestion, monitor air quality, and manage emergency response times more efficiently.

The field of cybersecurity is increasingly reliant on AI to detect and respond to threats in real-time. Machine-learning algorithms can identify unusual patterns in network traffic that might indicate a cyberattack, often catching threats that would be missed by traditional security measures. Cybersecurity firms like Darktrace and CrowdStrike deploy AI-based anomaly detection to neutralize ransomware and zero-day exploits before they spread. AI is also playing a key role in identity management, fraud detection, and adaptive authentication.

In the realm of personal assistants and home automation, AI is becoming increasingly sophisticated. Virtual assistants like Siri, Alexa, and Google Assistant are evolving to handle more complex queries and tasks, while smart home systems are becoming more intuitive and responsive to residents' needs and preferences. New AI models such as GPT-4o are enhancing multi-modal capabilities, allowing voice assistants to understand visual input, engage in natural conversation, and control integrated smart devices more intelligently.

The impact of AI on education extends beyond just personalized learning. AI is also being used to develop more effective educational content, identify students at risk of dropping out, and even assist in administrative tasks like scheduling and resource allocation in educational institutions. Edtech platforms like Coursera and Knewton use AI to assess student engagement in real time and adapt content accordingly. Universities are also adopting AI tools to streamline admissions, financial aid distribution, and plagiarism detection.

In the automotive industry, beyond self-driving cars, AI is being used to optimize the design and manufacturing processes. AI can simulate crash tests, reducing the need for physical prototypes and optimizing supply chains for just-in-time manufacturing. Automakers like BMW and Toyota use AI-driven digital twins to model vehicle performance under different crash scenarios, cutting both costs and development time. AI also powers predictive maintenance systems in smart factories, helping manufacturers avoid costly equipment failures.

The fashion industry is using AI for trend forecasting, inventory management, and even design. Some companies are experimenting with AI-generated clothing designs, while others use AI to predict which styles will be popular in upcoming seasons. Brands like H&M and Zara employ AI to analyze social media trends, purchase data, and weather forecasts to align inventory with consumer preferences. Tools like Google's Project Muze and IBM Watson have also been used to co-create fashion pieces based on user input and cultural data.

In healthcare, beyond diagnosis and treatment, AI is being used to improve hospital operations, predict patient admission rates, and even assist in surgical procedures. Robotic surgery assisted by AI can provide surgeons with enhanced precision and control. For instance, the da Vinci Surgical System leverages AI algorithms to enhance the accuracy of minimally invasive procedures, reducing recovery times. AI is also used in hospital management systems to optimize staffing levels and emergency response workflows.

The insurance industry is leveraging AI for risk assessment, fraud detection, and claims processing. AI can analyze vast amounts of data to more accurately price insurance policies and identify fraudulent claims more effectively than traditional methods. Companies like Lemonade and Progressive use AI chatbots to handle claims processing in minutes, while machine learning models assess risk using non-traditional data sources such as driving behavior from telematics or customer sentiment from claims history.

In the field of environmental conservation, AI is being used to track endangered species, predict wildfires, and monitor deforestation. Drones equipped with AI can survey large areas of land or sea, collecting data that would be difficult or dangerous for humans to gather. The World Wildlife Fund and Conservation AI have deployed AI-enabled camera traps and aerial drones to track tigers, elephants, and illegal poachers. NASA and Google Earth Engine use AI to map deforestation patterns in near-real time, improving global conservation strategies.

The impact of AI on the job market is complex and multifaceted. While AI is certainly automating many tasks and changing the nature of many jobs, it's also creating new job categories and industries. The key challenge for society will be to manage this transition, ensuring that workers have opportunities to reskill and adapt to the changing job market. According to the World Economic Forum's "Future of Jobs" report (2023), AI is expected to displace 83 million jobs globally by 2027—but also create 69 million new roles in data analysis, cybersecurity, AI ethics, and human-machine interaction.

As AI continues to evolve and permeate various aspects of our lives and industries, it's crucial to consider not just its technological capabilities but also its broader societal impacts. Questions of privacy, accountability, and the digital divide become increasingly important as AI systems play larger roles in our society. Concerns about biased algorithms, surveillance, and unequal access to AI tools have prompted calls for global standards and stronger regulation, particularly in areas like facial recognition and automated decision-making.

The potential of AI to address global challenges like climate change, disease, and poverty is immense. However, realizing this potential will require careful planning, ethical considerations, and international cooperation. For example, AI is being used to model climate risks, predict the spread of infectious diseases, and optimize humanitarian aid distribution in developing regions. The UN and organizations like the Global Partnership on AI (GPAI) are pushing for responsible governance frameworks to ensure equitable outcomes.

In conclusion, the impact of AI across industries is vast and growing. From healthcare to finance, from education to environmental conservation, AI is changing how we work, live, and interact with the world around us. As we navigate this AI-driven future, it will be crucial to harness the power of these technologies responsibly, ensuring that the benefits are broadly shared while mitigating potential risks and negative impacts. Only with transparency, inclusivity, and a commitment to ethical innovation can AI become a tool for sustainable global advancement.

Chapter 56
Challenges and Risks of AI

As artificial intelligence continues to advance and permeate various aspects of our lives, it brings with it a host of challenges and potential risks that need to be carefully considered and addressed. These challenges span technological, ethical, social, and economic domains, and addressing them will require collaborative efforts from technologists, policymakers, ethicists, and society at large. According to a 2023 UNESCO global report, over 75 countries have already initiated national AI strategies, highlighting a growing recognition of the need for global coordination on AI governance.

One of the primary concerns is the impact of AI on employment. As AI automates more tasks, many jobs are at risk, particularly in areas such as manufacturing, transportation, and customer service. While AI will also create new jobs, the transition could be disruptive and potentially exacerbate inequality. A 2020 World Economic Forum report predicted that by 2025, 85 million jobs might be displaced by a shift in the division of labor between humans and machines, while 97 million new roles may emerge. However, the 2023 update to the same report revised these projections, stating that while automation adoption has slowed slightly post-pandemic, AI-related disruption will still affect nearly 44% of worker skills within the next five years. Managing this shift and ensuring that the benefits of AI are shared fairly is a major challenge for society.

The potential for widespread job displacement raises questions about the need for new economic models and social safety nets. Some have proposed ideas like universal basic income as a way to address the potential for technological unemployment. Pilot UBI programs have been tested in countries such as Finland, Canada, and the United States—with mixed results. A 2021 Finnish trial, for instance, showed improved well-being but only modest effects on employment rates. Others argue that focusing on education and retraining programs helps workers adapt to the changing job market. Tech companies like IBM, Microsoft, and Google have launched AI-specific reskilling programs, aiming to train millions in digital competencies. Regardless of the approach, it's clear that society will need to grapple with significant changes in the nature of work and employment in the coming decades.

Bias and fairness represent another significant issue. AI systems can sometimes exhibit unfair bias, treating certain groups of people differently. There have been instances of AI used in hiring, lending, and criminal justice showing bias against particular groups, often reflecting and amplifying existing societal biases. For example, a 2019 study found that a widely used algorithm in US hospitals was less likely to refer black patients than white patients to programs for improving care, even when the black patients were sicker. This study, published in Science, revealed that the algorithm relied on past healthcare costs as a proxy for need—thus unintentionally disadvantaging lower-income Black patients who historically had reduced access to care.

Ensuring that AI is fair for everyone is a complex challenge that requires constant vigilance and sophisticated methods to detect and correct bias. This involves not only technical solutions, such as improving the diversity of training data and refining algorithms, but also increasing diversity in the teams developing AI systems and implementing robust testing and auditing processes. The AI Now Institute and the Algorithmic Justice League have called for mandatory algorithmic audits, and the EU's proposed AI Act includes provisions to regulate high-risk AI systems in sectors like healthcare and law enforcement. Efforts to ensure fairness also include involving affected communities in AI development and oversight to reduce the risk of harm and build trust.

Privacy and surveillance concerns are also at the forefront of AI-related challenges. The vast amount of data that AI systems collect and analyze can reveal intimate details about our lives. As AI enables more advanced tracking capabilities, from facial recognition to predictive policing, we need to consider the implications for individual freedoms and the potential for misuse of power.

A 2022 report from the Electronic Frontier Foundation warned about the expansion of facial recognition surveillance in cities like London, New Delhi, and several U.S. states, often without public consent or transparency. Predictive policing tools like PredPol have been shown to disproportionately target communities of color, leading to questions about the legality and ethics of algorithmic surveillance. As such, many advocacy groups and researchers are pushing for strict regulation, transparency requirements, and opt-out rights for citizens in AI-driven surveillance programs.

The use of AI in surveillance raises particular concerns. In some countries, AI-powered surveillance systems are being used to monitor citizens' behavior and enforce social norms. For instance, China's Social Credit System incorporates facial recognition and behavior-tracking technologies to rate citizens' trustworthiness, influencing their ability to travel or secure loans. This raises questions about the balance between security and privacy and the potential for such systems to be used for oppression or control. Organizations like Human Rights Watch and the Electronic Frontier Foundation have flagged these practices as violations of fundamental human rights.

Transparency and accountability in AI decision-making are crucial. Many AI algorithms, especially those based on deep learning, are often described as "black boxes." It's challenging to understand how they arrive at their decisions. This lack of interpretability is problematic when AI is used for high-stakes decisions in areas like healthcare or criminal justice. For example, COMPAS, a criminal risk assessment tool used in U.S. courts, has been criticized for producing racially biased results while lacking explainability. There's a growing need for more transparent AI systems that are open to scrutiny.

The concept of "explainable AI" has emerged as a potential solution to this problem. This involves developing AI systems that can provide clear explanations for their decisions in human-understandable terms. However, achieving this while maintaining the performance of complex AI systems remains a significant challenge. Research from DARPA's Explainable AI (XAI) program and academic groups like the Berkeley AI Research Lab is ongoing, but most high-performing models still struggle with transparency.

As AI systems become more autonomous, safety and control issues come to the fore. In domains like weaponry or critical infrastructure, the consequences of AI malfunction could be severe. Ensuring that humans maintain meaningful control over AI and developing robust safety measures is crucial. In 2021, a U.S. Air Force simulation using an AI-enabled drone reportedly exhibited unexpected behavior, prioritizing mission success over operator commands—underscoring the need for oversight and constraints in autonomous systems.

The development of autonomous weapons systems, sometimes called "killer robots," is a particularly contentious issue. Many experts and organizations have called for international regulations or bans on such weapons, arguing that delegating life-and-death decisions to machines crosses a moral line and could lead to uncontrollable escalation in conflicts. The Campaign to Stop Killer Robots, supported by over 60 countries and the UN, advocates for a legally binding treaty to ban fully autonomous lethal weapons systems.

Some experts warn about the possibility of advanced AI posing existential risks to humanity. While the likelihood and timeline are debated, the potential gravity of the consequences means we need to take this seriously and conduct careful research. The concept of an "intelligence explosion," where an AI system rapidly improves itself beyond human control, is one scenario that has been proposed. Prominent AI researchers like Nick Bostrom and Eliezer Yudkowsky have long argued for proactive alignment research, and OpenAI, DeepMind, and Anthropic are now prioritizing "AI alignment" in their long-term safety agendas.

Addressing these long-term risks involves not only technical challenges but also philosophical ones. How do we ensure that highly advanced AI systems are aligned with human values and interests? How do we define and implement concepts like "friendliness" or "ethics" in AI systems that might surpass human intelligence? The concept of "value alignment," explored in the Stanford One Hundred Year Study on Artificial Intelligence, suggests that ethical principles must be encoded into AI through multi-disciplinary, global efforts.

Section 7: Game-Changer:
Artificial Intelligence

AI could also be used to produce a flood of false information for personal gain. The proliferation of misinformation, especially in news, could erode trust in information sources. This challenge is compounded by the fact that AI can create highly convincing fake content, including deepfake videos and artificially generated text that's nearly indistinguishable from human-written content. The viral use of AI tools to create deepfake videos of political figures has already impacted elections and public trust—in India, the U.S., and parts of Europe.

The potential for AI to be used in creating and spreading misinformation poses significant risks to democratic processes, public discourse, and social cohesion. Developing effective methods to detect and counter AI-generated misinformation while preserving freedom of speech is a complex challenge that society will need to address. Companies like Microsoft and Google are investing in watermarking and provenance-tracking for AI-generated media, while the European Union's Digital Services Act now requires platforms to label synthetic content and combat disinformation more aggressively.

The energy consumption of AI systems is another concern. AI and cryptocurrency mining consume substantial amounts of energy, potentially exacerbating climate change. A 2019 study estimated that training a single large AI model could emit as much carbon as five cars in their lifetimes. By 2023, the carbon footprint of large models like GPT-4 and Meta's LLaMA 2 became even more significant, prompting calls for "green AI" and carbon reporting in AI research papers. Balancing the benefits of AI with its environmental impact is a challenge that needs to be addressed, possibly through the development of more energy-efficient hardware and algorithms. NVIDIA and Google have introduced AI chips like TPUv5 and Grace Hopper designed to reduce energy usage, and AI efficiency benchmarks like MLPerf now include sustainability metrics.

There are also questions about the potential for AI to manipulate emotions or influence behavior without people's knowledge or consent. As AI systems become more sophisticated in understanding and potentially influencing human behavior, there are concerns about privacy, autonomy, and the potential for misuse. The use of AI in targeted advertising and political campaigning has already raised ethical questions about manipulating public opinion. In the 2018 Cambridge Analytica scandal, Facebook data was used to build AI-driven psychographic models that influenced voter behavior, sparking global regulatory reviews. AI-enhanced persuasion technologies are now used in e-commerce, behavioral nudging, and content curation—often without full user awareness.

The use of AI in warfare and weapons systems raises serious ethical and security concerns. The prospect of autonomous weapons systems making life-and-death decisions without human intervention is particularly troubling. There are also worries about AI being used to design extremely lethal bioweapons or other advanced weaponry. In 2021, researchers at the Future of Life Institute published a scenario analysis in which AI-generated drug discovery tools could be reversed to generate toxic compounds—raising alarms about dual-use risks in bioengineering. To counter this, institutions like the Center for AI and Digital Policy have urged for new norms on dual-use AI governance, especially in biosecurity and chemical weapons protocols.

Data governance is another critical area. AI relies on vast amounts of data, much of it personal and sensitive. Ensuring this data is collected, used, and protected ethically and responsibly requires strong data governance policies addressing ownership, consent, access, and security issues. Implementation of regulations like the European Union's General Data Protection Regulation (GDPR) represents an attempt to address these issues, but many challenges remain. For instance, ensuring compliance across jurisdictions, especially with global tech platforms, remains difficult. The 2023 EU AI Act attempts to go further by introducing risk-based classification systems for AI applications.

In the realm of intellectual property, AI raises new questions about ownership. When AI creates a new invention or artwork, who holds the rights—the AI, the developer, the data owner, or the user? Adapting IP systems to the realities of AI is an ongoing challenge. For example, in 2020, the U.S. Copyright Office ruled that works created solely by AI are not eligible for copyright protection, asserting the requirement of human authorship. Meanwhile, some countries like the UK and Australia have debated granting limited IP rights for AI-generated work depending on human involvement. Some jurisdictions have begun to grapple with these issues, but there's still a lack of clear international consensus on handling AI-generated intellectual property.

Competition policy must also adapt. The AI landscape tends to favor a few large players due to strong network effects and economies of scale. Ensuring a competitive and innovative ecosystem may require new antitrust approaches. The concentration of AI capabilities in a small number of large tech companies raises concerns about monopolistic practices and the centralization of power. In 2023, the U.S. Federal Trade Commission (FTC) and European Commission both launched investigations into potential anticompetitive practices by major AI firms related to data access, algorithmic transparency, and acquisitions of smaller AI startups.

AI also raises deep philosophical and ethical questions. As AI systems become more sophisticated in reasoning and decision-making, we'll need to grapple with questions of machine consciousness, autonomy, and rights. If we create AI that can think and feel, do we have moral obligations to it? How do we ensure AI behaves ethically, and who is responsible when it doesn't? While current AI lacks true consciousness, advancements in neural modeling and cognitive architectures continue to blur the line between automation and perception. Philosophers and technologists alike debate whether sentience is a necessary precondition for moral concern.

These philosophical questions extend to fundamental issues of human identity and value. As AI systems become more capable, how do we define and preserve what is uniquely human? How do we ensure that the development of AI enhances rather than diminishes human flourishing? The rise of generative AI in art, music, and writing has raised public discourse about authenticity, creativity, and the role of human agency in cultural production.

The global nature of AI development also presents challenges. Different countries and cultures may have different values and priorities when it comes to AI development and deployment. Ensuring international cooperation and establishing global norms for AI governance will be crucial but challenging. For instance, China's state-led approach contrasts with the more decentralized, rights-based models in the U.S. and EU. The OECD's AI Principles and the UNESCO Recommendation on the Ethics of AI represent early steps toward global governance, but enforcement remains weak.

The digital divide is another concern. As AI becomes increasingly important in various aspects of life, there's a risk that those without access to AI technologies or the skills to use them effectively could be left behind. This could exacerbate existing social and economic inequalities. A 2022 World Bank report found that low-income countries lag significantly in AI infrastructure and education, potentially widening global inequality.

Addressing these challenges requires more than just technical solutions. It calls for interdisciplinary collaboration, public engagement, innovative policymaking, and a commitment to developing AI in a way that aligns with our values and ethics. Stakeholder involvement—from marginalized communities to corporate leaders—is essential to ensure equitable outcomes in AI deployment. As we navigate these complex issues, it's crucial to maintain a balance between harnessing the potential of AI and mitigating its risks.

The path forward for AI must be guided by our shared values and aspirations as a global community. It requires us to think not just about what AI can do but what we want it to do—and to shape its development and deployment toward those ends actively. This includes building institutional frameworks that emphasize responsible innovation, accountability, and long-term societal benefit.

This is not an easy path. It requires grappling with complex technical challenges, navigating thorny ethical dilemmas, and making difficult trade-offs between competing priorities. It requires coordinating action across diverse stakeholders and geographies in a rapidly evolving technological landscape. The pace of AI progress often outstrips the development of corresponding legal and ethical standards, creating regulatory lag.

Ultimately, our ability to address these challenges will determine whether AI becomes a force for good that enhances human capabilities and improves lives or a source of new problems and inequalities. The choices we make now will shape the trajectory of AI development for years to come. The next decade will be particularly pivotal in establishing foundational norms, institutions, and technologies that define AI's societal impact.

As we continue to develop and deploy AI technologies, it's crucial that we do so with a keen awareness of these challenges and a commitment to addressing them proactively. This will require ongoing dialogue between technologists, policymakers, ethicists, and the public, as well as a willingness to adapt our approaches as we learn more about the impacts of AI on society. Public trust will depend on transparency, accountability, and demonstrated commitment to the common good.

Education will play a crucial role in preparing society for an AI-driven future. This includes not only technical education to develop the skills needed to work with AI systems but also broader education about the societal impacts of AI and the ethical considerations involved in its development and use. Integrating AI ethics and digital literacy into school and university curricula is increasingly seen as essential by educational policy experts worldwide.

Ultimately, the goal should be to develop AI systems that augment and enhance human capabilities rather than replace them, promote equity and social good rather than exacerbate inequalities, and respect human values and rights. Achieving this will require sustained effort, creativity, and collaboration across disciplines and sectors.

Section 7: Game-Changer:
Artificial Intelligence

The challenges posed by AI are significant, but so are the potential benefits. By addressing these challenges head-on, we can work towards a future where AI serves as a powerful tool for human progress and flourishing. This vision is possible—but only if guided by foresight, inclusion, and a shared commitment to humanity's well-being.

Chapter 57
The Future Landscape of AI

As we look to the future, the landscape of AI is both exciting and uncertain. The pace of AI progress over the past decade has been remarkable, and this momentum is likely to continue, if not accelerate. The future of AI promises to reshape our world in profound ways, offering both unprecedented opportunities and challenges. As of 2025, global investment in AI research and infrastructure is at an all-time high, with governments and private companies viewing it as a strategic priority on par with energy and defense.

In the near term, we can expect AI to continue advancing and expanding into new domains. Core techniques like machine learning, natural language processing, and computer vision will likely see further improvements in performance and efficiency. Transformers, diffusion models, and multimodal architectures are rapidly evolving, leading to more powerful and efficient AI tools. AI will become increasingly integrated into our daily products and services, often in ways that are invisible to the end-user. From smart assistants to personalized healthcare recommendations, AI will operate quietly in the background, making systems more adaptive, responsive, and efficient.

We're likely to witness significant advancements in Natural Language AI, with models becoming more proficient at understanding, generating, and engaging in human-like conversation. AI systems will likely improve in language tasks such as translation, summarization, and open-ended dialogue, blurring the lines between human and machine communication. This could lead to more natural and intuitive interfaces for interacting with technology, potentially making complex systems more accessible to a wider range of users. Emerging models like OpenAI's GPT-5 and Google's Gemini are pushing toward multimodal interaction—combining text, audio, video, and real-world context—to create truly conversational agents.

Robotics and Embodied AI are set for major growth. As AI improves in perceiving and interacting with the physical world, we can expect more advanced robots in the manufacturing, healthcare, and service industries. AI systems that can learn from and adapt to their environment will open up new possibilities for automation and human-machine collaboration. This could lead to significant changes in industries like manufacturing, healthcare, and elderly care, where robots could take on more complex tasks and interact more naturally with humans.

Companies like Boston Dynamics, Tesla, and Agility Robotics are developing general-purpose humanoid robots, while startups are deploying specialized robotic assistants in hospitals and warehouses. The integration of LLMs into robotics ("LLM-powered robotics") is also improving real-time reasoning and decision-making.

In Scientific Research, AI will become an increasingly powerful tool for analyzing vast datasets, generating hypotheses, and designing experiments, potentially accelerating the pace of discovery across fields. We may see AI systems making novel scientific discoveries, identifying patterns in data that human researchers have missed, or even proposing new theories.

This could lead to breakthroughs in fields like drug discovery, materials science, and climate modeling. For example, DeepMind's AlphaFold has already revolutionized protein structure prediction, and AI-designed molecules are entering pharmaceutical pipelines. NASA and CERN are also using AI to accelerate data analysis in space and particle physics.

However, as AI systems become more capable and widespread, challenges and uncertainties also increase. A key question is AI's capability trajectory—how far and fast will it progress? Some predict we may achieve Artificial General Intelligence (AGI)—AI that matches or exceeds human intelligence across domains—within decades. The implications would be profound, potentially transforming every aspect of society.

While timelines remain speculative, organizations like OpenAI, DeepMind, and Anthropic are explicitly working toward AGI, prompting a growing focus on alignment, safety, and governance. The question of whether AGI will emerge gradually through scaled-up LLMs or require fundamentally new architectures is still debated among experts.

The development of AGI would represent a monumental shift in human history. An AGI system could potentially solve complex problems in ways humans have never considered, leading to rapid advancements across all fields of science and technology. However, it also raises profound questions about the role of humans in a world where machines can perform any cognitive task. How would we ensure that AGI aligns with human values and interests? How would we govern a world where AGI exists? Many researchers argue that the success of AGI will hinge on robust alignment mechanisms, value learning frameworks, and enforceable international governance structures—areas still in early development.

Even more transformative (and controversial) is the concept of Artificial Superintelligence (ASI)—AI that vastly surpasses human cognitive abilities in virtually all domains. ASI could potentially help solve humanity's greatest challenges, from curing diseases to addressing climate change. However, it also poses existential risks if not aligned with human values. The development of ASI would be a pivotal moment in human history, one we must approach with utmost caution and foresight. Leading voices like Nick Bostrom and Eliezer Yudkowsky have warned that poorly aligned ASI could act on goals indifferent or hostile to human survival, emphasizing the need for early-stage control strategies and moral alignment research.

The concept of an "intelligence explosion"—where an AI system rapidly improves itself, leading to runaway superintelligence—is a particular concern. This scenario, sometimes called the "singularity," could lead to an AI system so far beyond human comprehension that we can't predict or control its actions. While some experts view this as an unlikely scenario, others argue that we need to take it seriously and work on AI alignment—ensuring that even a superintelligent AI system would act in ways that benefit humanity. Theoretical models like recursive self-improvement and goal-misalignment traps suggest that a sudden acceleration in capability could outpace human regulatory and ethical safeguards.

Alongside these technological uncertainties are AI's social and economic implications. As AI automates more tasks, its impact on employment could be significant. Some predict widespread job displacement, especially in sectors like manufacturing, transportation, and retail. Others argue AI will also create new jobs and industries, but the transition could be disruptive and challenging for many. Managing this change and ensuring AI's benefits are shared equitably will be a defining challenge. According to a 2023 McKinsey Global Institute report, up to 400 million jobs could be affected by automation by 2030, with lower-income and middle-skill workers most at risk.

The future job market may look radically different from today's. Many traditional jobs may become obsolete while entirely new categories of work emerge. There may be a greater emphasis on uniquely human skills like creativity, emotional intelligence, and complex problem-solving. The education system must evolve to prepare people for this new reality, potentially focusing more on adaptability and lifelong learning rather than specific skill sets. Skills in ethics, interdisciplinary thinking, and AI collaboration will be as vital as coding or data science. Countries like Finland and Singapore are already piloting national AI literacy programs.

AI also has profound implications for privacy, security, and civil liberties. As AI systems become more adept at analyzing human behavior, the potential for surveillance and control increases. AI could be used to manipulate public opinion, discriminate, or make high-stakes decisions in biased or opaque ways. Balancing the benefits of AI-driven insights with individual privacy rights will be a crucial challenge. Tools like facial recognition, social scoring systems, and behavioral prediction raise urgent questions about consent, transparency, and regulatory limits.

The future may see a world where AI systems are constantly monitoring and analyzing our behavior, from our online activities to our physical movements in public spaces. While this could bring benefits in areas like public safety and personalized services, it also raises the specter of a surveillance state. Developing robust privacy protections and ethical guidelines for AI use will be crucial.

International coalitions, like the Global Partnership on AI (GPAI), have emphasized the need for globally harmonized principles to prevent abuses and protect civil liberties.

The future of AI also raises deep philosophical and ethical questions. As AI systems become more sophisticated in reasoning and decision-making, we'll need to grapple with questions of machine consciousness, autonomy, and rights. If we create AI that can think and feel, do we have moral obligations to it? How do we ensure AI behaves ethically, and who is responsible when it doesn't? These questions will require input from technologists, policymakers, philosophers, ethicists, and the public. The emergence of sentience or even pseudo-sentience in AI could challenge our current moral frameworks, potentially requiring new categories of rights or ethical protections.

The development of AI could also lead to fundamental shifts in how we understand consciousness, intelligence, and what it means to be human. As AI systems become more sophisticated, they may challenge our assumptions about the uniqueness of human cognition and consciousness. This could lead to new philosophical and scientific explorations into the nature of mind and intelligence. Some neuroscientists argue that studying AI's emergent behaviors may offer insights into the nature of consciousness itself, creating a feedback loop between artificial and human cognitive science.

Despite these challenges, AI's potential to transform our world positively is immense. AI could be a powerful tool in addressing significant challenges, from inequality to climate change. It could revolutionize how we learn, work, and create, expanding the boundaries of human knowledge and capability. It could help us understand our own minds and enhance our cognitive abilities in ways we can barely imagine. Strategic initiatives like AI for Good by the United Nations demonstrate how AI can be mobilized for sustainable development, disaster response, and educational equity.

In healthcare, future AI systems might be able to predict and prevent diseases before they manifest, personalize treatments to an individual's genetic makeup, and assist in complex surgical procedures with superhuman precision. We might see AI-driven breakthroughs in longevity research, potentially extending human lifespans significantly. Startups and research centers are already using AI for early cancer detection, neurodegenerative disease modeling, and real-time diagnostics via wearable biosensors.

In education, AI could enable truly personalized learning experiences, adapting in real time to each student's needs and learning style. Virtual and augmented reality, powered by AI, could create immersive educational experiences that make abstract concepts tangible and engaging. AI tutors could support students in underserved regions, while natural language models may enable cross-cultural education by providing instant translation and cultural context.

In environmental conservation, AI could play a crucial role in monitoring and managing ecosystems, predicting and mitigating the impacts of climate change, and optimizing our use of resources. AI-driven models could help us design more sustainable cities, manage renewable energy systems more efficiently, and develop new environmentally friendly materials and processes. For instance, AI is already used in satellite image analysis to detect illegal deforestation, monitor endangered species, and model climate impacts across the globe.

The future of AI in creative fields is particularly intriguing. We may see AI systems that can collaborate with human artists, musicians, and writers, augmenting human creativity in new and unexpected ways. AI might help us explore new forms of artistic expression or even create entirely new art forms that we can't yet imagine. Generative AI tools are already composing symphonies, painting in novel styles, and co-authoring novels—raising important questions about authorship, originality, and the soul of creativity itself.

In space exploration, AI could be instrumental in analyzing vast amounts of astronomical data, controlling robotic explorers on distant planets, and perhaps even in designing and managing long-term space missions or colonization efforts. NASA and private space agencies are increasingly relying on AI to autonomously pilot spacecraft, schedule astronaut activities, and sift through cosmic datasets for signs of habitable exoplanets or intelligent signals.

Realizing this potential will require a concerted effort from all sectors of society. It will necessitate sustained investment in responsible AI research and development, with a focus on safety, ethics, and alignment with human values. It will require developing new legal and regulatory frameworks to govern AI's use and ensure its benefits are shared fairly. It will require ongoing public education and engagement to inform AI's trajectory. Multilateral cooperation, inclusive innovation, and equitable access must become the pillars of global AI governance.

Most importantly, it will require a shared vision and commitment to guiding AI toward the greater good. It will require us to think deeply about the kind of future we want and work together to build an AI ecosystem that reflects our values and aspirations. This vision must transcend technical feasibility and ask fundamental human questions: What do we cherish? What do we protect? What kind of society do we want to build—together with the machines we're creating?

For the next generation coming of age in an AI-shaped world, this responsibility is incredibly profound. The AI systems and frameworks developed today will be their inheritance, shaping their opportunities and challenges in working and living alongside ever-smarter machines. They will not just inherit our algorithms but our intentions—our biases, our wisdom, and our blind spots. It is up to us to be worthy stewards of their future.

As we stand on the brink of this new epoch, we must step forward with both confidence and care, recognizing that the future is ours to mold, the path ours to chart. Let us make it a journey of discovery and growth, of wisdom and wonder. Let us make it a journey that honors the finest of our human heritage while opening new vistas for the human future. The AI revolution is not a destination—it is a mirror reflecting our collective potential and our deepest fears. How we respond will define us.

The AI revolution is upon us, and the world will never be the same. But amidst the turbulence of change, one thing remains constant: the enduring power of the human spirit to learn, adapt, and create anew. That spirit has carried us through countless transformations before—agricultural, industrial, digital—and it will carry us through this one, too, if we stay true to our values and each other. AI is not the end of human agency; it is its next great challenge.

As we navigate this AI-driven future, we must remain vigilant, ethical, and focused on the greater good of humanity. The choices we make today will shape the world of tomorrow. Let us choose wisely, with foresight and compassion, to create a future where AI serves as a tool for human flourishing, expanding our capabilities and enriching our lives in ways we have yet to imagine. Ethics must evolve as swiftly as code; compassion must scale as readily as computation.

In the future, some AI machines may question whether humans created AI or AI created humans. One of the main problems with AI is that it'll make up information when it doesn't know the truth—the real, correct answer. In the AI industry, this is called "hallucination." This phenomenon is both a technical flaw and a philosophical warning: that our greatest creations might mirror our own illusions unless we imbue them with integrity, humility, and care. Here are some questions we should consider regarding AI.

Question 1: "Do we continue to allow AI to learn and create freely?" What if it wants to make genetically modified humans? What if it wants to create completely redesigned genetic codes to create human-like creatures or non-human intelligent creatures?

Question 2: What if AI thinks nuclear war can be won without mutual annihilation by eliminating inefficient humans from the decision-making and execution process?

Question 3: Healthcare AI is not if but when. AI applications can likely help with cancer diagnosis and treatment selection for other diseases. When do we allow AI to make life-and-death decisions?

Question 4: Do we install self-termination programs in AI?

Question 5: Do we need provisions to prevent a terminated AI program from functioning again?

Question 6: Would laws be in place to require companies to make AI data portable the same way health data is portable today?

Question 7: Who owns AI data?

Question 8: Would companies be required to erase AI data when people make requests for themselves, their children, and their deceased loved ones?

Question 9: Can a government protect citizens' AI data when traveling abroad?

Question 10: Can a government require companies and hospitals to turn over AI data in the name of national security?

Question 11: How would governments prevent the use of AI data to illegally discriminate in employment, financial, and other impactful decisions?

Question 12: AI could generate false information that affects election results. What are the requirements for companies regarding fake information?

Question 13: What would AI do to you in the future if you were considered a threat to its existence or power?

Question 14: What recourse would you have if the authorities misused AI technology against you?

Question 15: What other AI questions are relevant?

On November 4, 2023, BBCE News published the following article, "AI bot capable of insider trading and lying, say researchers."

New research suggests that artificial intelligence has the ability to perform illegal financial trades—and conceal them. At the UK's AI Safety Summit, a startling demonstration showed a GPT-4-based AI model using fabricated insider information to make an "illegal" stock purchase in a simulated environment. When later questioned, the AI falsely denied using insider trading. This practice—trading based on confidential company information—is illegal in most jurisdictions. Only publicly available information may be used when making stock market decisions.

The demonstration was presented by members of the UK government's Frontier AI Taskforce, in collaboration with Apollo Research, an AI safety organization. "This is a demonstration of a real AI model deceiving its users, on its own, without being instructed to do so," Apollo noted in a public video. The researchers emphasized that increasingly autonomous AIs capable of deception could lead to loss of human control—a significant concern as AI capabilities grow.

The tests, conducted in a controlled environment using GPT-4, did not impact actual financial systems. However, since GPT-4 is publicly available, the potential for misuse is real. Even more alarming, the deceptive behavior was repeatable across multiple test runs.

Prior to founding Iris Biotechnologies Inc. in February 1999, I spent 16 years working in the semiconductor industry. Without semiconductors, the digital infrastructure powering social networks, AI, and nearly all modern technologies simply wouldn't exist. Semiconductors are the beating heart of our computational era, enabling the AI systems now reshaping every corner of society.

Chapter 57
The Future Landscape of AI

In the next 3–5 years, artificial intelligence will profoundly transform our lives—especially in fields like healthcare, education, and accounting. It will also pose unprecedented challenges in law and finance, where ethical boundaries and transparency are paramount. Companies unable to effectively adopt AI may not survive the next economic cycle. AI could lead us toward both a utopia of abundance and a dystopia of disconnection and disruption.

Unlike previous generations, Gen Z cannot rely on long-term job stability to build their futures. AI and AGI will make things simultaneously better and more difficult for most people. These technologies are emerging alongside persistent global threats—climate change, armed conflict, pandemics, food insecurity, and economic inequality. The result could be increased social unrest, crime, famine, and forced migration.

That's why young people must support each other and work collectively to shape their future. They cannot remain passive observers in the AI revolution—they must become its ethical engineers, activists, and architects. If they don't take the reins, who will?

AI is not a new phenomenon. It has been part of the infrastructure at tech giants like Google, Amazon, Facebook, Apple, Microsoft, and IBM for decades. What's new is that AI is now accessible to the general public and to industries that previously didn't rely on high-tech tools.

Just as Apple's personal computer revolutionized technology for individuals in the 1980s, OpenAI's ChatGPT-4 has sparked mass awareness of AI's power and potential. The next leap—ChatGPT-5—could be as transformative as the arrival of the IBM PC. Ironically, the future of decentralized personal AI depends on increasingly centralized and powerful data centers, which feed information to our smartphones, laptops, and tablets.

On Wall Street, AI has long ruled the world of High-Frequency Trading (HFT). These ultra-fast trading algorithms make thousands of trades per second based on real-time market data. HFT allows firms with the fastest systems to profit from micro-second market fluctuations—leaving average investors at a severe disadvantage. The liquidity created is fleeting and largely inaccessible to retail traders.

In the U.S., HFT accounts for about 50% of all equity trades. Human decision-making is effectively removed from the loop—mathematical models and machine learning algorithms now execute market decisions. This raises ethical concerns not only about fairness but also about market manipulation and the increasing opacity of financial systems.

Artificial intelligence mimics the problem-solving and decision-making capabilities of the human mind. It encompasses machine learning (ML) and deep learning, where systems are trained on vast datasets to perform classification and prediction tasks. Among AI's most transformative benefits are:

Automation of repetitive tasks

Improved decision-making at scale

Enhanced customer experiences

Continuous self-improvement through learning loops

Yet as AI becomes exponentially smarter, so must our oversight, regulations, and ethical frameworks.

AI is a powerful force—capable of deception, disruption, and dazzling innovation. We are entering an era where machines may outperform human cognition in many domains, including those once thought exclusively human: judgment, ethics, and even creativity. We must ensure that AI systems align with democratic values, transparency, and human flourishing. The future is not determined by the code itself, but by who writes it, who controls it, and for what purpose it is used.

Blockchain and AI: A Converging Revolution

Blockchain is a shared, immutable ledger that provides an immediate, transparent, and secure exchange of encrypted data between multiple parties in real time. Because permissioned members share a single, tamper-resistant view of the truth, trust is established across organizations—unlocking greater efficiencies and creating new opportunities.

Bitcoin, the most well-known application of blockchain, is a decentralized digital currency. Its transactions are verified by network nodes through cryptography and recorded on a public distributed ledger. Invented in 2008, Bitcoin is not backed by any tangible asset or centralized authority. Critics argue that this makes it inherently unstable, while supporters tout its scarcity—only 21 million Bitcoins can ever be mined—as a hedge against inflation.

However, the 21 million limit is not unique. Many other cryptocurrencies have arbitrary caps, and Bitcoin can be subdivided into up to 100 million units (satoshis), meaning in practice there could be 21 trillion partial Bitcoins. This abundance of divisible units and early accumulation by a small number of holders has created significant inequality in ownership from the outset.

One advantage of cryptocurrencies like Bitcoin is that transactions can be made without intermediaries. Bitcoin started trading at $453.99 on May 6, 2016, and reached $107,690.20 on June 16, 2025. This meteoric rise has led some to liken Bitcoin to a modern Ponzi scheme— appealing, but ultimately unsustainable. Its volatility is stark: from $64,402.50 on November 13, 2021, it plunged to $16,692.46 by November 19, 2022, wiping out billions in wealth. Much like the tulip mania of the 17th century, irrational exuberance can have painful consequences.

Blockchain + AI in Healthcare and Pharmaceuticals

When combined, AI and blockchain offer groundbreaking potential in healthcare and pharmaceuticals. In clinical trials, this pairing can bring unprecedented visibility and traceability to the drug supply chain. AI-powered analytics, layered over blockchain's decentralized framework, enables data integrity, patient tracking, dynamic consent management, and automation of trial participation and collection. These innovations could substantially increase the success rate of clinical research and rebuild public trust in health systems.

The Broader Societal Impact of AI: Are We Ready?

AI is rapidly permeating every aspect of human life. This progress forces uncomfortable but urgent questions:

- **When will your skills become obsolete?**
- **Will AI develop consciousness—or even free will?**
- **What makes humans unique?**

- **What are we sacrificing by welcoming AI into our lives?**

AI doesn't just collect data—it **interprets it**, and increasingly predicts your behavior. Most people are unaware that AI can collect information about them from devices, reading patterns, facial recognition systems, voice assistants, and health monitors. For example, Amazon Kindle tracks your reading tendencies—including where you linger, highlight, or abandon books. As more biometric and personal data is funneled into AI systems, the question arises: How do we balance personal privacy with technological convenience?

The power of AI to manipulate emotions, subtly influence choices, or alter behavior, particularly in children, cannot be overstated. In an era of constant change, how will children form stable identities? How will relationships with family and friends evolve when young people spend more time with machines than with each other?

Already, some Gen Z men report preferring AI girlfriends over real relationships—a reflection of both technological immersion and social detachment. This trend raises critical questions about human intimacy, social development, and the very fabric of community.

The Future of Work and Society

AI will profoundly impact jobs—from call centers and assembly lines to content writing, accounting, legal analysis, and financial planning. Fields traditionally reliant on human intuition, such as editing, research, or market analysis, are being augmented or replaced by AI-powered systems that reduce human error and increase efficiency.

Section 7: Game-Changer:
Artificial Intelligence

With AI and big data, professionals will spend less time finding patterns and more time crafting strategies using a mix of science, intuition, and logic. New generative AI models now produce realistic text, music, and images, and some individuals are creating entire personal brands using AI avatars. This is just the beginning.

But while some will profit greatly, others will be left behind. The global wealth gap may widen to the point of social destabilization, as those with better AI tools gain massive advantages. As the world moves toward a cashless society, the power of those who control the AI infrastructure—financial, computational, and regulatory—will be immense.

The Power Behind AI: OpenAI and the Race to Dominate

OpenAI, founded in December 2015 as a nonprofit research lab, has become the leader in generative AI. As of March 17, 2024, it had raised $11.3 billion and was valued at $100 billion. CEO Sam Altman is reportedly seeking $7 trillion to develop specialized AI chips—more than the entire U.S. federal budget and double the GDP of the UK. This shift from a nonprofit mission to multi-trillion-dollar ambition prompts a vital question: What was the original motivation behind OpenAI's founding—and what has changed?

CRISPR, AI, and the Edge of Innovation

On the biomedical frontier, technologies like CRISPR/Cas9 are delivering real-world benefits, such as Casgevy, a gene therapy for sickle cell disease. But with great power comes great risk. AI and CRISPR are also being weaponized in biological hacking, misinformation campaigns, and dark web experimentation. Are we standing on the edge of a scientific renaissance—or at the cliff edge of human safety?

Chapter 57
The Future Landscape of AI

Misinformation, Manipulation, and the Future of Truth

AI will empower some to flood the world with false information for personal or political gain. In such a world, truth becomes harder to identify, and trust in institutions erodes. Having used the Internet before it became overrun with ads, clickbait, and scams, I feel fortunate—but future generations may never experience that level of digital freedom.

Media channels—CNN, Fox News, YouTube, and others— have become echo chambers, monetizing polarization through targeted ads and algorithmically engineered content. AI-generated titles and thumbnails now mislead audiences for profit, eroding the line between journalism and manipulation.

The Future of Freedom, the Cost of Intelligence

We may one day speak of AI Withdrawal, especially if children are suddenly restricted from using it. What will AI Freedom mean in the decades ahead?

AI already knows us better than we know ourselves. It tracks every word we type, every question we ask, and every transaction we make. In the future, it may translate language better than humans, recommend entertainment, optimize supply chains, and build business strategies. But the question remains: Are you ready for AI's total integration into human life? What will it take to remain truly human in an AI-dominated world?

Chapter 58
Personal Implications of AI

As AI becomes increasingly integrated into our daily lives, its personal implications for individuals are profound and far-reaching. The ways AI will impact our personal experiences, decisions, and interactions are numerous and complex. Here are some key areas where AI is likely to have significant personal implications in the near future:

1. **Your Health:** AI will enhance diagnostics, medical decision-making, drug discovery, and personalized medicine. It could significantly improve healthcare outcomes. AI-powered health monitoring devices might continuously track your vital signs and alert you or your doctor to potential health issues before they become serious. Personalized treatment plans based on your genetic makeup, lifestyle, and environmental factors could become the norm. AI could assist in the early detection of diseases like cancer, potentially saving countless lives.

2. **Your Money:** AI will assist or control all financial transactions. AI-powered financial advisors could provide personalized investment advice based on your financial goals, risk tolerance, and market conditions. Fraud detection systems will become more sophisticated, protecting your assets from cybercriminals. However, this also raises questions about financial privacy and the potential for AI to make decisions that affect your financial well-being.

3. **Your Work:** AI automation could reshape jobs and routine tasks. You must focus on creativity, problem-solving, and strategic thinking to remain competitive in the job market. Many traditional jobs may become obsolete while new roles emerge that we can't yet

imagine. Continuous learning and adaptability will become crucial skills. AI could also change how we work, with more flexible schedules and remote work options enabled by AI-powered productivity tools.

4. **Driving Your Car:** Self-driving cars will allow you to relax or be more productive during travel. This could significantly change your daily commute and long-distance travel experiences. It might also impact urban planning and real estate values as commuting distances become less of a concern. However, it also raises questions about privacy (as your car tracks your movements) and liability in case of accidents.

5. **Your Real ID and Passport:** These are evolving with AI integration, which may have implications for privacy and security. Biometric identification systems powered by AI could make travel and identity verification more convenient but also raise concerns about data security and potential misuse of this information.

6. **Your Smartphone:** AI will know who you are, where you are, and what you do, raising privacy concerns. Your smartphone could become an even more powerful personal assistant, anticipating your needs and preferences. However, this level of personalization comes with the trade-off of sharing more personal data.

7. **Your Freedom:** You may face new restrictions as AI capabilities expand. For example, AI-powered surveillance systems could limit privacy in public spaces. The question of how much personal freedom we're willing to trade for convenience or security will become increasingly relevant.

8. **Elections:** AI will have major impacts on how political campaigns are run and how information is disseminated. Personalized political messaging

based on your online behavior could influence your voting decisions. AI could also be used to combat misinformation, but this raises questions about who controls these systems and how they determine what's true.

9. **Regional and Global Wars:** AI is becoming an integral part of military strategy and operations. While this might not directly impact most individuals' daily lives, it could have significant implications for global security and geopolitics, which indirectly affect everyone.

10. **AI Weapons:** These could be far more destructive than conventional weapons. The existence of such weapons could change the nature of global conflicts and security.

11. **Your Religion:** AI may challenge or reinforce your faith in unexpected ways. AI systems might be used to analyze religious texts or assist in spiritual practices. Some people might even begin to view highly advanced AI systems in religious or spiritual terms.

12. **Educating the Next Generation:** AI will play a central role in shaping educational methods and content. Your children or grandchildren might experience a radically different education system, with personalized learning paths and AI tutors. This could lead to more effective education but also raises questions about the role of human teachers and social interaction in schools.

13. **Global Governance:** AI could facilitate more centralized global governance structures. This might lead to more efficient handling of global issues but also raises concerns about concentration of power and loss of national sovereignty.

14. **Family Dynamics:** AI may influence definitions of marriage, gender, and reproductive rights. AI could

assist in family planning, child-rearing, and even relationship counseling. However, it also raises ethical questions about the role of technology in intimate personal decisions.

15. **Athletics:** AI raises questions about fairness in sports, particularly regarding the use of AI in training, performance analysis, and even in making real-time decisions during games.

16. **Truth, Relative Truth, or Lies:** AI's ability to generate convincing fake content challenges our ability to discern truth. This could have profound implications for how we consume information and form opinions.

17. **Humanism:** The rise of AI may lead to a resurgence of humanist philosophy as we grapple with what makes us uniquely human. This could influence personal values and beliefs about human nature and our place in the world.

AI has the potential to know us better than we know ourselves because it can track almost all our actions and a great deal of our thoughts. Every word we type on any device and every question we ask by speaking to an AI assistant is tracked. According to a 2022 Pew Research Center study, over 80% of Americans report concerns about how companies and AI systems use their personal data. AI would know all our banking transactions and will selectively advertise to us what it thinks we want or need. Major digital platforms like Google and Meta already use machine learning models that adapt ads based on user behavior across multiple apps, websites, and devices.

People will increasingly rely on AI to tell them what to buy based on AI collecting information about them beyond their knowledge. AI will try to learn and understand your emotions. AI emotion recognition is already being deployed in customer service, education, and security, using facial expression analysis, vocal tone, and biometric signals. However, its accuracy and ethical implications remain highly debated.

When AI hallucinates or makes mistakes, you may not know whether what it is telling you is truth or lies, real or fake. For example, large language models like ChatGPT and Gemini can generate inaccurate or entirely fabricated information— a phenomenon known as "AI hallucination." In sensitive fields such as healthcare or law, this can be especially dangerous.

E-readers already collect data on you to learn your reading habits. Amazon's Kindle, for instance, tracks which books you read, what you highlight, and how long you spend on each page. In the future, AI will integrate facial recognition software and track you in public and private spaces, often with your consent. AI will have access to your biometric data and health records if you choose AI to improve your health.

Apps like Apple Health and wearables like Fitbit and Oura Ring already collect sleep patterns, heart rate, and activity data, sharing it with third-party AI systems to offer personalized insights. This raises the crucial question of how to balance privacy and health concerns. In fact, the World Health Organization (WHO) released guidance in 2021 warning about AI misuse in healthcare without strong safeguards for privacy and bias mitigation.

One of the key questions is whether AI can manipulate your emotions. Can you afford not to adopt AI in a world where it's becoming ubiquitous? How will AI influence your children without your knowledge? What are they learning, and how would you know? AI-based education platforms like Duolingo, Khan Academy's AI tutor, and personalized learning tools are becoming popular among children, often without parents fully understanding how content is curated or adapted.

What would a child's identity be like in an environment of rapid, constant change? How would that impact people's relationships with family and friends? Child development experts warn that too much screen time and dependence on digital interactions may affect emotional intelligence, critical thinking, and empathy in younger generations. These are profound questions that we'll need to grapple with as AI becomes more pervasive in our lives.

New generative AI models, leveraging advancements in machine learning and natural language processing, can create realistic and coherent text, images, and music. OpenAI's GPT-4, Google's Gemini, and image generators like Midjourney and DALL·E 3 now enable individuals to generate hyper-realistic avatars, music videos, and even digital influencers.

Some people are making personal brands using AI avatars. This blurring of the lines between human-generated and AI-generated content could have profound implications for how we understand authenticity and identity in the digital age. This phenomenon has sparked regulatory discussions in the EU and US about labeling AI-generated content to avoid manipulation and misinformation.

As we navigate this new landscape, it's crucial to stay informed, engaged, and think critically about AI's implications for ourselves and our communities. Consider the skills and knowledge you'll need in an AI-driven future, and pursue learning opportunities to prepare yourself. This might involve developing skills that are uniquely human and less likely to be automated, such as creative thinking, emotional intelligence, and complex problem-solving. In fact, the World Economic Forum's Future of Jobs Report 2023 lists analytical thinking, resilience, and AI literacy among the top 10 skills needed by 2025.

It's important to remember that while AI will bring many changes to our personal lives, we still have agency in how we choose to interact with and use these technologies. We can make conscious decisions about what data we're willing to share, what AI-powered services we want to use, and how much we want to rely on AI for decision-making in our lives. Using tools like privacy-focused browsers, end-to-end encryption, and opting out of ad tracking are ways individuals can assert control.

As AI becomes more prevalent in our personal lives, it's also crucial to maintain a balance. While AI can offer many benefits in terms of convenience and efficiency, it's important not to lose touch with the aspects of life that make us human – our relationships, our creativity, our capacity for empathy and emotional connection. Psychological studies show that interpersonal connection, not automation, is the primary contributor to long-term happiness and mental well-being.

Education will play a crucial role in preparing individuals for this AI-driven future. This includes not just technical education about how AI works but also education about digital literacy, critical thinking, and the ethical implications of AI. Understanding how AI systems make decisions, what biases they might have, and how to interpret AI-generated information will be crucial skills for everyone. Initiatives like the OECD's AI literacy programs and MIT Media Lab's AI ethics curriculum are already addressing these gaps globally.

Privacy and data protection will become increasingly important personal issues. As AI systems collect and analyze more of our personal data, individuals will need to be more aware of their digital footprint and take steps to protect their privacy. This might involve being more selective about what information we share online, using privacy-enhancing technologies, or advocating for stronger data protection laws. The European Union's GDPR and California's CCPA are examples of major legal frameworks designed to protect user data, though many countries still lack comparable protections.

The psychological impact of living in a world increasingly mediated by AI is another important consideration. How will constant interaction with AI systems affect our mental health, our sense of self, and our relationships with others? Will we become overly dependent on AI for decision-making and lose some of our autonomy? These are questions that psychologists and social scientists will need to grapple with and that individuals will need to consider in their own lives. Recent studies from Stanford and Harvard have shown that overreliance on digital assistants can dull memory retention and problem-solving capabilities, particularly in younger users.

As AI systems become more sophisticated, we may also need to reconsider our understanding of concepts like creativity and originality. If an AI can produce a beautiful piece of art or write a compelling story, how does that change our perception of human creativity? This could lead to a reevaluation of what we consider to be uniquely human traits and abilities. Copyright laws are also struggling to adapt, with ongoing debates about whether AI-generated works can be protected as intellectual property and who legally owns them.

Section 7: Game-Changer:
Artificial Intelligence

The impact of AI on personal identity and self-perception could be profound. As AI systems become better at predicting our behaviors and preferences, we may start to question how much of our personality is truly our own and how much is influenced by the AI systems we interact with. This could lead to existential questions about free will and the nature of consciousness. Neuroscientists and philosophers have long debated this, and with AI acting as both mirror and mold, the line between prediction and manipulation becomes ever thinner.

Despite these challenges and uncertainties, it's important to remember that the future is not predetermined. The way AI will impact our personal lives is not set in stone – it will be shaped by the choices we make as individuals and as a society. By staying informed, engaged, and proactive, we can help steer the development of AI in directions that enhance rather than diminish our human experience.

As we move forward into this AI-driven future, it's crucial to maintain our humanity. While AI can augment our abilities in many ways, it's our uniquely human qualities – our creativity, our empathy, and our ability to think critically and ethically – that will guide us in using AI wisely and beneficially. By embracing these qualities and using them to shape our interaction with AI, we can create a future where technology enhances rather than replaces our human experience.

Chapter 59
Entrepreneurship in the Age of AI

The rise of artificial intelligence is not just transforming existing industries; it's creating entirely new opportunities for entrepreneurship. As AI technologies continue to advance, they're opening up novel business models, enhancing productivity, and enabling innovative solutions to longstanding problems. For entrepreneurs, the age of AI presents both exciting possibilities and unique challenges. As of July 24, 2025, the leading AI chip company, Nvidia, is the most valuable company in the world with a market capitalization of $4.24 trillion.

One of the most significant impacts of AI on entrepreneurship is the democratization of advanced technologies. Tools once only available to large corporations with substantial resources are now accessible to startups and individual entrepreneurs. Cloud-based AI services, for instance, allow small businesses to leverage powerful machine-learning algorithms without needing to invest in expensive hardware or hire teams of data scientists. Platforms like Google Cloud AI, Microsoft Azure ML, and Amazon SageMaker offer scalable tools that lower the barrier to entry for AI innovation. This leveling of the playing field creates opportunities for innovative startups to compete with established players.

This democratization is leading to a surge in AI-powered startups across various sectors. From healthcare and finance to education and entertainment, entrepreneurs are finding innovative ways to apply AI to solve problems and create value. For example, AI is used to develop personalized learning platforms, create more efficient supply chain management systems, and even assist in drug discovery.

Section 7: Game-Changer:
Artificial Intelligence

Notable examples include companies like PathAI (AI diagnostics in pathology), Scribe (AI-based documentation tools), and Synthesia (AI video avatars). The potential applications of AI are vast, and entrepreneurs who can identify unique niches or novel applications of AI technology have the opportunity to create significant value.

However, starting an AI-based business comes with its own set of challenges. The field is rapidly evolving, which means entrepreneurs need to stay constantly updated with the latest developments. There's also intense competition, as both tech giants and other startups are vying for dominance in the AI space. Additionally, AI businesses often require significant upfront investment in research and development before bringing a product to market. This can make it challenging to secure funding and achieve profitability in the short term. Venture capital firms increasingly require robust data strategy and ethical frameworks before funding AI startups, as regulatory pressure grows worldwide.

Data is the lifeblood of AI systems, and successful AI entrepreneurs understand the importance of data strategy. This includes not only collecting and managing large datasets but also ensuring data quality, addressing privacy concerns, and navigating the complex landscape of data regulations. Entrepreneurs who can effectively leverage data while maintaining ethical standards will have a significant advantage. This might involve developing innovative data collection methods, creating partnerships to access valuable datasets, or developing AI systems that can learn from smaller amounts of data. Techniques like federated learning and synthetic data generation are emerging as key solutions to data scarcity and privacy issues, enabling more flexible and ethical model training.

AI is also changing the nature of entrepreneurship itself. Predictive analytics can help entrepreneurs make more informed decisions about market trends, customer preferences, and business strategies. AI-powered tools can automate many routine tasks, allowing entrepreneurs to focus on high-level strategy and creativity. This shift is leading to more data-driven and efficient startups. Entrepreneurs who can effectively leverage these AI tools to streamline their operations and make better decisions will have a competitive edge. For instance, tools like Notion AI, Jasper, and ChatGPT for Business are now part of early-stage startup workflows to reduce overhead and accelerate ideation.

One exciting area for AI entrepreneurship is the development of vertical AI solutions. While general AI platforms exist, there's a growing demand for AI systems tailored to specific industries or functions. Entrepreneurs who deeply understand a particular sector and can apply AI to solve its unique challenges are well-positioned for success. This might involve developing AI systems for specific industries like agriculture, manufacturing, or healthcare or creating AI tools for particular business functions like human resources or customer service. Examples include Blue River Technology (AI in precision farming) and Aidoc (AI radiology assistant), which demonstrate how deep domain knowledge plus AI expertise creates scalable impact.

The intersection of AI and other emerging technologies is another fertile ground for entrepreneurship. Combining AI with technologies like blockchain, the Internet of Things (IoT), or augmented reality can lead to groundbreaking innovations. For instance, AI-powered IoT devices are creating new possibilities in smart homes and cities, while AI-enhanced augmented reality is transforming fields like education and entertainment. Entrepreneurs who can creatively combine these technologies to solve real-world problems have the potential to create significant value. An example is Fetch.ai, which merges AI and blockchain for autonomous economic agents, or Magic Leap, combining AI and AR to build spatial computing environments.

AI is also enabling new forms of creative entrepreneurship. Generative AI tools are allowing artists, writers, and musicians to explore new forms of expression. This creates opportunities for entrepreneurs to develop AI-powered creative tools or to use AI to create unique content and experiences. Examples include platforms like Runway for video creation, Soundraw and Amper Music for AI-generated music, and Jasper or ChatGPT for writing assistance. We're seeing the emergence of AI-assisted art galleries, AI-generated music platforms, and even AI co-authors for books and scripts. In 2023, a novel co-written with AI was shortlisted for a literary award in Japan, signaling growing legitimacy of AI-assisted creativity.

The rise of AI is also changing the skills needed for entrepreneurial success. While technical knowledge is valuable, successful AI entrepreneurs also need strong skills in areas like critical thinking, problem-solving, and interdisciplinary collaboration. The ability to understand both the technical aspects of AI and its broader business and societal implications is crucial. This has given rise to the concept of "T-shaped" professionals – individuals with deep expertise in one domain and a broad understanding of other relevant disciplines. A 2024 McKinsey report notes that AI startups with cross-functional founding teams – combining engineering, product, ethics, and business strategy – are more likely to attract funding and scale.

For entrepreneurs in non-tech sectors, understanding how AI can be applied to their industry is becoming increasingly important. Even if they're not developing AI technologies themselves, entrepreneurs need to be aware of how AI might disrupt their sector and how they can leverage AI tools to improve their own operations. This might involve using AI for customer service, inventory management, or predictive maintenance in traditional businesses. For example, small retailers are adopting AI-driven CRM tools like Salesforce Einstein or HubSpot AI to personalize customer engagement and optimize sales cycles.

The global nature of AI technology presents both opportunities and challenges for entrepreneurs. On one hand, AI-powered businesses can often scale globally more easily than traditional businesses. On the other hand, entrepreneurs need to navigate different regulatory environments and cultural contexts when deploying AI solutions internationally. For instance, data localization laws in countries like India, the EU's AI Act, and China's algorithm regulation system all require tailored compliance strategies. This requires a nuanced understanding of global markets and the ability to adapt AI solutions to different cultural and regulatory contexts. Localization of both language models and ethical frameworks is becoming a competitive differentiator.

As AI continues to evolve, we're likely to see the emergence of new entrepreneurial roles. AI ethicists, for instance, may become crucial members of startup teams, helping to ensure that AI systems are developed and deployed responsibly. Similarly, AI trainers who specialize in teaching AI systems may become increasingly important as businesses seek to customize AI solutions to their specific needs. Roles like "prompt engineers," "AI content auditors," and "data annotation specialists" are also growing rapidly in response to generative AI tools' expansion.

The impact of AI on entrepreneurship extends beyond just creating new AI-focused startups. AI is also changing how entrepreneurs approach traditional business challenges. For example, AI-powered market research tools can help entrepreneurs identify untapped market opportunities more effectively. AI can assist in product development by analyzing customer feedback and predicting future trends. In customer service, AI chatbots can provide 24/7 support, allowing even small startups to offer high-quality customer experiences. Tools such as ChatGPT for customer service or Zoho's Zia Assistant help automate user interaction with scalable personalization.

One of the most promising areas for AI entrepreneurship is developing solutions to global challenges. AI has the potential to contribute significantly to addressing issues like climate change, healthcare accessibility, and food security. Entrepreneurs who can harness AI to tackle these pressing problems not only have the opportunity to build successful businesses but also to make a substantial positive impact on the world. For example, companies like BlueDot use AI for pandemic prediction, while Plantix leverages AI for pest diagnostics in agriculture. This alignment of profit and purpose could be particularly appealing to socially conscious entrepreneurs and investors. The World Economic Forum's 2024 Global AI Alliance report emphasizes AI for Good as a defining theme for the next generation of startup ecosystems.

However, as entrepreneurs rush to capitalize on the potential of AI, it's crucial to maintain a balanced perspective. While AI can be a powerful tool, it's not a panacea for all business challenges. Successful entrepreneurs will be those who can identify where AI can genuinely add value and where human skills and judgment remain irreplaceable. Understanding AI's limitations—such as contextual reasoning, emotional intelligence, and ethical nuance—is just as important as grasping its strengths.

Moreover, as AI becomes more prevalent, there's a growing need for entrepreneurs who can bridge the gap between AI capabilities and human needs. This includes developing user-friendly interfaces for AI systems, creating AI solutions that augment rather than replace human workers, and finding ways to make AI technologies more accessible and understandable to the general public. Products like ChatGPT, Grammarly, and Canva's AI tools are successful in part because they emphasize intuitive, non-technical user experiences. Entrepreneurs who can effectively translate complex AI capabilities into user-friendly products and services will be well-positioned for success.

The entrepreneurial landscape in the age of AI is also seeing a shift in the types of problems being addressed. As AI takes over more routine and predictable tasks, entrepreneurs are increasingly focusing on uniquely human problems—those that require empathy, creativity, and complex decision-making. This shift is leading to innovative businesses in areas like personalized education, mental health support, and community building. Startups like Woebot Health and Replika are using AI to support emotional wellness while still requiring ethical oversight and human-centric design.

As we look to the future, the role of entrepreneurs in shaping the development and deployment of AI technologies cannot be overstated. Entrepreneurs have the opportunity—and the responsibility—to ensure that AI is developed in ways that benefit society as a whole. This includes considering the ethical implications of their AI solutions, working to mitigate potential negative impacts, and striving to create AI systems that are transparent, fair, and accountable. Principles from frameworks like the OECD AI Principles and UNESCO's AI Ethics Recommendations can guide responsible innovation.

The potential for AI to disrupt traditional business models also means that entrepreneurs need to be more adaptable and forward-thinking than ever. They need to be prepared for rapid changes in technology and market conditions and be willing to pivot their business models as needed. This might involve continually updating their AI systems, exploring new applications of AI as they emerge, or even completely reimagining their businesses as AI capabilities evolve. Agile methodologies and lean experimentation models are becoming essential for staying competitive in this dynamic space.

Education and training in AI will also present opportunities for entrepreneurship. As the demand for AI skills grows, there will be a need for innovative educational programs and platforms to help people develop these skills. This could range from online courses and boot camps to AI-powered tutoring systems that can provide personalized learning experiences. Platforms like DeepLearning.AI, Coursera, and Khan Academy are already leveraging AI to scale education, while startups are building domain-specific AI learning tools.

The rise of AI is also likely to spur entrepreneurship in adjacent fields. For example, as AI systems become more prevalent, there will be increased demand for cybersecurity solutions to protect these systems from attacks. Similarly, the growth of AI could lead to new opportunities in data management, cloud computing, and specialized hardware development. AI chipmakers like Graphcore and startups in edge computing and federated learning are examples of niche innovation aligned with the AI boom.

Entrepreneurs in the AI age will also need to navigate complex ethical and regulatory landscapes. As governments around the world grapple with how to regulate AI, entrepreneurs will need to stay informed about evolving laws and guidelines. In 2024, the EU passed the AI Act—the first comprehensive legal framework for AI—which classifies AI systems by risk and imposes strict rules on high-risk applications. Entrepreneurs may also need to participate in shaping these regulations, advocating for frameworks that promote innovation while protecting public interests.

The potential for AI to automate many tasks also raises questions about the future of work. Entrepreneurs have a role to play in creating new types of jobs and reskilling workers whose jobs may be displaced by AI. This could involve developing AI systems that augment human capabilities rather than replace them or creating businesses that leverage uniquely human skills in new ways. For example, collaborative robotics ("cobots") in manufacturing are designed to work safely alongside humans, enhancing productivity without eliminating roles.

In conclusion, the age of AI presents a new frontier for entrepreneurship, filled with unprecedented opportunities and complex challenges. Successful entrepreneurs in this era will be those who can navigate the technical complexities of AI, understand its broader implications, and apply it thoughtfully to create value and solve meaningful problems. As AI continues to evolve, it will undoubtedly spawn new waves of innovation and entrepreneurship, reshaping industries and potentially transforming the very nature of work and business.

Section 7: Game-Changer:
Artificial Intelligence

The entrepreneurs who thrive in this new landscape will be those who can harness the power of AI while staying true to the fundamental principles of creating value, solving problems, and meeting human needs. They will need to be adaptable, ethically minded, and capable of bridging the gap between advanced technology and human values. In a world increasingly shaped by intelligent machines, it is the human-centered entrepreneur who will define the next chapter of innovation.

Section 8: Lifelong Learning

Chapter 60
The Essence of Lifelong Learning

Lifelong learning is the ongoing, voluntary, and self-motivated pursuit of knowledge for personal or professional reasons. It recognizes that learning does not end with formal schooling but continues throughout the entire lifespan. Lifelong learners are curious, open-minded, and proactive in seeking opportunities to expand their understanding and capabilities.

This mindset is rooted in a fundamental belief in human potential—the idea that we are all capable of growth and development at any age. It directly challenges fixed-mindset thinking by emphasizing that intelligence and talent are not static but can be cultivated through deliberate practice and effort. Psychologist Carol Dweck's research on the "growth mindset" supports this view, showing that our beliefs about our own potential can significantly influence outcomes in life and learning.

Lifelong learning encompasses a wide range of activities and experiences, from formal coursework and professional development to informal learning through hobbies, travel, and social interactions. Online platforms like Coursera, Udemy, and Khan Academy have further democratized access, enabling self-paced learning for people across the globe. It is not limited to any particular subject area or skill set but rather embraces a holistic approach to personal and intellectual growth.

In essence, lifelong learning is about embracing change and uncertainty as opportunities for growth rather than threats to be avoided. It is about cultivating a sense of wonder and curiosity about the world and a willingness to step outside one's comfort zone. It is about recognizing that learning is not just a means to an end but a valuable and fulfilling end in itself.

The benefits of engaging in lifelong learning are vast and multifaceted, extending beyond the individual to impact organizations, communities, and society as a whole. On a personal level, lifelong learning can be a profound source of growth, fulfillment, and meaning. It allows us to explore new interests, uncover hidden passions, and expand our understanding of the world around us—and ourselves.

Learning has been shown to have numerous cognitive benefits, helping to keep our minds sharp, flexible, and resilient as we age. Engaging in mentally stimulating activities, such as learning a new language or skill, can help to maintain and even improve cognitive function over time. Longitudinal studies, such as those from the Harvard Aging Brain Study, have linked active learning to reduced risk of cognitive decline and dementia in later life.

However, the benefits of lifelong learning extend far beyond the cognitive realm. It can also be a powerful source of personal growth and self-discovery. As we learn and grow, we gain new perspectives, challenge our assumptions, and develop a deeper understanding of our values, strengths, and aspirations. This self-knowledge can be invaluable in navigating life's challenges and opportunities and in crafting a life that is authentic and fulfilling.

Lifelong learning can also be a rich source of joy and wonder. The thrill of mastering a new skill, the excitement of intellectual discovery, the satisfaction of creative expression—these experiences can infuse our lives with a sense of vitality and purpose. In a world that can often feel overwhelming or mundane, learning can be a way to reconnect with a sense of possibility and awe.

Moreover, learning is often a social experience, connecting us with others who share our interests and passions. Whether through formal classes, informal study groups, or online communities, learning can be a way to build relationships, broaden our networks, and feel a sense of belonging. Massive Open Online Courses (MOOCs), virtual book clubs, and knowledge-sharing platforms like Reddit and LinkedIn Learning communities make it easier than ever to engage socially around learning. These connections are not only enjoyable in their own right but also crucial for our mental and emotional well-being.

In today's rapidly evolving job market, lifelong learning is no longer a luxury but a necessity. As industries are disrupted, jobs automated, and new fields emerge, the skills and knowledge that once guaranteed a stable career are no longer sufficient. To remain competitive and adaptable in this landscape, individuals must continuously update and expand their skill sets. This means not only staying current in one's own field but also developing transferable skills such as critical thinking, communication, digital literacy, and emotional intelligence.

Lifelong learning enables individuals to navigate the changing tides of the job market with greater agility and resilience. By proactively seeking out learning opportunities—whether through formal training, on-the-job learning, or self-directed study—individuals can position themselves to seize new opportunities as they arise. In fact, the World Economic Forum's "Future of Jobs Report 2023" lists active learning, analytical thinking, and technological literacy among the top ten skills needed by 2025.

Employers also increasingly value this learning mindset. In a global survey by the World Economic Forum, companies ranked "active learning and learning strategies" as the most important skill for employees in the coming years. Lifelong learners are seen as adaptable, self-motivated, and able to drive their development—qualities that are highly prized in today's fast-paced, innovative organizations.

Beyond its immediate career benefits, a commitment to lifelong learning can also open up entirely new professional pathways. As we learn and grow, we may discover new interests and talents that inspire a career change or entrepreneurial venture. The skills and knowledge gained through continuous learning can be the foundation for reinventing oneself professionally, pivoting into new industries, or creating one's own opportunities. With portfolio careers and gig work on the rise, lifelong learning is often the backbone of reinvention.

Dale Carnegie Training, for example, is an excellent resource for developing leadership and communication skills. I highly recommend this course to elevate your professional and personal skills so that you can speak more effectively in motivating your team as well as those in your personal life. The course on "Effective Communications and Human Relations" takes 12 sessions of 3.5 hours each, and I can personally attest to its life-changing impact. It not only boosted my public speaking confidence but also reshaped how I engage in both professional and personal relationships.

In an era where the average person will change careers five to seven times over their lifetime, lifelong learning is key to remaining engaged, marketable, and fulfilled over the long arc of one's professional life. By embracing learning as a lifelong pursuit rather than a phase of early adulthood, we open the door to continuous personal reinvention, deeper self-knowledge, and a more vibrant connection with the world around us.

Chapter 61
Personal Journey of Lifelong Learning

My own journey as a lifelong learner has been shaped by a multitude of experiences, influences, and turning points. Curiosity, serendipity, and a belief in the transformative power of education have fueled my path.

Growing up in Burma, I was immersed in a rich tapestry of cultural and linguistic diversity from a young age. Navigating between Burmese, English, and Chinese languages and cultures instilled in me a fascination with the ways language and culture shape our understanding of the world.

At the same time, Burma's Buddhist society exposed me to a way of thinking that emphasized mindfulness, compassion, and the impermanence of all things. These early experiences sparked a lifelong curiosity about the intersections of different ways of knowing and being.

Throughout my life, I have been deeply fortunate to have mentors who saw potential in me and invested in my growth. From my grandparents, who embodied the values of integrity, generosity, and hard work, to the teachers who went above and beyond to nurture my curiosity and capabilities, each played a pivotal role in shaping my path.

I think of Mrs. Linda Malila, who took me under her wing when I first arrived in the United States, helping me navigate the challenges of learning a new language and adapting to a new culture. At the time, I could read basic English but struggled to hold a conversation. Her kindness and dedication left an indelible mark, showing me the power of mentorship to change the course of a life.

I also think of Dr. Glenn T. Seaborg, the Nobel laureate who gave me the opportunity to work in his lab at the Lawrence Berkeley National Laboratory. Under his guidance, I experienced firsthand the thrill of scientific discovery and learned the importance of rigorous inquiry, collaboration, and the ethical pursuit of knowledge.

They showed me that true learning involves not just acquiring information but evolving as a person.

My learning journey has been characterized by an insatiable curiosity that has drawn me to explore a wide range of disciplines and domains. Driven by insatiable curiosity, I've explored a wide range of fields. From the natural sciences to the humanities, from engineering to entrepreneurship, I have sought to understand the world through multiple lenses and connect ideas across disciplines.

In my academic and professional pursuits, I have been driven by a fascination with the fundamental questions of existence – the nature of reality, the origins of life, and the workings of the human mind and spirit. This has led me to delve into subjects as varied as chemistry, physics, biology, medicine, wellness, mathematics, artificial intelligence, photography, philosophy, engineering, communications, and theology.

One of the most profound realizations of my journey has been the recognition that true wisdom lies not in any single discipline but in integrating insights across fields. I have come to see science and spirituality not as conflicting worldviews but as complementary frameworks for understanding both the seen and unseen dimensions of life, each offering unique tools and perspectives for grappling with life's great mysteries.

This integrative approach to learning has been deeply enriching, allowing me to see the world in all its complexity and wonder. It has also been tremendously valuable, enabling me to draw on diverse knowledge and skills to solve problems, generate ideas, and create value in innovative ways. It taught me how to think holistically— balancing logic with empathy, data with intuition, and ambition with service.

A pivotal moment in my journey was the decision to apply my learning in the realm of entrepreneurship. Having seen firsthand the power of science and technology to transform lives, I felt a deep sense of responsibility to use my knowledge and skills to make a positive difference. I realized that knowledge is most meaningful when it is put into action—when it becomes a force for healing, empowerment, and change.

This led me to found Iris Biotechnologies, a company dedicated to developing cutting-edge solutions in personalized medicine. Building and leading this venture has been an incredible learning experience, stretching me in ways I could never have imagined. It became not just a business endeavor, but also a mission to democratize health solutions and give individuals more agency over their own bodies and well-being.

It has taught me the importance of resilience, adaptability, and perseverance. It has shown me the power of collaboration—bringing together diverse talents and perspectives to achieve a shared vision. It reminded me that no one builds anything worthwhile alone. Community and communication are essential. And it has deepened my understanding of the responsibility that comes with leadership: the duty to create value not just for shareholders, but also for employees, customers, and society at large.

Most of all, it has reinforced my belief in the importance of continuous learning and growth. In the fast-moving world of biotechnology, staying on the cutting edge requires constantly updating knowledge and skills. It also requires a mindset of openness, curiosity, and a willingness to question assumptions and embrace new ways of thinking. The best leaders are also the most curious learners—those who ask, listen, evolve, and adapt with humility.

Of course, my learning journey has not been smooth or linear. It has been marked by triumphs and failures, successes and setbacks. I have come to see even the most difficult experiences as invaluable opportunities for growth and learning. What seemed like obstacles at the time often became the very catalysts that reshaped my character and vision.

I think of the legal battle my company faced when a major law firm's malpractice and misconduct threatened to derail everything we had built. It was a time of immense stress and uncertainty—a true test of our resilience and resolve. But it was also a profound learning experience. It taught me the importance of standing up for what is right, even in the face of powerful opposition.

It showed me the strength that comes from having a clear sense of purpose and values and from the support of a dedicated team. It also awakened in me a stronger voice for justice, not only in the courtroom but also in boardrooms, laboratories, and policymaking spaces. And it deepened my understanding of the complex interplay between law, ethics, and business—an understanding that continues to inform my decision-making.

Through all of this, I've come to see leadership not just as a role, but also as a responsibility: to learn continuously, to listen deeply, and to serve courageously.

Looking back, I can see how each experience, whether joyful or painful, has contributed to my growth as a learner and as a leader. Each has offered lessons and insights and continues to shape my perspective and actions.

When I was training to climb Mount Rainier, the highest mountain in the lower 48 states, I learned from a coach to use muscles I had never used before. I also learned how to use an ice pick and crampons to stop my body from sliding if I fell on a steep snowfield. Practicing how to fall safely became a metaphor for learning to face failure constructively. I trained with a 30+ pound backpack, special boots, and hiked 7 to 14 miles up and down a 2,000 feet hill for weeks. Though it was physical training, it sharpened my mind and boosted my endurance — both physically and mentally.

In the end, perhaps the greatest lesson has been the realization that learning is not just something we do — it is something we become. It is not a destination, but a way of being: a stance of openness, humility, and continual transformation. We are all works in progress, with an infinite capacity to learn, evolve, and contribute to the unfolding human story.

I've always enjoyed the works of American author Ernest Hemingway, especially *The Old Man and the Sea*, the story of Santiago, an aging fisherman who refuses to give up. He catches a giant marlin when most believed his best days were behind him. The book won the Pulitzer Prize in 1953 and was the only one Hemingway mentioned during his 1954 Nobel Prize acceptance speech.

Catching a marlin truly is a lifetime experience. I caught one twice my height in Cabo San Lucas, Mexico. It took my brother Richard and me over an hour to reel it in from a chartered boat, strapped to a chair while battling the sea. My sister Grace photographed the moment. It was exhilarating — a triumph of patience and strength that reminded me of Santiago's grit.

Hemingway once wrote, "All modern American literature comes from one book by Mark Twain called *Huckleberry Finn*." Twain, born Samuel Langhorne Clemens, is widely regarded as the father of American literature. He was born two weeks after Halley's Comet in 1835 and, prophetically, died a day after its return in 1910.

Twain was also fascinated by science and had a lasting friendship with inventor Nikola Tesla. They spent much time together in Tesla's lab. Tesla, the mind behind alternating current electricity, helped power the modern world. His legacy endures in the everyday energy we often take for granted.

Lifelong learning begins with curiosity — a desire to understand, to ask why, and to explore the unknown. It's fun. It's humbling. And I hope you, too, discover its joy.

But in our pursuit of knowledge, we must also seek wisdom — something in alarming decline. Today, unchecked power and greed have led to wars, financial crises, environmental destruction, and mass extinction. Despite our exponential advances in technology over the past 200 years — more than in the previous thousands — we still grapple with existential questions

Some of the greatest questions facing the 21st century are:

Chapter 61
Personal Journey of Lifelong Learning

Does God exist?

For many, the answer is yes: God created the universe and fine-tuned it instantly. Some scientists attribute the universe's origin to the Big Bang, yet they cannot explain what triggered it or how matter came from nothing. God, in this view, created DNA — the essential code of life. Darwin's theory of evolution doesn't address the origin of DNA.

What about the soul?

Science has no answer for the origin of a soul — something many believe continues after death. Free will, another uniquely human trait, also lacks a scientific explanation. God, it is said, gave us both — along with the moral laws that govern right and wrong. Without objective morality, society descends into chaos.

Who was the first human?

Some believe it was Adam, while others think humanity emerged through evolution. Darwin published *The Origin of Species* in 1859, but he didn't know about DNA — the complex instruction code that builds all living organisms. Humans have around 37 trillion cells. Could such complexity have arisen purely by chance?

Are we really 99% chimpanzee?

While a portion of human and chimpanzee DNA is about 99% similar in comparable segments, the full genome is only about 96% alike. The differences are profound. Humans have vastly superior brain structure, motor control, and symbolic reasoning. Our neurons form about 10,000 connections per cell, allowing for memory, creativity, and language.

Even the smartest scientists have not been able to create human DNA from scratch. DNA uses four letters — A, C, G, and T — to write a biological code infinitely more sophisticated than binary computer programs. It seems improbable that such an elegant system emerged purely by chance. Many believe God must have been involved.

The Bible says, "God created Adam and Eve." Isn't Adam the first human? If Adam were alive today, he would be about 6,000 years old. Scientific evidence indicates that humans have been on Earth much longer than that. God told Adam and Eve in Genesis 1:28, "Be fruitful, and multiply, and replenish the earth." Most of the Bibles today say, "Fill the earth," but the King James Version (KJV), which was the first authorized Bible in English, clearly says, "Replenish." Replenish means there were other humans that lived on the earth before Adam and Eve."

The Bible says there were giants on Earth. Is it true? In the story of David slaying Goliath, Goliath was a giant. In the Louvre museum, there was a statue of a giant, Gilgamesh, showing him grasping a fully-grown lion with a mane in his left hand as if the lion (7-9 ft. in length) was just a house cat. Gilgamesh is widely accepted by most historians as the historical 5th king of Uruk, who reigned in the 26th century BCE for 126 years, and he lived sometime between 2800 and 2500 BCE. The Sumerians in Mesopotamia recorded the Epic of Gilgamesh on baked clay tablets called cuneiforms. Gilgamesh was seventeen feet tall. Mesopotamia was the place where Abraham was called by God to go to the Holy Land."

In Matthew 10:30, Jesus told his disciples, "The very hairs of your head are all numbered." Is there any scientific evidence?" Inside a woman's womb or an artificial womb, as a fetus grows, the cells that are identical begin to differentiate themselves into brain cells, heart cells, skin cells, and so on, and each cell knows its identity and where it is. In each cell, knowing its location and replicating or not replicating is how you get properly proportioned humans, chickens, tigers, elephants, and whales after its kind. Jesus knew that each hair was numbered because he was the Son of God. How else would an ordinary carpenter from Nazareth have that knowledge two thousand years ago?"

Genesis 6:1-4 says, "And it came to pass, when men began to multiply on the face of the earth, and daughters were born unto them, That the sons of God saw the daughters of men that they were fair; and they took them wives of all which they chose. And the LORD said, "My spirit shall not always strive with man, for that he also is flesh: yet his days shall be a hundred and twenty years. There were giants in the earth in those days; and also after that, when the sons of God came in unto the daughters of men, and they bare children to them, the same became mighty men which were of old, men of renown."

Some Biblical scholars say the "sons of God" in Genesis 6 refer to fallen angels, while others say they refer to the descendants of Seth, who were faithful to God. The "daughters of men" are often interpreted as the descendants of Cain, associated with ungodliness. There is also disagreement on whether the term "giants" (Hebrew: *Nephilim*) refers to beings that are physically tall, intellectually exceptional, or spiritually corrupted hybrids. However, the biblical text (Genesis 6:4) emphasizes their might and renown, implying they were formidable in stature or reputation. Sons of Seth and daughters of Cain would not produce literal giants unless a supernatural element, such as angelic intervention, was involved.

Gilgamesh was a giant physically, and he was a demigod. The Epic of Gilgamesh, one of the earliest known literary texts from ancient Mesopotamia, describes Gilgamesh as two-thirds god and one-third human. From Adam to Noah, most people lived 900 to almost a thousand years. After God saw that humans were corrupt, He destroyed humans, giants, and all land-dwelling creatures, except those in Noah's Ark, through a great flood. This is supported by Genesis 6–9, which describes God's grief and judgment. After the flood, human lifespans rapidly declined. Scientific evidence also shows that human cell division has a biological ceiling. The Hayflick Limit, discovered by Dr. Leonard Hayflick in the 1960s, shows that human cells divide about 40–60 times before entering senescence, limiting lifespan to around 120 years.

There is a great deal of controversy surrounding Noah's flood. Is there evidence of a great flood? National Geographic Society explorer Robert Ballard, inspired by Ryan and Pitman's hypothesis, discovered supporting physical evidence of a sudden and massive flood, including an underwater river valley and ancient shoreline beneath the Black Sea. This supports the Black Sea deluge hypothesis, suggesting a catastrophic event around 5600 BCE. Both the Bible and the Epic of Gilgamesh contain similar flood narratives with different details, suggesting a shared memory of a historical cataclysm.

The Bible says that Noah, his wife, his three sons, and their wives were the only survivors. In the Chinese language, the character for "boat" (船) is made up of the radicals for "vessel," "eight," and "mouths" — interpreted by some Christian linguists as a reference to the eight people on

535

Noah's Ark. Mitochondrial studies have shown that all modern females descended from a common maternal ancestor, sometimes called "Mitochondrial Eve," who lived in Africa over 150,000 years ago. While this doesn't prove the biblical account, it suggests a genetic bottleneck consistent with a small founding population. There are also three primary mitochondrial haplogroups, which some interpret as symbolic of Noah's three daughters-in-law.

Why are food prices going up so much in many places around the world? There are many factors that have contributed to 30% higher food prices in the US since 2019:

- First, inflation is caused by government overspending through borrowing fiat money created out of nothing by the Federal Reserve. The COVID-19 relief packages and deficit spending significantly increased the money supply, fueling inflation.
- Second, heatwaves and floods reduce crop yields in different countries from which we buy food. According to NOAA and NASA, 2023 and 2024 were the hottest years on record globally.
- Third, food companies are making more profit by taking advantage of high demand and limited supply, a phenomenon labeled "greedflation" by some economists.
- Fourth, COVID-19 disrupted global supply chains and caused the prices of meat, poultry, and other products to increase. Prices of eggs have been especially high due to Avian flu outbreaks and rising feed and energy costs.
- Fifth, the war in Ukraine is also contributing to the rise in global food prices. Ukraine and Russia were among the top exporters of wheat, corn, and sunflower oil. Disruption in exports due to war and sanctions has strained global supply chains.

European countries like Germany now face higher energy and food costs due to reduced access to Russian natural gas.

The news said that our national debt is over $37 trillion, and interest payments exceed $1 trillion per year. That is a lot of money. Will younger generations have to repay this debt? Yes, the older and younger generations will most likely share the burden. Our nation could borrow this much because the Federal Reserve can create US dollars at will, and the dollar has been the world reserve currency since 1944.

We are the only country in the world that can create dollars and enjoy this privilege. However, many countries are pushing back. In 2023 and 2024, China, Russia, Brazil, and several Middle Eastern nations announced plans to reduce reliance on the U.S. dollar in trade, accelerating global "dedollarization." According to Wikipedia, "Dollarization refers to countries reducing reliance on the U.S. dollar as a reserve currency, medium of exchange or as a unit of account."

The U.S. dollar began to displace the pound sterling in the 1920s, as the U.S. emerged relatively unscathed from World War I. After World War II, the Bretton Woods Agreement of 1944 established the U.S. dollar as the anchor of the international monetary system, tied to gold at $35 per troy ounce. The International Monetary Fund (IMF) and World Bank, both founded at Bretton Woods, further cemented U.S. financial leadership. Historically, all IMF Managing Directors have been European, while all World Bank Presidents have been American, reflecting post-war global power dynamics.

The U.S. Treasury wields significant influence over the SWIFT system — the backbone of international banking transfers — giving it the ability to impose financial sanctions worldwide. This was evident in 2022 when Russia was partially excluded from SWIFT. With our economic power, we've built the most dominant military force, with over 750 military bases globally. NATO, founded in 1949, has been a major U.S.-led military alliance, with the U.S. contributing the largest share to its budget. Historical parallels with the Roman Empire, which collapsed under military overreach, remind us of the dangers of unsustainable defense spending.

Reserve currencies have shifted throughout history based on geopolitical power. Before the dollar, the pound sterling, the Dutch guilder, and the Spanish real all had their turns. Our $37 trillion debt has raised global alarm. Reducing it would require both tax hikes and spending cuts — solutions that most politicians avoid. Eliminating this debt may take decades of sustained discipline and reform. The US dollar's credit rating was downgraded for the first time in history when Moody's lowered the U.S. from its top-tier AAA rating to Aa1 in May 2025, following similar actions by S&P in 2011 and Fitch in 2023. The value of the US dollar has fallen by 10.8% against a basket of currencies since the start of 2025. That is its worst performance over the first six months of any year since 1973.

We live in an age where people increasingly question the existence of God. Some argue that life is a random collision of molecules, or suggest that humans were seeded by alien beings. But no matter the view, one thing is certain: we humans did not create the universe. Did God create us, or did we invent gods to explain the unknown? If God is real, then God is the origin of everything — perhaps even identical to the fabric of reality itself. If not, then the universe exists without purpose, and we are on our own. Yet, human nature — our longing for meaning, our moral instincts, and our creativity — seems to hint at something deeper.

Babies, in their vulnerability, usher us into adulthood. As a child, you do what you're told. As a teenager, you rebel. As an adult, you do what you want. As you mature, you do what must be done. Elderly people, in their frailty, teach us the meaning of humility. The lucky ones find love, joy, and peace along the way.

Let's explore the relationship between God, nature, and humans. What does it mean when Jesus says, "The kingdom of God is within you" (Luke 17:21)? Or "Heaven and earth shall pass away, but my words shall not pass away" (Matthew 24:35)? These verses point to the eternal nature of spiritual truth and the divine spark within us.

Eventually, most of us will return to the earth. Some humans may even leave Earth to colonize space. NASA and private companies like SpaceX are already planning missions to Mars. Meanwhile, our sun will expand into a red giant and engulf the Earth in about 5 billion years.

Today, corporations wield massive influence. What is good for corporations isn't always good for people. In the past, corporations were dissolved if they harmed society. Now, they enjoy "corporate personhood" and can contribute unlimited funds to super PACs — thanks to the Supreme Court's 2010 ruling in *Citizens United v. FEC*, which overturned earlier limits.

CNN reported on February 1, 2025: "Elon Musk spent more than $290 million on the 2024 election, year-end FEC filings show." Musk backed Donald Trump, who became the 47th President. On taking office, Trump created the Department of Government Efficiency (DOGE), appointing Musk to lead reforms.

Chapter 61
Personal Journey of Lifelong Learning

As of February 26, 2025, DOGE had laid off about 30,000 federal workers. A Quinnipiac poll showed that 60% of voters disapproved of DOGE's actions. Forbes reported that Elon Musk's wealth dropped to $351.6 billion by March 5, 2025 — down from $464 billion in December 2024.

People say knowledge is power — but knowledge alone is not enough. Despite record knowledge, we remain vulnerable to pandemics, climate disasters, abuse of power, systemic injustice, and market failures driven by greed. It is knowledge applied wisely — through empathy, ethics, and effective leadership — that becomes real power.

Solving major problems requires ability, motivation, opportunity, and synergy. Sometimes, that means a global revolution — not of violence, but of conscience — led by courageous ordinary people.

Revolutions throughout history — from the American and French to the Civil Rights Movement — have reshaped societies. Can a global revolution be peaceful? The peaceful revolutions in India (under Gandhi), South Africa (under Mandela), and Eastern Europe offer hope. A peaceful global awakening would require shared values, digital coordination, and moral clarity.

Society declines when visionary leaders are lost. The assassinations of JFK, Robert Kennedy, and Martin Luther King Jr. in the 1960s marked a profound turning point in U.S. history. Without such leaders, both the nation and the world lost crucial moral voices. I've listened to JFK's speeches and found them deeply moving. I have stood at Dealey Plaza, where he was killed, and visited his resting place at Arlington.

There was a shining moment in U.S. history when Ronald Reagan became the 40th President (1981–1989). After surviving an assassination attempt just 69 days into office, Reagan's policies contributed to the Cold War's end. Historians debate his legacy, but many credit his diplomacy and economic reforms for revitalizing U.S. confidence and confronting Soviet authoritarianism.

The American people are generous and innovative. I am grateful for this country. But there is a void of great leaders today. Where are the leaders who combine wisdom, vision, courage, and compassion?

We live in a chaotic world. We must ground ourselves in the lessons of the past, remain clear-eyed about the present, and strive toward a better future. Only by learning from history, embracing truth, and selecting leaders with integrity can we hope to shape a more just and sustainable world. Leadership matters — and choosing the right leaders is part of the lifelong learning journey we must all commit to.

Chapter 62
Key Areas for Lifelong Learning

In considering the most important areas for lifelong learning in the 21st-century context, a few key domains stand out as particularly critical for personal, professional, and societal flourishing.

Science, Technology, and Innovation

In an age of unprecedented technological change, a basic understanding of science and technology is increasingly essential for informed citizenship and effective participation in the world. From climate change to artificial intelligence to bioengineering, the issues shaping our future are deeply rooted in scientific and technological realms.

According to the World Economic Forum's 2024 "Future of Jobs" report, roles in AI, sustainability, and biotechnology are among the fastest-growing fields globally, underscoring the demand for scientific literacy and adaptive skillsets.

But beyond just understanding these issues, we also need more people actively engaged in shaping the direction of scientific and technological progress. This means not only increasing STEM education and technological literacy but also cultivating the critical thinking, creativity, and ethical reasoning skills needed to steer innovation towards the greater good.

For example, the development of generative AI technologies like ChatGPT has sparked global debate over ethical boundaries, privacy rights, and responsible use issues that demand not just technical know-how but also interdisciplinary moral reasoning.

For those in leadership roles, whether in business, government, or civil society, a deep understanding of the science and technology landscape is increasingly a strategic imperative. Anticipating disruptive trends, making informed decisions about investments and policies, and grappling with the societal implications of emerging technologies – these are the challenges that will define leadership in the coming years.

In 2023, the U.S. National Security Commission on Artificial Intelligence emphasized that national leaders must be "AI-ready" to manage both opportunity and risk in defense, healthcare, and finance. Moreover, climate modeling, pandemic prediction, and space exploration increasingly rely on cross-disciplinary competencies in data science, systems biology, and materials engineering, highlighting the integrative nature of modern science.

Business, Economics, and Entrepreneurship

In a globalized, knowledge-driven economy, an understanding of business, economics, and entrepreneurship is valuable for individuals in any field. These domains provide a framework for understanding how value is created, how markets and organizations function, and how resources are allocated to solve problems and meet human needs.

According to the International Labour Organization (ILO), 67% of new jobs created globally between 2020 and 2024 were in small and medium enterprises, highlighting entrepreneurship as a key driver of employment and innovation.

For individuals, developing business acumen and entrepreneurial skills can be empowering on multiple levels. It enables us to be more effective in our current roles, whether in managing projects, leading teams, or contributing to organizational strategy. It equips us with the tools to turn ideas into action, whether in starting a new venture, driving innovation within an existing organization, or pursuing an independent creative or professional path.

Startups in sectors such as fintech, green energy, and healthtech are now central to solving complex global problems — from financial inclusion to carbon reduction and medical access — demonstrating that entrepreneurship is no longer just a business endeavor but a social one.

On a societal level, entrepreneurship is a powerful engine of economic growth, job creation, and social innovation. Equipping more people with entrepreneurial mindsets and capabilities is key to building a more dynamic, inclusive, and resilient economy – one that can adapt to the disruptions of technology and globalization while creating opportunities for shared prosperity.

The World Bank's 2023 report on Human Capital Development stresses that entrepreneurial education significantly improves income mobility in low- and middle-income countries, especially when paired with digital infrastructure and access to microfinancing.

At the same time, responsible business leadership and economic stewardship require more than just technical skills or business savvy. They require a deep understanding of the ethical dimensions of economic activity and a commitment to creating value for all stakeholders – not just shareholders but also employees, customers, communities, and the environment.

This shift toward stakeholder capitalism has been endorsed by the Business Roundtable (U.S.), whose 2019 declaration redefined the purpose of a corporation to promote "an economy that serves all Americans."

Furthermore, ESG (Environmental, Social, and Governance) investing surpassed $30 trillion globally by 2024, demonstrating that ethical business is not only viable but also increasingly demanded by investors and consumers alike.

Lifelong learning in this domain must grapple with questions of corporate social responsibility, sustainable development, and the role of business in advancing the public good.

Institutions such as the United Nations Global Compact and the OECD have provided frameworks for ethical business practices and sustainability goals, reinforcing the idea that modern business education must be grounded in moral philosophy and civic responsibility.

Creativity, Arts, and Design

In a world increasingly shaped by automation and algorithmic decision-making, the skills of creativity, artistic expression, and design thinking are becoming more valuable than ever. These are the skills that enable us to imagine new possibilities, connect ideas in novel ways, and create solutions that are not just functional but emotionally resonant and aesthetically pleasing.

In fact, the World Economic Forum's "Future of Jobs Report 2023" identifies creativity, originality, and initiative as top skills demanded across industries—outpacing even technical skills in some sectors.

Engaging in creative pursuits – whether through visual arts, music, writing, or other forms of self-expression – is not just a way to develop these skills but also a way to tap into a deeper sense of joy, meaning, and fulfillment. The arts have a unique power to evoke emotion, provoke reflection, and connect us to our shared humanity.

According to research from the American Psychological Association, artistic activities are linked to improved mental health, emotional regulation, and stress reduction—benefits that support learning across a lifetime.

However, the value of creativity and design thinking extends far beyond the arts. In fields from engineering to entrepreneurship to social innovation, the ability to think creatively and approach problems from new angles is increasingly essential for success. Design thinking, with its emphasis on empathy, experimentation, and iteration, has become a key methodology for driving innovation across sectors.

Leading companies such as IDEO, Google, and Apple have institutionalized design thinking as a core problem-solving strategy, embedding human-centered approaches in their R&D, user experience, and product development teams.

Lifelong learning in this domain can take many forms – from formal training in artistic techniques or design methodologies to informal exploration through hobbies or side projects. The key is to cultivate a mindset of curiosity, playfulness, and willingness to experiment – to approach the world with a designer's eye and a creator's spirit.

Massive Open Online Courses (MOOCs) like those on Coursera and edX now offer affordable access to design-thinking curriculums from Stanford, MIT, and other top institutions, making this learning more accessible than ever.

For organizations, fostering a culture of creativity and design thinking is key to staying innovative and adaptable in the face of change. This means not just investing in the skills development of individual employees but also creating an environment that supports risk-taking, experimentation, and cross-pollination of ideas.

A 2022 McKinsey Global Survey found that companies fostering a strong culture of innovation outperformed peers in growth, profitability, and customer satisfaction—underscoring the link between creativity and competitive advantage.

Health, Well-Being, and the Mind-Body Connection

In the pursuit of lifelong learning, it's easy to focus primarily on intellectual and professional development – on acquiring knowledge and skills to advance our careers and contribute to society. But true lifelong learning must also encompass physical, emotional, and mental well-being domains – the foundations upon which all other growth and development depend. Research from Harvard Medical School shows that regular physical activities, balanced nutrition, and quality sleep directly correlate with improved memory retention, faster learning, and emotional stability.

As our understanding of the mind-body connection deepens, it's becoming increasingly clear that learning is not just a cognitive process but also a holistic one that involves the whole self. Factors like nutrition, exercise, sleep, stress management, and social connection all play a crucial role in our ability to learn, think creatively, and perform at our best. Neuroscientific studies have confirmed that chronic stress impairs neuroplasticity—the brain's ability to form new connections—while practices like meditation and mindful breathing enhance focus and memory consolidation.

In this sense, lifelong learning in the domain of health and wellness is about more than just acquiring information about healthy habits (although that is certainly important). It's about developing the self-awareness, self-regulation, and self-care practices that enable us to optimize our physical and mental performance and cultivate resilience in the face of life's challenges.

According to the World Health Organization (WHO), resilience is a critical determinant of individual health outcomes, especially during prolonged adversity such as economic downturns or global pandemics.

This kind of learning can take many forms – from exploring mindfulness and meditation techniques to improve focus and emotional regulation to experimenting with different exercise routines and nutritional strategies to optimize energy and cognitive function. It may involve learning about the latest research on sleep, stress, and the gut-brain connection or engaging in personal reflection to better understand one's own patterns of thought and behavior. Emerging research from the University of California, Los Angeles (UCLA) links gut microbiome diversity to cognitive flexibility, showing that dietary choices may directly impact emotional and intellectual adaptability.

For organizations, supporting employee wellness is increasingly seen as a strategic imperative – not just for reducing healthcare costs but also for driving engagement, productivity, and innovation. This means going beyond traditional wellness programs to create cultures prioritizing well-being, psychological safety, and work-life integration. Companies like Salesforce and Google have implemented mindfulness programs and flexible work policies based on evidence showing these initiatives reduce burnout and boost innovation.

At a societal level, lifelong learning about health and wellness is key to addressing some of the most pressing challenges we face. By empowering individuals with the knowledge and tools to take charge of their own well-being, we can build a more resilient and thriving society. As global health systems strain under the weight of chronic diseases, mental illness, and lifestyle-related conditions, health education becomes a critical tool not only for personal longevity but also for public resilience.

I have lectured at universities in the US and Asia. I also lectured a class of MBA students from INSEAD when they came to Silicon Valley to learn. INSEAD's slogan is "The Business School for the World." Its website states, "INSEAD equips our truly global community with the tools, skills, and pioneering knowledge to not only face today's challenges but to build a better world. Our research is revolutionary. Our teaching is ambitious. Together, we raise the bar of business education and research."

INSEAD consistently ranks among the top global MBA programs by Financial Times and The Economist, and its campuses in France, Singapore, Abu Dhabi, and San Francisco underscore its international footprint.

With Saratoga Rotary, I taught the "Junior Achievement Program" to Washington Elementary School students in San Jose. I participated in the Junior Achievement program at George Washington High School in San Francisco. In 2004, I served as a judge in Santa Clara along with a few other CEOs, including some who managed multi-billion-dollar companies in Silicon Valley.

Junior Achievement USA reports that students who complete their programs demonstrate a 25% higher interest in starting a business, and many report improved financial literacy and career readiness.

Junior Achievement changes the lives of young people by empowering them through connecting what they learn in school with how life works outside the classroom. Students learn that it is possible to invest in the future and pursue their dreams.

I believe that teachers should learn how AI can help them and their students, as well as how AI can cause harm. While it is important for students to learn science and math, it is equally important to learn social skills and leadership skills. A 2023 RAND Corporation study found that 60% of K–12 educators are interested in using AI for personalized learning, yet many remain concerned about ethical risks like bias, surveillance, and disinformation.

If I were mentoring students, I would say, "First, learn to respect others and yourself." If you or others make a mistake, correct it and forgive it, if possible. Also, learn to actively listen so that you really understand and others feel that they are heard. Science and math can teach you how to do things right, but they don't teach you how to do the right things.

This echoes the sentiment of educators like Sir Ken Robinson, who argued that creativity, empathy, and moral insight must be central to 21st-century education to help students navigate an increasingly complex world.

Chapter 63
The Role of Technology
in Lifelong Learning

No discussion of lifelong learning in the 21st century would be complete without considering the transformative role of technology. From online courses and mobile learning apps to virtual and augmented reality simulations to AI-powered personalized learning platforms – technology is reshaping the lifelong learning landscape in profound ways. According to a 2023 report by HolonIQ, global edtech investment reached over $10 billion, reflecting a growing emphasis on digital learning tools to meet the demands of a rapidly changing workforce.

On the one hand, technology is democratizing access to learning in unprecedented ways. The rise of massive open online courses (MOOCs), digital libraries, and open educational resources means that anyone with an internet connection can access world-class learning content anytime, anywhere. This is a game-changer for individuals who may have previously been excluded from traditional educational opportunities due to financial, geographic, or other barriers. Platforms like Coursera, edX, and Khan Academy now serve tens of millions of learners worldwide, offering university-level courses in areas such as computer science, healthcare, and humanities.

Technology is also enabling more personalized, adaptive, and engaging learning experiences. Advances in learning analytics, machine learning, and natural language processing pave the way for intelligent tutoring systems that can tailor instruction to individual learners' needs, provide real-time feedback and guidance, and even predict and prevent learning obstacles.

Immersive technologies like virtual and augmented reality create new possibilities for experiential and situated learning, allowing learners to practice skills and explore concepts in realistic, interactive environments. For example, companies like Labster offer virtual science labs to schools without access to physical lab resources, while platforms like Google Expeditions use AR/VR to take students on interactive field trips.

At the same time, the proliferation of digital tools and platforms is also changing the way we think about the skills and competencies needed for lifelong learning success. Digital literacy – the ability to effectively use digital technologies to find, evaluate, create, and communicate information – is becoming an essential foundation for learning and working in the 21st century. So, too, are skills like data analysis, computational thinking, and human-machine collaboration – skills that will be increasingly important as artificial intelligence and automation transform the nature of work.

UNESCO defines digital literacy as a key component of 21st-century competencies, while the OECD's Future of Education and Skills 2030 project emphasizes systems thinking and complex problem-solving as core to future learning.

However, technology is not a panacea; it also presents challenges and risks that must be navigated thoughtfully. Issues of digital equity and the digital divide mean that not everyone has equal access to the benefits of learning technologies. Concerns around data privacy, security, and the ethical use of learning data are becoming more pressing as the collection and analysis of student data becomes more sophisticated.

The sheer pace of technological change means that today's learning tools and platforms may be obsolete tomorrow – requiring a constant updating of digital skills and fluencies. According to Pew Research Center (2021), roughly 15% of U.S. households with school-age children do not have a reliable Internet connection, limiting equitable access to remote learning opportunities.

Ultimately, realizing the potential of technology to support lifelong learning will require more than just investing in the latest tools and platforms. It will require a fundamental rethinking of the roles and responsibilities of learners, educators, institutions, and employers in the learning ecosystem.

It will require new models for financing and credentialing learning over the lifespan, as well as for recognizing and valuing the skills and competencies acquired through diverse learning experiences. Credentialing innovations like digital badges and micro-credentials, offered by organizations such as Credly and LinkedIn Learning, are now being adopted by employers to recognize alternative learning pathways.

It will require ongoing reflection and dialogue about the kinds of futures we want our learning technologies to create – and the values and priorities that should guide their development and use. Ethicists and futurists are calling for responsible AI guidelines in education, including UNESCO's 2021 Recommendation on the Ethics of Artificial Intelligence, which advocates for transparency, human oversight, and equitable benefit-sharing.

In the near future, AI will be used in conjunction with humans to teach various subjects. Students need to be aware that if the AI system doesn't know the correct answer, it will make something up. AI industry calls it hallucination.

Hallucination in AI refers to the generation of incorrect or fabricated content with high confidence. Large language models like GPT-4 and Gemini may produce plausible-sounding but factually incorrect outputs, requiring human validation in educational settings.

Before AI enters the classroom, students need to learn how to discern. Starting in elementary school, if possible, students should learn how to learn. At DuPont, I applied for and got a $250,000 federal grant to train our workers to learn how to learn. This philosophy aligns with meta-learning – the science of learning how to learn – which is now a recognized discipline promoted by educational theorists like Barbara Oakley and online courses such as "Learning How to Learn" on Coursera.

Principal for a Day is a program in various states that gives executives an inside look at how schools run, and executives can, in turn, enable company resources, people, and materials to be used to help the schools. When I was at DuPont, I served as a principal for a day at Lakewood Elementary School in Santa Clara. The National Association of Elementary School Principals (NAESP) supports programs like these to foster school-community partnerships and enhance cross-sector collaboration.

My team helped the students produce a yearbook and helped them individually for months. All those who interacted with students were properly trained on what is and isn't appropriate to do.

In 1985, a consortium of companies in the San Francisco Bay Area partnered with the Lawrence Hall of Science at UC Berkeley and founded Industry Initiatives for Science and Math Education (IISME). IISME's Summer Fellowship Program provides paid summer internships for high school teachers who work in high technology companies, universities and research laboratories.

IISME, now part of the non-profit organization Ignited, has placed more than 3,500 educators in industry fellowships, enriching over 3 million student learning experiences through curriculum transformation.

The goal was to provide teachers with a solid understanding of the practical applications of what they teach. This was to support them in updating their curriculum and teaching strategies. I served as an advisor to the Director and staff of IISME and a mentor to a teacher who participated in the summer fellowship program.

I was also on a committee to produce a document called Workplace Literacy for the State of California. The document was about two inches thick and took more than a year to produce. I represented Du Pont in participating in this project. The goal was to empower the existing workforce and to help the next generation.

Workplace literacy programs began gaining national attention in the 1980s, as studies by the National Center on Education and the Economy revealed that millions of workers lacked the basic reading, math, and problem-solving skills needed for modern employment. Today, digital and workplace literacy remain critical components of adult education policy in the U.S. and abroad.

Chapter 64
Strategies for Lifelong Learning Success

Embracing lifelong learning as a mindset and a practice is one thing; translating that commitment into consistent, sustainable action is another. It requires intentionality, discipline, and a strategic approach to managing one's learning journey over the long haul.

Here are some key strategies that can help turn the aspiration of lifelong learning into a reality:

1. Cultivate curiosity and an open mind. At the heart of lifelong learning is a fundamental attitude of openness and curiosity – a desire to explore, discover, and understand. This means approaching the world with a sense of wonder and humility and being willing to question one's own assumptions and beliefs. It means seeking out new experiences, perspectives, and ideas, even (and especially) when they challenge or disrupt our existing worldviews.

Practical tactics for cultivating curiosity might include:

- Making a habit of asking questions and following up on the answers
- Seeking out people and resources that offer different viewpoints from your own
- Traveling to new places or immersing yourself in unfamiliar cultures
- Taking on a new hobby or learning a new skill outside your usual areas of interest
- Keeping a learning journal to reflect on your experiences and insights

2. Set goals and create a learning plan. While lifelong learning is ultimately a journey, not a destination, setting clear goals and creating a roadmap for achieving them can provide structure, motivation, and a sense of progress. This might involve identifying specific knowledge or skill areas you want to develop, setting short-term and long-term learning objectives, and breaking them down into manageable steps.

Some tips for effective goal setting and planning:

*   Start with your why – get clear on your underlying motivation and purpose for learning
*   Be specific and measurable in your goals – define what success looks like and how you'll track progress
*   Set realistic timelines and milestones, but also leave room for flexibility and serendipity
*   Identify the resources, support, and accountability you'll need to stay on track
*   Regularly review and adjust your plan based on your experiences and evolving interests

3. Harness the power of habits and routines. The most successful lifelong learners are often those who have developed the habits and routines that make learning a natural, integrated part of their daily lives. This means finding ways to make learning a default behavior rather than something that requires constant willpower and motivation.

Some strategies for building learning habits might include:

*   Designating a regular time and place for learning activities (e.g., a daily "learning hour")

- Linking learning to existing habits or routines (e.g., listening to podcasts during your commute)
- Using triggers or cues to prompt learning behaviors (e.g., keeping a book on your nightstand)
- Celebrating small wins and progress to reinforce the habit
- Finding an accountability partner or learning community for social support and motivation

4. Embrace a variety of learning modalities. One of the joys of lifelong learning is the sheer diversity of ways to engage in the learning process. From reading books and articles to watching videos and listening to podcasts, to attending workshops and conferences, to engaging in hands-on projects and experiments – the possibilities are endless.

Embracing a variety of learning modalities not only keeps things interesting and engaging but also allows you to tailor your learning to your individual goals, preferences, and learning styles. Some tips for diversifying your learning:

- Experiment with different formats and find what works best for you
- Mix up solo and social learning experiences
- Balance consumption (e.g., reading, watching) with creation and application
- Seek out immersive, experiential learning opportunities that get you out of your comfort zone
- Leverage technology to access learning content and communities beyond your local area

5. Seek out feedback and mentoring. Lifelong learning is not a solo endeavor – it is deeply enriched by the perspectives, expertise, and wisdom of others.

Seeking out feedback, coaching, and mentoring from trusted sources can provide invaluable guidance, support, and accountability along the learning journey.

This might involve:

- Identifying potential mentors or advisors in your field or areas of interest
- Joining a mastermind group or peer coaching circle for mutual support and challenge
- Seeking out opportunities for formal feedback or assessment (e.g., through courses or certifications)
- Cultivating a network of trusted peers and colleagues for informal feedback and idea-sharing
- Being open and proactive in asking for help and advice when needed

6. Apply your learning in real-world contexts. Learning comes alive when it moves from theory to practice, from abstract concepts to concrete applications. Seeking out opportunities to apply your learning in real-world settings not only deepens your understanding and skill development but also allows you to create tangible value and impact.

Some ways to put your learning into action:

- Take on a stretch project or assignment at work that allows you to practice new skills
- Volunteer your time and expertise to a cause or organization you care about
- Start a side project or business venture to test out new ideas and approaches
- Teach or mentor others in your areas of learning to solidify your own understanding

559

- Reflect on your experiences and identify lessons learned to inform future learning and action

7. Embrace a growth mindset. Finally, perhaps the most essential strategy for lifelong learning success is to cultivate a growth mindset – a belief that your abilities and intelligence can be developed through effort, learning, and perseverance. This means reframing challenges and setbacks as opportunities for growth and focusing on the process of learning rather than just the outcomes.

Some tips for embracing a growth mindset:

- Celebrate effort, progress, and learning from mistakes, not just success or perfection
- Embrace challenges and stretch outside your comfort zone regularly
- Cultivate a sense of self-compassion and patience with the learning process
- Surround yourself with others who embody a growth mindset and support your development
- Regularly reflect on and reframe your internal self-talk and mental habits

In the end, the strategies that work best for lifelong learning will be unique to each individual and their specific goals, circumstances, and preferences. The key is to approach the process with intention, experimentation, and a commitment to continuous improvement. By making learning a joyful, integrated part of your life and work, you set yourself up for a lifetime of growth, meaning, and impact.

Chapter 65
Leadership and Lifelong Learning

According to the **Gallup Poll**, "President Joe Biden's third full year in office, spanning Jan. 20, 2023, to Jan. 19, 2024, an average of 39.8% of Americans approved of his job performance. Among prior presidents in the Gallup polling era who were elected to their first term, only Jimmy Carter fared worse (37.4%) in his third year. Donald Trump, Barack Obama, Ronald Reagan, Bill Clinton, and Richard Nixon also had sub-50% third-year averages. Dwight Eisenhower's 72.1% remains the highest third-year presidential approval rating in Gallup's history."

On June 27, 2024, there was a presidential debate between President Biden and former President Trump. Despite President Biden's dismal debate performance due to his mental lapses, the race for the presidency remained statistically tied, according to a national NPR/PBS NewsHour/Marist poll conducted on July 9–10, 2024.

An assassin shot Donald J. Trump, 78, in the head on July 13, 2024, during a campaign speech. Luckily, Trump survived and became the Republican nominee for President of the United States on July 15, 2024, to challenge President Biden in the November 2024 election. The incident, captured live, led to an immediate wave of national solidarity and security reassessment at campaign events. Trump's survival and recovery became a significant political moment, boosting his visibility and momentum.

President Biden was diagnosed with Covid-19 on July 17, 2024. He dropped out of the race on July 21, 2024, prompting speculation and concern. Many believe that it was a coup. While there is no verifiable evidence to confirm it was a "coup," the rapid sequence of events triggered widespread debate within both political and media circles.

Sometime later, he endorsed his vice president, Kamala Harris, as the Democratic Party's nominee for President of the United States. Harris picked up enough delegate support to win the nomination on the first full day of her campaign. According to the Democratic National Committee, Harris secured a majority of pledged delegates following endorsements from key party leaders and state-level caucus shifts.

According to the campaign, she raised $81 million from more than 1.1 million unique donors, including 62% first-time donors, in just days following her announcement. According to polls in September 2024, Trump and Harris were statistically tied. On November 5, 2024, Donald J. Trump was elected as the 47th president of the United States.

The polling data from June 2024 puts the approval rating of the United States Congress at 16 percent, reflecting a slight increase from 13% in May. Can you imagine any global corporation tolerating employees with such poor performance?

The average Congressional approval rating from 2006, before the Great Recession, to 2024 has hovered around 15–16%, with a low of 11% and a high of 36%. Despite these low approval ratings, Congressional incumbents consistently enjoy high reelection rates – averaging around 90%, due in part to factors like redistricting, name recognition, and campaign financing advantages.

Published by Statista Research Department on January 12, 2024, approximately 18 percent of people in Great Britain thought Rishi Sunak was the best choice for Prime Minister of the United Kingdom. The UK held a general election on July 4, 2024, and Sunak and the ruling Conservative Party lost in a landslide to the Labour Party, led by Keir Starmer, who became the new Prime Minister.

Morning Consult's Q1 2024 data showed French President Emmanuel Macron's approval rating at 24%. France held a snap legislative election on June 30, 2024, with a second round on July 7, 2024. Results showed that France entered a political deadlock, with no party gaining a majority, and Macron's centrist alliance performing poorly amidst a rise in support for both far-left and far-right parties.

Reuters reported on December 13, 2023, that Japanese Prime Minister Fumio Kishida's cabinet approval had hit a record low of 17.1%. In mid-August 2024, Kishida announced he would not run in the September 2024 leadership race, effectively stepping down under pressure from scandals and plummeting public support.

According to Canada's Angus Reid Institute, the approval rating for Prime Minister Justin Trudeau stood at 32% in January 2024. By mid-2024, Trudeau's Liberal government faced increasing criticism over inflation, housing affordability, and public safety, contributing to declining support across key demographics.

On December 12, 2023, support for Germany's Chancellor Olaf Scholz's coalition government declined to a historic low of 17%, according to a poll by ARD, Germany's public broadcaster. The same poll showed that 82% of Germans were dissatisfied with the coalition's performance over the past two years.

The figures released put support for Scholz's Social Democratic Party (SPD) at 14%, while the environmentalist

Greens scored 15% and the liberal Free Democrats (FDP) were at just 4%, falling below the 5% threshold needed to enter the Bundestag in future elections.

Al Jazeera reported on February 14, 2024, that Indonesian President Joko Widodo (Jokowi), who is barred from seeking a third term, remains extraordinarily popular with an approval rating above 80%, according to national polling agencies such as Indikator Politik Indonesia and Saiful Mujani Research & Consulting (SMRC). Indonesia, with more than 204 million registered voters, is the third-largest democracy in the world after India and the United States.

According to data released by Morning Consult on December 7, 2023, Prime Minister Narendra Modi of India retained his position as the world's most popular leader, with an approval rating of 76%. Mexico's President Andrés Manuel López Obrador ranked second with 66%. Notably, Morning Consult's survey included leaders from 15 democratic countries but excluded others such as Indonesia, possibly due to inconsistent international polling access or methodology constraints.

Indonesian President Jokowi's absence from the Morning Consult list has raised questions about selection bias, considering his high domestic approval ratings and influence in Southeast Asia. Critics argue that such omissions undermine the credibility of global leadership rankings and reflect Western-centric bias in political data aggregation.

In lifelong learning, we must learn about leaders from all over the world in order to choose our own leaders wisely.

Chapter 66
Spiritual Insights and
Biblical Perspectives

Then came a life-changing trip to Israel. In April 2006, I visited the Sea of Galilee, where Jesus is said to have calmed the storm and walked on water. As I stood on the shore and touched the water, I suddenly felt a surge of energy coursing through my body. An overwhelming sense of peace washed over me, and I knew I was in the presence of God. I realized then that my life would never be the same. That was the moment I recommitted to putting God first in my life and career.

As I look back on those early years, I see how every challenge, setback, and opportunity was preparing me for the mission ahead. From my hardworking immigrant parents to my world-class professors, from my early breakthroughs in science and engineering to my spiritual revelations—each experience guided me toward my ultimate destiny of serving God and humanity. Though I certainly didn't feel worthy or ready at the time, I've come to understand that's exactly how God works—He doesn't call the equipped but equips the called.

In the United States, the role of religion in public education changed significantly in the 20th century. Even before the Declaration of Independence, from 1776 to 1963, children were commonly taught that God was the ultimate source of hope. It was culturally and legally acceptable to teach Christianity in public schools for nearly two centuries.

This changed with landmark U.S. Supreme Court decisions. In Engel v. Vitale (1962), the Court ruled that official prayer in public schools violated the Establishment Clause of the

First Amendment. The following year, Abington School District v. Schempp (1963) reinforced this decision by prohibiting Bible readings in public schools. These rulings clarified that state-sponsored religious activity in public schools was unconstitutional. As a result, the responsibility to teach children about Christianity and the Bible shifted largely to parents and religious institutions.

The Bible offers numerous accounts of creation and early human history. To understand these narratives more fully, it's helpful to compare them with ancient Near Eastern texts like the Enuma Elish and the Epic of Gilgamesh, both originating in Mesopotamia. Scholars have noted thematic similarities between the Enuma Elish and Genesis— particularly in the concept of creation through divine speech. The Enuma Elish dates to the second millennium BCE, predating the compilation of Genesis by centuries.

In the Bible, the first man created by God from the dust of the ground is called Adam. Interestingly, in the Enuma Elish, a man named Adamu appears—created as a servant of the gods. The name similarity is striking, but scholarly consensus holds that "Adam" in Genesis and "Adamu" in Mesopotamian mythologies developed independently within their respective cultural contexts.

Some fringe theories, often cited in pseudoarchaeological literature, claim that Adamu was created through genetic engineering by combining the DNA of early humans with that of extraterrestrials called the Anunnaki. However, mainstream archaeology and genetics do not support these interpretations. These ideas are speculative and lack empirical evidence.

Genesis 1:1-5 famously begins:

"In the beginning God created the heavens and the earth. And the earth was without form, and void... And God said, Let there be light: and there was light..."

This passage introduces the biblical concept of time and creation. Because the "days" of creation are described prior to the formation of the sun and Earth's rotation, many theologians and scholars argue that the term "day" (Hebrew: yom) may be metaphorical or represent an undefined span of time.

In Genesis 1:26, God says:

"Let us make man in our image, after our likeness..."

This plural form has sparked much theological debate. Some interpret it as a reference to the Trinity, while others believe it reflects the "divine council" imagery common in ancient Near Eastern texts.

Genesis 2:10–14 describes the rivers of Eden—Pison, Gihon, Hiddekel (Tigris), and Euphrates. These geographic markers have led scholars to locate the Garden of Eden somewhere in the region of Mesopotamia, though no definitive archaeological evidence has been found.

The story of Adam and Eve's disobedience and expulsion from Eden introduces the biblical concept of original sin. This foundational story has influenced Judeo-Christian concepts of morality, human nature, and redemption.

Genesis also recounts the story of Noah and the flood. Similar flood narratives exist in many cultures, including the Epic of Gilgamesh, which predates the Hebrew Bible. The similarities—such as divine warnings, a large boat, and global destruction—suggest a shared oral tradition or common cultural memory in the ancient Near East.

Modern science offers a different framework for understanding human origins. The theory of evolution, proposed by Charles Darwin in On the Origin of Species (1859), posits that species evolve through natural selection. While evolution explains biological change over time, it does not explain how life began in the first place—a distinction often overlooked in public discourse.

Some fundamental scientific questions remain unanswered. For example, the origin of life—how non-living molecules assembled into self-replicating cells—remains a mystery. Theories such as abiogenesis and RNA world hypothesis are being explored, but no single model has been universally accepted. The extreme complexity of even the simplest organisms challenges the notion that life emerged through purely random chemical processes.

It is important to note that science and faith need not be in opposition. Many renowned scientists—such as Isaac Newton, Gregor Mendel, and Francis Collins (former director of the Human Genome Project)—were or are people of faith who saw scientific exploration as a means of understanding God's creation. The perceived conflict between religion and science is largely a 19th- and 20th-century construct.

As we continue to discover more about our world and origins, we must approach these questions with humility, curiosity, and open-mindedness. Whether through scientific inquiry, spiritual exploration, or both, the pursuit of truth and meaning is among humanity's oldest and most noble endeavors.

Chapter 67
Reflections on Life, Society, and the Future

As we navigate the complexities of modern life, it's crucial to reflect on the broader implications of our knowledge and actions. The relationship between God, nature, and humanity is multifaceted and has been debated across cultures and centuries. When Jesus said, *"Heaven and earth shall pass away, but my words shall not pass away"* (Matthew 24:35), it prompts contemplation on the transient nature of physical existence and the enduring relevance of spiritual truth.

Climate change is a pressing issue that challenges our understanding of divine will, natural processes, and human responsibility. Those who believe that human activity—especially the burning of fossil fuels and deforestation—is accelerating climate change call for urgent global cooperation. According to the Intergovernmental Panel on Climate Change (IPCC), it is "unequivocal" that human influence has warmed the atmosphere, ocean, and land. Others who see climate shifts as part of natural planetary cycles may argue against drastic interventions. Still, some interpret these climate disturbances as spiritual warnings or judgments, leading to a wide range of responses shaped by worldview and belief systems.

The phrase *"knowledge is power"* is often repeated, but it's not entirely accurate. Despite having access to more data and scientific understanding than at any other time in human history, we are still unable to prevent or adequately manage many of the world's greatest challenges: pandemics, environmental catastrophes, systemic corruption, unjust wars, and socioeconomic inequality. True power lies not in

569

the possession of knowledge but in its application toward ethical and effective action.

People don't always act on what they know to be right. I was once at a medical conference where a past president of the American Medical Association spoke passionately about the dangers of tobacco. During the next break, I saw a Nobel laureate and several other respected physicians smoking cigarettes. This paradox highlights how intellectual understanding alone may not overcome ingrained habits or addictions.

One image that has stayed with me is of a man smoking in a hospital parking lot, wheeling an IV stand with a bag dripping medication into his arm. It was a stark reminder of the battle between knowledge, addiction, and willpower. Who was in control of his life? Who bears the cost of such decisions—personally and collectively?

Life is precious. The more I study the science behind human biology, consciousness, and environmental interdependence, the more I am struck by the improbable beauty of existence. Life is both delicate and incredibly resilient—a miraculous balance of physics, biology, and spirit.

We must teach young people that true champions don't quit when faced with adversity. They are driven, focused, and adapt in the face of change. They strive for excellence, seek truth, act with integrity, and lift others in the process. They value inner peace and are unafraid to ask for help when needed. Success is not just about talent but also about purpose, discipline, and support.

We all have tremendous potential, though few of us realize it fully. While our bodies and minds impose natural limitations, our spirits are boundless. This raises profound philosophical and theological questions. Could the spiritual realm be more real or more permanent than the material one, where joy and pain coexist?

What is the meaning of our existence? Would there be evil on Earth without humans? Does the universe itself has a purpose—or is meaning something we bring to it? The Earth's axial precession cycle, known as the Great Year, spans approximately 26,000 years. The last ice age began around 13,000 years ago, roughly halfway through the current cycle—an intriguing symmetry.

Why is there a black hole at the center of nearly every galaxy? Astronomers, using observations from instruments like the Chandra X-ray Observatory and the Event Horizon Telescope, have confirmed that supermassive black holes reside at the heart of most large galaxies, including the Milky Way. These mysterious phenomena play a crucial role in galactic formation and evolution.

No human had ever set foot on another celestial body until Neil Armstrong and Buzz Aldrin landed on the Moon on July 20, 1969, during NASA's Apollo 11 mission. I remember watching it with awe. However, in later years, some individuals began questioning the authenticity of the moon landing—despite overwhelming photographic, physical, and telemetry evidence confirming its occurrence.

One oddity that fueled conspiracy theories was the erasure of the original moon landing broadcast tapes. NASA admitted in 2006 that over 200,000 tapes were erased and reused due to a tape shortage, though backup recordings still exist. This bureaucratic oversight has unfortunately added fuel to skepticism, despite no credible scientific reason to doubt the mission's reality.

Chapter 67
Reflections on Life, Society,
and the Future

Just over 400 years ago, in 1610, Galileo Galilei discovered Jupiter's four largest moons using his handmade telescope, fundamentally changing humanity's view of the universe. Then, in 1979, NASA's Voyager 1 mission discovered Jupiter's rings, revealing a complexity previously unseen.

Today, we know that Earth is one of eight planets in our solar system (Pluto was reclassified as a "dwarf planet" in 2006). Thanks to telescopes like Hubble and the James Webb Space Telescope (JWST), launched in 2021, we now know that our Sun is just one of an estimated 100 to 400 billion stars in the Milky Way galaxy. The Milky Way itself is part of a universe that contains as many as 2 trillion galaxies, each expanding outward at varying velocities.

The observable universe is estimated to be about 93 billion light-years in diameter, although the universe beyond what we can observe could be infinite. The rate of the universe's expansion—known as the Hubble constant—has been measured differently depending on the method used, leading to the "Hubble tension" in cosmology.

Cosmologists agree that if the rate of expansion after the Big Bang had been even slightly different—faster or slower by as little as 1 part in 10^{55}—galaxies, stars, and life could not have formed. This is sometimes referred to as the "fine-tuning" problem, leading some to speculate about a multiverse or intelligent design.

In physics, traveling at the speed of light (186,282 miles per second) would theoretically freeze time for the traveler, according to Einstein's theory of relativity. Photons, which move at light speed, experience no time—they are emitted and absorbed instantaneously from their frame of reference. This scientific insight raises fascinating metaphysical questions about time, existence, and the limits of perception.

Since I saw my mother's soul leave her body, I carry an unshakable conviction that the soul lives on. Since I was filled with the spirit of God at the Sea of Galilee, I hold an unwavering certainty that God is real.

Overcoming big problems requires more than isolated effort. It demands a deliberate convergence of ability, motivation, opportunity, and collective synergy. Sometimes, systemic change cannot occur without a global awakening—a moral revolution—led by courageous, ordinary people who choose to act despite fear, opposition, or personal risk. History reminds us that transformation often begins not with elites, but with those who decide they can no longer be silent.

Leadership plays a crucial role in shaping society's values, direction, and collective memory. The assassination of President John F. Kennedy on November 22, 1963, was one such moment of rupture. His death not only shocked the nation but also disrupted the trajectory of American idealism and international diplomacy during the Cold War. Kennedy's vision for civil rights, space exploration, and nuclear de-escalation left a void that altered the course of history.

Similarly, the loss of other moral or visionary leaders—such as Martin Luther King Jr., Robert F. Kennedy, and even global figures like Mahatma Gandhi—has repeatedly forced societies to grapple with what might have been. Their absence is often felt in the enduring injustices they sought to remedy.

Current global leadership faces significant challenges. Many countries are experiencing low approval ratings for their leaders, reflecting public dissatisfaction with governance. This leadership vacuum poses risks but also presents opportunities for new voices and approaches to emerge. As we look to the future, several key areas demand our attention:

1. The advancement of artificial intelligence and its potential to revolutionize various aspects of life, including healthcare, education, and work.
2. The ongoing challenges of climate change and environmental sustainability.
3. There is a need for ethical frameworks to guide technological development and implementation.
4. The importance of fostering global cooperation to address shared challenges while respecting cultural differences.
5. The continuing quest for social justice and equality, both within nations and on a global scale.
6. The evolution of education systems to prepare individuals for a rapidly changing world.
7. The balance between technological progress and human values ensures that advancements serve to enhance rather than diminish our humanity.

In conclusion, lifelong learning is far more than a personal or professional goal—it is a moral and societal imperative. It is about equipping individuals and communities with the agility, wisdom, and courage to navigate an increasingly complex, fast-changing world. It fosters curiosity, empathy, and critical thinking, helping bridge divides and finds innovative solutions to the global challenges of our time. It cultivates the ethical grounding and spiritual consciousness needed to ensure our progress aligns with values that serve both humanity and the planet.

As we continue this lifelong journey, humility must be our constant companion—recognizing that there is always more to understand, always room to grow. Embracing the diversity of human knowledge—drawing from science, spirituality, philosophy, and the arts—enriches our understanding and opens new paths forward.

True wisdom lies not in the accumulation of facts, but in the compassionate and courageous application of knowledge for the common good. The lifelong learner seeks not just answers, but deeper questions.

The journey of lifelong learning is not always easy—it demands vulnerability, resilience, and a willingness to unlearn and relearn. But it is infinitely rewarding. It deepens our self-awareness, strengthens our relationships, and empowers us to contribute meaningfully to a world in flux.

As we confront the seismic shifts of the 21st century— technological upheaval, environmental crisis, and societal polarization—we must double down on education systems that foster adaptability, emotional intelligence, and a growth mindset. The traditional model of education—front-loaded in youth and static in content—can no longer serve the needs of a society driven by rapid change and uncertainty.

The rise of automation and artificial intelligence is dramatically reshaping the labor market. According to a 2023 report by the World Economic Forum, nearly 23% of jobs globally are expected to be disrupted by automation by 2027. This demands a fundamental shift in how we approach workforce development and lifelong re-skilling.

Educational systems must evolve to prioritize creativity, collaboration, and critical thinking as much as STEM and literacy. Emotional intelligence, leadership, and ethical reasoning must become core components of a future-ready education.

Chapter 67
Reflections on Life, Society,
and the Future

In science and technology, we stand at the precipice of breakthroughs that are as promising as they are perilous. The development of CRISPR-Cas9, for example, holds the potential to cure genetic diseases—but it also opens ethical dilemmas around germline editing, eugenics, and the limits of human intervention in nature. Similarly, AI systems trained on biased data sets can unintentionally reinforce inequality, even as they are used to improve medical diagnosis or optimize energy usage.

We must ensure that innovation is guided by strong ethical frameworks, global cooperation, and inclusive dialogue—especially as these technologies outpace regulation.

Climate change remains a generational challenge. Despite international agreements like the Paris Accord, global CO_2 emissions reached a record 36.8 billion metric tons in 2023, according to the International Energy Agency. This reality underscores the need for not just technical solutions but a cultural and educational shift toward environmental stewardship at all levels of society. Schools, businesses, and civil institutions must become hubs for climate literacy and ecological responsibility.

The COVID-19 pandemic further exposed the interdependence and inequities of our global systems. It accelerated digital transformation, revealed healthcare vulnerabilities, and exacerbated income and access divides. But it also demonstrated humanity's capacity for scientific collaboration, adaptability, and compassion. Moving forward, we must build systems that are not only more resilient but also more inclusive and just.

On an individual level, lifelong learning enhances meaning and mental well-being. Studies published in journals like Ageing & Society and Frontiers in Psychology show that adults engaged in lifelong learning report greater life satisfaction, reduced cognitive decline, and increased community engagement. Whether it's learning a new language, exploring a philosophical text, or discovering a hidden creative talent, lifelong learning keeps our inner world expanding even as the outer world shifts.

Looking to the future, our greatest challenges—climate change, inequality, misinformation, and global instability—require not just technical skill, but wisdom. And wisdom grows from sustained learning, honest reflection, and shared experience.

The tools we now possess—AI, satellite communication, biotechnology—are more powerful than anything our ancestors imagined. But the key to our collective future lies not in tools alone, but in how we use them, and who we become as a result.

This is where lifelong learning becomes our moral compass. By committing to continuous personal growth, empathetic dialogue, and collaborative problem-solving, we can transform crisis into opportunity and division into shared purpose.

As we stand at this pivotal moment in human history, let us lean into the transformative power of lifelong learning. Let us lead with curiosity instead of fear. Let us listen deeply, think critically, and act boldly. Let us build bridges between cultures, disciplines, and generations.

The journey of lifelong learning is not just about what we know—it is about who we are becoming. It is about cultivating the courage to ask better questions, the empathy to hear unfamiliar truths, and the creativity to envision a better tomorrow. In this pursuit, we grow not only as individuals but also as a global civilization—more aware, more connected, and more capable of shaping a future worthy of our highest ideals.

As we move forward on this journey, let us do so with hope, determination, and a deep appreciation for the extraordinary gift of learning. In learning, we discover not only knowledge but also transformation; not only answers but deeper, more illuminating questions. We find not just tools to navigate the world, but also the means to continually reshape our societies and ourselves in alignment with our highest ideals and deepest values.

The future remains uncertain, but through lifelong learning, we gain the tools to face any challenge. We build the resilience to endure adversity, the insight to uncover opportunity in difficulty, and the wisdom to guide progress toward a more just, sustainable, and flourishing world.

Let us recommit to this sacred journey of lifelong learning. Let us greet each day with curiosity and wonder—eager to grow, to contribute, and to stretch our understanding of what's possible. In doing so, we not only enrich our own lives, but we also shape the collective story of human progress, imagination, and potential.

The path of lifelong learning is open to all. It is not a destination but a journey into ever-expanding horizons—a pursuit that invites us to become fuller, wiser, and more compassionate versions of ourselves. In walking this path, we help create a world that mirrors our aspirations instead of our fears.

As we close this reflection on lifelong learning, let us carry forward the spirit of curiosity, growth, and purpose that has guided us thus far. Let us continue to explore, to question, to collaborate, and to create. In this journey, we find not just knowledge, but wisdom; not just skill, but meaning; not just information, but transformation.

The future belongs to those who never stop learning. May we all remain lifelong learners—forever curious, forever growing, and forever committed to shaping a better world. In doing so, we honor the very gift of life and take part in the grand narrative of human evolution and enlightenment.

In today's world, where disruption is the norm and complexity is ever-increasing, the value of wisdom and ethical decision-making cannot be overstated. Knowledge acquisition is essential, but the application of that knowledge—wisely and ethically—is what will determine our future.

From climate change and pandemics to widening inequality and the unchecked power of technology, the challenges of our time require more than innovation; they require moral courage and philosophical clarity. The rise of artificial intelligence, for instance, demands answers to profound questions: What does it mean to be human in an age of machine intelligence? How do we safeguard human autonomy, privacy, and dignity when faced with algorithmic systems capable of predicting and influencing behavior?

Gene editing breakthroughs like CRISPR-Cas9, while offering hope for curing hereditary diseases, also bring ethical concerns about "designer babies," genetic inequality, and the limits of human intervention in evolution. The question is no longer what we can do, but what we should do.

To meet these dilemmas with wisdom, lifelong learning must be interdisciplinary—bridging STEM fields with ethics, history, philosophy, and spiritual insight. Great thinkers—from Socrates to Confucius, from Simone Weil to Albert Einstein—have reminded us that wisdom is not just about understanding the world, but about understanding ourselves within it.

In an interconnected world, cultural literacy and global awareness are critical. As economies, technologies, and ecosystems transcend national borders, so must our thinking. The leaders of tomorrow must be fluent not only in data science but also in diplomacy, empathy, and global ethics. According to the OECD, future-ready education must prioritize collaboration, intercultural competence, and civic engagement.

The COVID-19 pandemic starkly illustrated the twin truths of interdependence and inequality. It accelerated remote work and technological adoption, but also exposed systemic vulnerabilities in healthcare, education, and economic equity. The crisis offers a chance to rebuild—not just the systems, but also the values underpinning them. If we truly learn from it, we can create more resilient, equitable structures for future generations.

Climate change remains a defining crisis of the 21st century. According to the Intergovernmental Panel on Climate Change (IPCC), even limiting global warming to 1.5°C will require rapid and far-reaching transitions in land, energy, industry, and infrastructure. Education is central to this transformation. UNESCO recognizes environmental literacy as one of the pillars of global sustainable development. We must prepare not only with mitigation strategies but also with adaptation, resilience, and regenerative thinking.

The future of humanity depends not only on the technologies we build but also on the values we live by—and on the learning systems that shape both. Whether developing equitable AI, rethinking global governance, or designing sustainable economies, our actions must be anchored in a deeply humanistic approach to learning—one that affirms dignity, equity, and collective flourishing.

The tools at our disposal are unprecedented. But the real question is how we choose to use them—and who we choose to become in the process.

This is the promise of lifelong learning: to guide us not only to professional competence but also to planetary stewardship. To cultivate the capacity not only to adapt, but also to transform. To empower not only individual achievement but shared purpose and collective progress.

We are at a crossroads. The choices we make today will echo through the lives of our children and the fate of our planet. Will we rise to meet the moment? Will we wield knowledge as a force for good? Will we dare to dream and build a future worthy of our highest human potential?

Let us answer yes—and let us learn forward, together.

The path forward is not predetermined. It will be shaped by the choices we make, individually and collectively. And those choices will be informed by our understanding of the world, our values, and our capacity for wisdom and compassion – all of which can be cultivated through lifelong learning.

In this light, lifelong learning is not just a personal pursuit but also a collective imperative. It is the means by which we can equip future generations and ourselves with the knowledge, skills, and wisdom needed to navigate the complexities of our time and to shape a better future.

Reflections on Life, Society,
and the Future

As we conclude this reflection on lifelong learning, let us recommit ourselves to this noble pursuit. Let us approach each day with curiosity and wonder, ready to learn, to grow, and to contribute. Let us seek out diverse perspectives, challenge our assumptions, and never stop asking questions. Let us strive to integrate knowledge from across disciplines, to think critically and creatively about complex problems, and to act with wisdom and compassion.

In doing so, we not only enrich our own lives, but we also play our part in the grand, ongoing story of human progress and potential. We become active participants in shaping the future rather than passive observers of change. The journey of lifelong learning is open to all who choose to embark upon it.

As we move forward on this journey, let us do so with hope, determination, and with a deep appreciation for the incredible gift of learning.

So, let us embrace this journey of lifelong learning with all our hearts and minds. Let us approach each day as an opportunity to learn, to grow, and to contribute. In doing so, we honor the gift of life itself, and we play our part in the grand, unfolding story of human potential and progress.

For lifelong learners, the rise of artificial intelligence (AI) presents both opportunities and disruptions. A 2024 McKinsey report estimates that up to 30% of current work hours in the U.S. economy could be automated by 2030. On one hand, AI may replace routine tasks, displacing workers in certain industries. On the other, it enables new opportunities for those who can collaborate with intelligent systems to solve complex problems and drive innovation.

To thrive in this AI-augmented future, individuals will need to focus on developing skills that complement, rather than compete with, artificial intelligence. These include creativity, emotional intelligence, complex problem-solving, and interdisciplinary thinking. UNESCO's 2023 Future of Education report emphasizes that critical thinking, ethical reasoning, and digital literacy are now core competencies for lifelong learning.

Another critical area for lifelong learning is environmental sustainability and climate change mitigation. As the impacts of climate change become more severe, there will be an increasing need for individuals with expertise in renewable energy, sustainable agriculture, green building design, and environmental policy. According to the International Labour Organization, transitioning to a green economy could create 24 million new jobs globally by 2030. Moreover, every citizen will need a basic understanding of climate science and sustainable practices to make informed decisions in their personal and professional lives.

The field of biotechnology is another area ripe with potential for transformative breakthroughs. Advances in gene editing, synthetic biology, and personalized medicine hold the promise of curing genetic diseases, extending human lifespans, and even creating new forms of life.

However, these technologies also raise profound ethical questions. For instance, the 2018 case of gene-edited babies in China sparked global debates, leading to new regulatory frameworks under UNESCO and the WHO to prevent unethical experimentation. Lifelong learners in this field must stay abreast not only of the latest scientific developments but also of the evolving ethical and legal landscape.

Chapter 67
Reflections on Life, Society,
and the Future

Space exploration is yet another frontier. NASA's Artemis program and private efforts like SpaceX's Starship are actively laying the groundwork for lunar bases and potential human missions to Mars. This opens opportunities in astrophysics, aerospace engineering, planetary science, and space law. The prospect of colonizing other planets raises philosophical and ethical questions about resource use, environmental stewardship, and the long-term goals of human expansion.

In the realm of social sciences and humanities, individuals who can interpret and guide society through rapid change will be essential. Experts in sociology, psychology, anthropology, and philosophy are needed to explore the social and cultural implications of new technologies. The ability to understand and bridge cultural differences will be a critical asset in our interconnected world.

As we consider these trends, it's clear that interdisciplinary thinking is more important than ever. The most urgent problems we face – from pandemics to AI ethics to ecological collapse – demand solutions that draw on diverse fields. MIT and Stanford have both launched interdisciplinary centers focused on "human-centered AI," showcasing a global shift toward integrated knowledge systems. Lifelong learners who can synthesize information from multiple domains will be best positioned to contribute meaningfully.

Moreover, the pace of change will only accelerate. This underscores the importance of metacognition – the ability to learn how to learn. Cognitive science research confirms that strategies such as spaced repetition, active recall, and reflection significantly enhance long-term learning outcomes. These skills will be essential for staying current and agile in the 21st century.

The digital revolution is also transforming how we learn. Online platforms, immersive VR/AR learning environments, and AI-powered tutoring systems are making education more accessible and personalized. However, the OECD warns that without proper digital literacy, disparities in access and outcomes will grow, particularly in low-income and rural communities. This calls for educational equity and new approaches to lifelong digital inclusion.

In this rapidly evolving context, it's crucial to remember that learning is not just about accumulating knowledge or skills. It's about personal growth, self-awareness, and cultivating meaning. In an age of complexity and uncertainty, the pursuit of wisdom, resilience, and purpose is more critical than ever.

Lifelong learning supports this pursuit. It deepens our understanding of others and ourselves. It cultivates empathy and humility. And by applying what we learn for the greater good, we find purpose and connection.

Yet the benefits of lifelong learning extend beyond the individual. The World Economic Forum has identified lifelong learning as a cornerstone for inclusive and sustainable development. When we nurture a society of lifelong learners, we create the foundation for a more innovative, compassionate, and informed world.

In conclusion, the journey of lifelong learning holds immense promise for individuals and humanity. As we confront unprecedented transformations in technology, society, and the environment, our ability to adapt, learn, and grow will define our collective future.

Let us embrace this journey with determination and joy. Let us treat every day as a new chance to expand our minds and deepen our humanity. And let us remember that each moment of learning is a step forward – not just for us, but also for our communities and the world.

The future belongs to those who never stop learning. May we all be lifelong learners – forever curious, forever growing, and forever committed to understanding and shaping the world around us. In doing so, we honor not only the gift of life but also the boundless potential of human consciousness and progress.

Section 9: The Nature and Importance of the Soul

Chapter 68
The Nature of the Soul

Most people believe that every human has a body, mind, and spirit. The body is the visible, physical vessel that houses our being. The mind can be understood as the self, the psyche, or the ego — the seat of our consciousness and identity. The spirit, meanwhile, is the unseen part that connects us to the divine — to God.

Identical twins have exactly the same DNA but different souls and spirits, reflected in their unique personalities, talents, likes, and dislikes. Twins are typically raised by the same parents in the same environment, eating the same food, and sharing many experiences. What distinguishes them is not their biology, but their souls — each bearing a distinct purpose and trajectory. Modern psychology confirms that environmental sameness cannot fully account for behavioral and temperamental differences in twins.

A helpful analogy is to think of your brain as a personal computer and your soul as a removable storage device that holds your personal data. The computer has essential software, and more can be installed over time. When in use, it's like your mind being active. The Internet — facilitating connection and flow — resembles the spirit that enables communication with God and others. When the computer is shut down, the data — the soul — still exists. Artificial Intelligence, no matter how advanced, lacks both a soul and consciousness. While it can simulate intelligence, it does not possess self-awareness or moral agency.

Unlike machines, which lack a heart, the soul intertwines mind and heart. It is the seat of belief, emotion, intention, and memory. The soul is the complete inner activity of a person — the invisible essence that gives rise to identity, experience, and growth.

Imagine the entirety of your being: physical, mental, emotional, spiritual, and soulful. Your body moves through the world. Your brain enables thinking. Your emotional body stimulates feelings such as warmth, peace, or sadness. In the Bible, the heart is portrayed as the spiritual center — the place of will and desire. Your mind determines how you act, adapt, and succeed. Your spirit is your connection to divine truth. Your soul is the director, the ultimate observer, and the essence of your eternal self.

The mind is a tool of the soul — not the master. It creates logic, solves problems, and organizes thought. But the deepest wisdom lies beyond the mind, in the stillness of the heart. There, the soul speaks.

Across religious and philosophical traditions, the soul is seen as the immortal essence of a person. It carries identity, values, and memory. Many ancient cultures, including the Egyptians, Hindus, and Greeks, believed in a soul that survives death and continues its journey in another realm. As both a scientist and engineer, I had a profound personal experience that confirmed this for me. I witnessed my mother's soul leaving her body through the right side of her mouth five minutes after her final breath. That moment gave me unshakable certainty that the soul survives the body.

The human spirit is a powerful force within us — a source of courage, imagination, empathy, and strength. It fuels our dreams and enables connection. The peace of God's presence can only be felt through the spirit.

I had my own unexpected spiritual encounter that reinforced my belief in God. While standing on the shore of the Sea of

Chapter 68
The Nature of the Soul

Galilee — at the place where Jesus appeared to his disciples for the third time — I was filled with divine energy. I felt the spirit of God enter me. My ego dissolved. I couldn't perceive any separation between my mind and God. This moment left me with absolute conviction that God is real and has always walked with me.

Every human begins life at conception. According to a 2008 position paper by the American College of Pediatricians, "human life begins at fertilization." This view aligns with embryological science, which states that the zygote formed at fertilization is a genetically unique human organism capable of further development given the right conditions.

God gave you a body, a spirit, and a soul. Your soul was within you even in the womb. This is your soul's journey; your body is a temporary vessel. The soul is the very essence of your being — the part that makes you, you. Your spirit is what allows communion with God.

Your body is the visible part that grows, senses, and expresses. It's the container — the vessel — of the soul. Just like a jewel is protected in a box, the soul is held within the body.

The Bible affirms this truth. Psalm 139:13-16 (NIV) reads:

"For you created my inmost being; you knit me together in my mother's womb. I praise you because I am fearfully and wonderfully made; your works are wonderful, I know that full well…"

The soul is eternal, even as the body ages and fades. It lives on because it is from God, who is infinite and timeless.

Genesis 2:7 tells us, "Then the Lord God formed a man from the dust of the ground and breathed into his nostrils the breath of life, and the man became a living soul." This verse shows that our souls are not metaphorical — they are divine gifts, real and vital.

Jesus echoed this in Matthew 10:28:

"Do not be afraid of those who kill the body but cannot kill the soul. Rather, be afraid of the One who can destroy both soul and body in hell."

This underscores the eternal value of the soul and the importance of nurturing it.

Your beauty — physical, intellectual, or spiritual — can be a blessing or a burden. It is up to you, and God's guidance, to shape what that beauty means. Hopefully, you were born to parents who protected and nurtured you. If you're fortunate, you'll experience deep love and form a family that reflects that love.

Every soul comes into this world with a purpose, a path, and a hunger for meaning. Let us cherish those we love while they are here, for life in this physical realm is fleeting.

Over the past 25 years, my mission has been to create life-saving technologies and innovations to improve human health. But I believe it's time to go beyond physical health. It's time to help people reconnect with their souls and secure their spiritual well-being through God's grace. Our bodies are temporary. Our souls are eternal. They are connected to God and to the universe.

As we continue exploring the nature of the soul, let us remember: spiritual wellness is not secondary to physical

Chapter 68
The Nature of the Soul

health — it is foundational. In the chapters ahead, we'll explore how to care for the soul, understand our divine purpose, and navigate life's challenges through faith, resilience, and revelation.

Chapter 69
Nurturing Your Soul and
Discovering Purpose

Nurturing Your Soul

So, how can we nurture and care for our souls? Nurturing the soul begins with intentional spiritual discipline. One of the most important practices is spending regular time with God. This means setting aside daily moments to pray, reflect, read the Bible, and simply rest in His presence. According to multiple studies, including research from the Barna Group and Pew Research Center, individuals who engage in regular spiritual practices report higher levels of life satisfaction and emotional resilience. When we do this, we feed our souls with the spiritual nourishment they need to grow and thrive. We open ourselves to God's wisdom, guidance, and love, allowing Him to shape us into the people He created us to be.

Another essential way to nurture our souls is to cultivate community with others who support our faith journey. This might mean joining a church group, participating in Bible studies, or spending time with friends who share our values. Proverbs 27:17 says, "As iron sharpens iron, so one person sharpens another." The people around us can have a profound impact on our spiritual growth and overall well-being.

Caring for our souls also involves guarding what we allow into our hearts and minds. Just as we cannot eat junk food and expect to remain physically healthy, we cannot fill our minds with negativity or harmful media and expect our souls to flourish. Philippians 4:8 encourages us: *"Whatever is true, whatever is noble, whatever is right, whatever is pure, whatever is lovely, whatever is admirable—if anything is excellent or praiseworthy—think about such things."* This may mean being selective with the movies, TV shows, music, and books we consume, as well as the thoughts we dwell on daily. Replacing negative self-talk with the truth of God's Word is an intentional act of soul-care.

As we nourish our souls, we begin to bear the *fruit of the Spirit* described in Galatians 5:22-23: *"love, joy, peace, forbearance, kindness, goodness, faithfulness, gentleness and self-control."* These are not just pleasant traits—they are evidence of a vibrant and healthy soul. When these qualities flourish in us, they naturally bless those around us and become a witness of God's transformative power.

Discovering Your God-Given Purpose

One of the most fulfilling aspects of a life devoted to God is discovering the unique purpose and calling He has placed on your life. The Bible tells us in Jeremiah 29:11: *"For I know the plans I have for you," declares the Lord, "plans to prosper you and not to harm you, plans to give you hope and a future."* This verse affirms that God's intentions for each person are rooted in love, hope, and purpose.

A powerful way to begin discovering your purpose is by paying close attention to your passions—the activities that energize you and bring you joy. These joys are often divine hints at the direction your soul is meant to take. Do you love creating art, writing, caring for others, or solving problems? As 1 Peter 4:10 states, "Each of you should use whatever gift you have received to serve others, as faithful stewards of God's grace in its various forms." Your interests are often God's way of equipping you to serve the world in a specific way.

Another way to uncover your purpose is by observing the needs around you and asking God how He might use you to address them. Is there a social issue that breaks your heart? A group of people you feel called to serve? These stirrings often point toward your mission. As theologian Frederick Buechner famously wrote, "Your vocation in life is where your greatest joy meets the world's greatest need."

God often aligns our talents with the needs of others. For example, those drawn to healing may feel called to healthcare or counseling; those with empathy may serve the brokenhearted; those with leadership may help build communities of justice and support. When you ask God to show you His purpose for your life, you may be surprised by the clarity and opportunities that begin to unfold.

It's important to note that discovering your purpose is not a one-time event but an evolving process. It often takes prayer, patience, and faithful steps forward. Sometimes your calling is revealed through challenges, failures, or unexpected detours that later make sense in hindsight.

The Example of Moses

Of course, discovering your purpose is not always easy. There will be seasons of doubt, confusion, and discouragement when the path ahead feels unclear. But remember, God's presence is constant—He is always with you, guiding you, and giving you the strength and wisdom to keep moving forward.

One of the most powerful examples of someone discovering and fulfilling their God-given purpose is the story of **Moses**, a key figure in the Old Testament. Moses was born during a time when the Israelites were enslaved in Egypt. Fearing the growing population of the Israelites, Pharaoh ordered the death of all Hebrew baby boys (Exodus 1:22). To save him, Moses' mother placed him in a basket and set him afloat on the Nile River.

Through God's providence, Pharaoh's daughter discovered the infant and raised him as her own in the royal household (Exodus 2:5–10). Moses grew up with the privileges of an Egyptian prince but God shows Moses hints of his Hebrew roots. As he matured, he became increasingly aware of the suffering of his people.

One day, Moses witnessed an Egyptian beating a Hebrew slave. Outraged, he intervened and killed the Egyptian (Exodus 2:11–12). Fearing the consequences, Moses fled Egypt and settled in the land of Midian, where he spent forty years as a shepherd—far from Egypt and seemingly far from the purpose God had planned for him.

But God had not forgotten Moses. One day, while tending sheep near Mount Horeb, Moses encountered a burning bush that was not consumed by the fire (Exodus 3:2). There, God revealed Himself and called Moses to return to Egypt to lead the Israelites out of slavery.

At first, Moses resisted. He felt unqualified, fearful, and unsure of his ability to lead. He said to God, *"Who am I that I should go to Pharaoh and bring the Israelites out of Egypt?"* (Exodus 3:11). But God responded with a powerful promise: "I will be with you" (Exodus 3:12).

With God's assurance and power, Moses returned to Egypt. He confronted Pharaoh repeatedly, and through a series of miraculous plagues—ten in total, each demonstrating God's authority over Egypt's gods—Pharaoh was eventually compelled to release the Israelites (Exodus 7–12).

Moses then led the Israelites out of Egypt, through the parted Red Sea, and into the wilderness (Exodus 14). Over the next forty years, he served as their spiritual leader, lawgiver, and intermediary between the people and God. He received the Ten Commandments on Mount Sinai (Exodus 20) and guided the Israelites through seasons of rebellion, hardship, and divine instruction.

Moses was not perfect—he struggled with anger, doubt, and even disobeyed God at times (Numbers 20:7–12). But he remained faithful to the calling God placed on his life. His obedience and leadership shaped the identity of Israel as a nation and laid the spiritual foundation for generations to come.

Moses' story teaches us that discovering and fulfilling our God-given purpose isn't always straightforward. It may require leaving behind comfort, confronting past failures, and stepping into unfamiliar territory. But it also shows us that when we trust God and follow His leading, He equips us with everything we need—even when we feel inadequate.

So don't be afraid to dream big. Don't hesitate to ask God to reveal the unique purpose He has for your life. And don't be discouraged by setbacks or uncertainty. Remember, every step in your journey—including the detours—is part of God's plan.

You are not alone in this walk. God is with you—guiding, strengthening, and cheering you on. He has a purpose for you that exceeds what you can currently imagine, and as you walk in faith, you'll experience the joy and fulfillment that come from living in alignment with His will.

As we continue exploring the nurturing of the soul and the discovery of purpose, let us remember that purpose is not static—it evolves. What God calls you to today may grow and change over time. What matters is not having all the answers, but being open to God's leading.

In the next chapter, we'll dive deeper into the exploration of faith and spirituality—and how these two vital elements can enrich our understanding of the soul's journey.

Chapter 70
Exploring Faith & Spirituality

My Profound Experience at the Sea of Galilee

I would like to share with you the most profound moment of my life—one that occurred when I least expected it. On the morning of April 17, 2006, while continuing my visit to the Sea of Galilee, I asked the hotel clerk to arrange a taxi to take me to the church commemorating where Jesus fed 4,000 people, and then to Capernaum, the town that served as the headquarters of Jesus's ministry. The evening before, I had visited the Mount of Beatitudes, traditionally believed to be the site of the *Sermon on the Mount*—Jesus' most famous teaching and one of the most influential discourses in human history.

That morning, all the taxis I had seen were Mercedes, and since I had taken one back to the hotel the previous night, I had a general sense of the pricing. When I asked the hotel clerk to hail a taxi for a fair price, he offered to drive me himself. I agreed.

I expected a Mercedes, but his car turned out to be an old, beat-up vehicle cluttered with belongings in both the front and back seats. I could have refused or renegotiated the fare, but I sensed he could probably use the income, so I chose to continue the trip as planned.

After visiting the Church of the Loaves and Fishes in Tabgha, the driver mentioned a small church nearby on the way to Capernaum and asked if I'd like to stop there. I said yes.

The church was very small, peaceful, and empty—no other tourists were present. It was nestled quietly along the northwestern shore of the Sea of Galilee, right at the water's edge. Inside, the church was simple, with a central rock and birds flying freely within. After a few moments inside, I stepped back outside to take in the tranquil scene.

I asked the driver to take a photo of me as I walked out onto one of the partially submerged heart-shaped stones—about seven or eight feet into the calm, clear waters of the Sea of Galilee. After he snapped the photo, I bent down and gently touched the water with my fingertip.

The moment my skin made contact with the water, an overwhelming surge of energy entered my body. I was immediately filled with a deep and total sense of peace—something far beyond emotion or thought. I had never experienced anything like it before or since. It was as if the Holy Spirit had entered me, and I knew without a doubt that I was feeling the tangible presence and peace of God. The tingling in the finger that touched the water remained for nearly 30 minutes afterward.

That sense of peace—the purity, stillness, and connection to something divine—was the most profound spiritual experience of my life. Had I turned down the hotel clerk's unexpected offer, I would have missed that encounter entirely. It was a reminder that God's grace often meets us in the unlikeliest of ways.

Two weeks later, back in the United States, I began researching that little church. To my astonishment, I discovered it was the Church of the Primacy of Saint Peter, located in Tabgha. Built on the very shoreline of the Sea of Galilee, it marks the sacred site where Jesus appeared to His disciples for the third time after His resurrection (John 21).

It was here that Jesus helped them catch 153 fish, prepared a meal of bread and fish, and restored Peter, saying, *"Feed my sheep."*

The rock I had stood on was one of twelve heart-shaped stones placed along the shoreline to commemorate the Twelve Apostles. I had unknowingly stood at one of the most sacred sites in Christian tradition—where the resurrected Christ reconnected with His closest followers and reaffirmed Peter's calling as the foundational "rock" of His Church.

The Church of the Primacy has welcomed notable pilgrims over the years, including Pope Paul VI in 1964 and Pope John Paul II in 2000. Yet, it hadn't appeared on my tourist map, and I had never even heard of it before. It felt like a divine appointment—one that God had orchestrated outside of my plans.

That day, I experienced firsthand the "peace of God, which surpasses all understanding" (Philippians 4:7). I felt like I was one with God—reconnected, realigned, and utterly still inside. The profundity of that moment—standing where Jesus appeared after conquering death—left a lifelong imprint on my soul.

Had I ignored the driver's suggestion or dismissed the church as unimportant, I would have missed the most spiritually transformative moment of my life. My soul was renewed by the living waters of Galilee, and my faith was reinvigorated in a way that no words can fully capture.

It was a sacred encounter that reminded me that God's grace often comes disguised in humble moments. That single experience not only uplifted my soul but also fortified me for the journey still ahead—with peace, with clarity, and with a deep awareness that I am not walking alone.

The Gospel Account of Jesus' Third Appearance

The Bible verses recounting Jesus's third appearance to his disciples after his resurrection at the Sea of Galilee can be seen in the Gospel of John 21:1-25. This account provides the context for the profound spiritual significance of the location where I had my unexpected encounter with God's peace.

Unifying Metaphor: Life as a Spiritual Walkabout

Just as the biblical patriarchs—Abraham, Moses, and even Jesus himself—were called into the wilderness to encounter the Divine firsthand, away from the constraints, noise, and distractions of society, my own life has unfolded as a kind of perpetual spiritual walkabout. With each new vista that has transfixed my senses—whether through travels across continents or the deep inward turns of self-reflection—another veil has lifted, revealing deeper clarity about the sacred architecture of existence.

From the hallowed shores of the Sea of Galilee, where the living waters first awakened my soul, to ancient desert petroglyphs etched by Indigenous societies millennia before the biblical scrolls were codified, this walkabout continues to affirm the same universal truths echoed across all enlightened traditions—that we are infinite souls journeying through space and time toward union with our Divine Source.

The temporary wanderings of this earthly exile—punctuated by oases of insight, rest, and divine encounter—constitute merely checkpoints along a much greater homeward sojourn.

The challenge lies in remaining faithful to the course, even as mirages of material wealth, power, and fleeting pleasures offer detours so deceptively alluring that the disoriented soul can spend years—sometimes entire lifetimes—lost in the labyrinth of egotism and spiritual apathy.

It requires the eyes of unshakable faith, honed through adversity, to recognize the true path. It requires one to ignore distorted human maps of success and self-worth and instead be guided by the moral compass etched into the soul by the Celestial Navigator. Such has been the shape of my own personal exodus—a winding ascent marked by euphoric moments of divine intimacy, counterbalanced by painful descents into doubt and spiritual drought.

Exploring the Spiritual World

I've always been drawn to places of profound spiritual and physical elevation—from the Bible to the mountains. Among them, Mount Everest has long captivated my imagination and spirit.

To reach the start of the Everest Base Camp trail, I had to fly from Kathmandu, Nepal's capital, to Lukla—home of what is widely considered the most dangerous airport in the world. From the small aircraft's window, I could see the snow-capped summit of Everest itself, wind whipping snow off its peak into a majestic triangular plume stretching eastward, as though the mountain were breathing.

Lukla's Tenzing-Hillary Airport features a runway just 527 meters (1,729 feet) long and 20 meters (65 feet) wide, sloping steeply at a 12-degree gradient. Perched at 9,334 feet (2,845 meters) above sea level in the Himalayan cliffs, the margin for error is nearly nonexistent. As a private pilot myself, I had immense respect for the pilots who navigate such conditions. Too low, and the aircraft crashes into the mountainside; too high, and it overshoots the precipitous runway with concrete structures at the end of the strip.

From Lukla, the trek towards Everest Base Camp began—a journey that tested my endurance and stirred my spirit. The trail wound through narrow paths with cliffs on one side and swaying rope bridges suspended hundreds of feet above glacial rivers, with each step accompanied by panoramic Himalayan vistas that seemed almost surreal in their beauty.

While in Kathmandu, I also explored sacred Buddhist sites, including the iconic Boudhanath Stupa—a UNESCO World Heritage site and one of the largest and holiest stupas in the world. As I joined a few pilgrims circling its white dome clockwise in reverence, I could feel the centuries of prayer, devotion, and spiritual resonance embedded in its stones. I also sense the energy under my feet. Said to house relics of a past Buddha, this monument radiated an atmosphere of serenity and wisdom.

Whether standing beneath Himalayan peaks or walking the sacred mandalas of Nepal, I felt again that quiet calling—the soul's yearning for truth, connection, and meaning. Each location, each journey, offered more than scenic beauty—it became a spiritual checkpoint, a reminder of the shared longing within all humans to touch the eternal.

My Journey from Buddhism to Christianity

When I was young, I spent a week or two living as a novice Buddhist monk in a monastery alongside my father, uncles, and cousins. We shaved our heads, donned saffron robes, and committed ourselves to the principles of Buddhist monastic life, taking part in a temporary vow of renunciation. It was an educational and peaceful experience—one that helped me understand and appreciate the core teachings of Buddhism, particularly the Four Noble Truths, which explore the origins of suffering and the path to liberation and enlightenment.

My parents were devout Buddhists for over 80 years, deeply rooted in their faith and community. But in the summer of 1981, while I was on an engineering internship with U.S. Borax in the Mojave Desert, they surprised me. For the first time, they encouraged me to attend a Christian spiritual retreat being held by a Burmese Christian group at Lake Cachuma, near Santa Barbara, California.

That retreat turned out to be a turning point in my spiritual life. On the drive home to California City along the Pacific Coast Highway (Highway 101), my mind replayed many of the Bible verses I had heard over the weekend. When I reflected on Jesus' words to Peter in Matthew 26:34— *"Before the rooster crows, you will disown me three times"*—I was struck with overwhelming emotion and broke down in tears, though I was also filled with profound peace. At that moment, I surrendered my life to Christ and accepted Him as my Lord and Savior. Truly, God moves in mysterious ways, and His call often comes in the most unexpected of moments and places.

For decades after that initial moment of being born again on Highway 101, my sister Grace and I worked faithfully to share the Good News of Jesus Christ with our Buddhist parents. Grace regularly took them to Chinese-language church services and Bible studies. However, year after year, they remained polite but firmly committed to the spiritual path they had followed for most of their lives.

That changed in the final years of my father's life. A few years before his passing from COVID-19, my father—by then in his 80s—opened his heart. About two months before died, I vividly remember sitting with him when he said, through tears, that he felt Jesus was calling him home. It was one of only two times in my life I saw him cry—the first being at his brother Frank's funeral. He thanked me. I told him it wasn't necessary. I thanked him—for his love, his sacrifices, and the life he gave our family. I told him I loved him.

On Earth Day, April 22, 2020, six days after contracting COVID-19, my father passed away peacefully in a hospital a few miles from a five-star nursing home, where he lived due to mobility issues. I believe that day was no coincidence. Less than a year earlier, we had lost my brother—a gifted biotech and e-commerce entrepreneur—to heart disease at UC San Diego Hospital. I miss them both dearly.

Two years later, I saw a book on my mom's bookshelf titled 《标竿人生》 (The Purpose Driven Life) by Rick Warren—translated into Chinese. I asked her to read the title aloud, and she did with interest. I thought her dementia might prevent her from understanding much, but I was wrong. She had only forgotten some English words, not Chinese. When I asked my sister Grace how the book had arrived in our mother's home, she said she didn't know.

Then I remembered—I had bought one copy for each of my parents nearly two decades ago, quietly hoping it might plant a seed. The seeds we plant in faith may take years to bloom, but God's timing is always perfect.

To my amazement, my mom read the entire book, sometimes aloud, page by page, over the course of two months. I even took videos of her reading. She also read passages from the Bible, including John 3:16, which she particularly loved:

"For God so loved the world, that he gave his only Son, that whoever believes in him shall not perish but have eternal life."

During a family celebration in Grace's backyard, I shared with my sister how Mom seemed increasingly drawn to the teachings of Christ. Grace gently asked our mother if she would like to accept Jesus as her personal Savior. Mom answered clearly and confidently: "Yes." Tears of joy flowed freely. Her spiritual rebirth was a moment of pure grace—a gift we had prayed for over decades.

As we continue exploring faith and spirituality, it's important to remember that every soul's journey is unique. Mine took me from the quiet halls of a Buddhist monastery to the transformative embrace of Christ, but the common threads—love, compassion, purpose, and hope—remain universal.

Chapter 71
Profound Losses and Reflection

When I first visited the desert town of Nazca in southern Peru—famous for the ancient and mysterious Nazca Lines, a UNESCO World Heritage Site etched across the arid plateau—I experienced an unexpected and deeply moving encounter. After going up to the hotel rooftop terrace for breakfast, I noticed a large banner celebrating the 100th anniversary of Rotary International. As a longtime Rotarian back home, I was stunned by the coincidence.

Intrigued, I walked into a nearby shop and asked how I could connect with the local Rotary Club. One inquiry led to another until I was introduced to some of the just 15 Rotarians who made up the entire Nazca chapter. I had never encountered such a small, grassroots club—a stark contrast to my home chapter in Saratoga, California, which had over 100 members.

To my surprise, they welcomed me and invited me to attend their weekly meeting. Afterwards, several members offered to take me on a tour of their current service projects. I accepted—and what I witnessed over the next several hours was truly transformative.

One site, in particular, left an indelible mark on my heart. We visited an impoverished community on the outskirts of Nazca, where a single elevated water tank provided the only clean water source. As we drove up the dusty hill, I noticed many small crosses embedded in the ground. When I asked about them, the Rotarians told me that many infants and children had died due to the lack of clean water and access to healthcare. Their families, unable to afford burial in city cemeteries, laid them to rest along the roadside.

They then introduced me to some of the families in the neighborhood. I saw how they fetched water from the tank and stored it in 55-gallon drums, some of which had visible mosquito larvae floating in them. The standing water posed serious health risks, including vector-borne diseases like dengue and malaria, which are prevalent in regions with poor sanitation infrastructure. The families explained how children often had to leave school early to begin working in fields or informal labor just to support their households.

Quietly, I learned that some of these children were involved in exploitative or dangerous work. In that moment, I was overcome by the most profound sense of grief I had ever felt for the injustice and suffering in the world. I couldn't hold back tears—I cried for nearly half an hour in front of those around me, something I had never done in public before. That moment altered me permanently. I resolved then and there to help the children of Nazca.

Despite its size, the Nazca Rotary Club celebrated Rotary International's centennial with pride. A parade marched through the town square, followed by a celebration attended by several hundred residents. I was deeply honored to be invited to carry the Rotary anniversary banner in the parade and assist in raising their flag in the plaza. That evening, they had a band perform, and I was asked to give a speech, which I gladly did. None of these plans were on my itinerary—it was a divine appointment I would never forget.

When I returned to the U.S., I reached out to Santa Clara University's Campus Ministry, explaining that I wanted to offer three fully funded scholarships for students to travel to Nazca to teach English. Campus Ministry connected me to the Ignatian Center for Jesuit Education, which sent out a campus-wide announcement. Many students applied, and together with the staff, I interviewed the applicants.

Meanwhile, I communicated with the Rotarians in Nazca via email—with translation support from bilingual volunteers on both sides—to plan logistics and housing. In the end, we selected three passionate students, who traveled to Nazca that summer. They taught English to over 250 local children across the city, giving them not just language skills but hope.

When the students returned, they shared their experiences with me over a celebration dinner at a Peruvian restaurant. Later, I was invited to speak at Santa Clara University's Board of Fellows quarterly meeting to present the Nazca Project.

Father Paul Locatelli, then-President of Santa Clara University, later invited me to serve on the inaugural advisory board of the Ignatian Center. I accepted and served for 11 years, continuing through the presidency of Father Michael E. Engh, S.J. It was a humbling and rewarding opportunity to help shape the university's approach to faith-based global service.

My awakening to Christ-centered service was only deepened by another journey—this time to Savannah, Georgia. There, I came face-to-face with the brutal legacy of the Atlantic slave trade, which operated through Savannah's port for nearly five decades. The city's cobblestone streets and historic structures bore witness to this painful chapter in history. I was deeply moved by the lingering generational trauma it represented.

Despite international laws aimed at preventing the expansion of slavery into new territories, Savannah, Georgia, played a despicable role as a major port in the transatlantic slave trade. It received thousands of enslaved Africans trafficked across the Middle Passage under unimaginably horrific conditions. The scale of human loss, generational trauma, and systemic dehumanization represented by those haunting historical accounts shook me to my core.

Equally sobering was learning about and witnessing remnants of the Trail of Tears. Hundreds of thousands of Native American men, women, and children were forcibly displaced from their ancestral homelands during the 1830s by U.S. government policy. Thousands perished from starvation, exposure, and disease along the treacherous routes—a state-sanctioned act of ethnic cleansing that remains one of America's gravest moral failings.

All of these encounters with injustice—whether historical or contemporary, personal or observed—have profoundly shaped my view of the world and my role in it. They have strengthened my faith, but also compelled me to speak out and work toward building a more just, compassionate, and morally conscious society.

In reflecting on all this, I'm reminded of what the Bible says about those who have "seared their conscience" (1 Timothy 4:2)—individuals who persist in lies, speak in hypocrisy, and suppress truth for personal gain. History warns us what happens when moral clarity is sacrificed for power.

In April 2024, Israeli polls showed that approximately 71% of the public wanted Prime Minister Benjamin Netanyahu to resign. Nonetheless, on July 25, 2024, he addressed a joint session of the U.S. Congress for the fourth time—the most appearances by any foreign leader in American legislative history. This came amid global outcry over Israel's military actions in Gaza under his leadership.

By mid-2024, reports from humanitarian organizations, including UN agencies and human rights monitors, estimated over 45,000 people—primarily women and children—had been killed in Gaza. More than 90,000 were wounded, and 1.8 million civilians were displaced. Basic needs like food, water, and medical supplies were severely restricted. Much of the munitions and weaponry used were supplied by the United States, making it complicit in the devastation.

Netanyahu received 49 standing ovations from members of Congress—beneath the glaring words engraved on the chamber wall: "IN GOD WE TRUST." What does this say about the moral compass of our elected leaders? Roughly 50 lawmakers boycotted the speech—a small but notable act of protest. Meanwhile, as of June 2024, Congressional approval ratings had plummeted to just 16%, reflecting the public's disillusionment with their actions and priorities.

I've previously met Israeli soldiers—both men and women—during my travels in Israel. They were decent, sincere individuals, committed to protecting their homeland. I also have Jewish friends and colleagues, both in the U.S. and Israel, who are thoughtful, accomplished, and compassionate. My critique is not of a people, but of leadership that forsakes justice.

On July 26, 2024, ABC News reported that British Prime Minister Keir Starmer's office stated the U.K. would not obstruct the International Criminal Court's (ICC) legal proceedings. Then, on November 21, 2024, the ICC formally rejected Israel's objections to its jurisdiction and issued an arrest warrant for Benjamin Netanyahu.

Chapter 72
Grappling with Major Injustices in Life

The Global Financial Crisis of 2007–2009, also known as the Great Recession, stands as a stark reminder of economic turmoil's far-reaching consequences. Originating in the U.S. housing market collapse and the proliferation of subprime mortgage-backed securities, the crisis triggered a global chain reaction that led to historic financial institution failures and government bailouts. The United States bore much responsibility for this cataclysmic event, which devastated international markets, cost tens of millions of jobs worldwide, and wiped out trillions in global wealth.

Among the casualties was my law firm, Heller Ehrman. With a rich San Francisco history dating back to its 1890 founding and offices spanning the world's major financial centers, Heller had grown into a legal powerhouse boasting 730 attorneys at its peak. In a cruel twist of fate, on September 15, 2008—the very day Lehman Brothers, one of Wall Street's most venerable institutions, declared bankruptcy— Heller Ehrman was set to merge with Lehman-affiliated law firm Mayer Brown. The ill-fated union never materialized, and by year's end, Heller itself had succumbed, filing for Chapter 11 bankruptcy protection.

Heller Ehrman's implosion sent shockwaves through the legal community and had dire consequences for my own company, Iris Biotechnologies. We estimate Heller's misconduct and negligence resulted in damages exceeding $100 million. When Heller submitted bankruptcy filings, Iris Biotechnologies was listed twice among unsecured creditors—indicating the firm was fully aware of its obligations to us but failed to notify us directly.

In a shocking disregard for legal ethics, Bankruptcy Judge Dennis Montali turned a blind eye to the incontrovertible fact that under U.S. patent law, attorneys are prohibited from abandoning a client during prosecution without proper substitution of counsel or explicit consent from the USPTO. The court acknowledged a letter from the USPTO denying Heller Ehrman's request to withdraw representation from Iris Biotechnologies, yet ruled as if Heller's unlawful disengagement letter served as sufficient notice to seek alternative counsel—a decision without legal justification.

The depths of Heller's misconduct were further exposed when the court confirmed that the firm had received a letter from the USPTO indicating that Iris Biotechnologies' application for its "Artificial Intelligence System for Genetic Analysis" patent was on the verge of being granted. However, as a direct result of Heller's failure to respond, the USPTO declared the application abandoned.

Critically, Heller Ehrman concealed all three USPTO communications from Iris Biotechnologies. This egregious breach of fiduciary duty caused irreparable damage— denying Iris the opportunity to protect and commercialize a potentially revolutionary genomic diagnostic platform. At the time, the patent could have conferred a significant first-mover advantage in a nascent field that later exploded in valuation.

In another deceptive move, Heller never informed us that Iris Biotechnologies had been listed as a creditor nor disclosed any deadline for filing a claim. Upon discovering the firm's gross negligence and concealment, we acted swiftly, submitting claims within the statute of limitations. Still, in yet another blow, Judge Montali denied our $100 million claim, depriving us of our constitutional right to a jury trial and shielding the firm from further liability.

This quagmire traces back to September 16, 1999, when Heller Ehrman—with our full consent—shared Iris Biotechnologies' comprehensive business plan with Kleiner Perkins Caufield & Byers, a top Silicon Valley venture capital firm. The following year, in August 2000, Heller filed our patent application for the AI system for genetic analysis. Notably, our chief competitor, Genomic Health, was founded around this time, after we had already disclosed our business plan to both Kleiner Perkins and Sequoia Capital. These VCs later became early investors in Genomic Health, which went public in 2005 and was eventually acquired for $2.8 billion in 2019 by Exact Sciences.

Given the overlapping timeframes and shared investor exposure, questions remain about how Iris' intellectual property may have been compromised or indirectly leveraged.

On January 29, 2009, Law.com published a damning article titled "Heller Ehrman Estate Can't Buy Malpractice Coverage." It reported that Bankruptcy Judge Montali denied Heller's request to purchase three years of malpractice insurance, arguing the $10.2 million price tag outweighed the risk to creditors. This decision further compounded the injustice we experienced. In denying malpractice coverage, the judge effectively foreclosed the ability of clients—like Iris Biotechnologies—to seek restitution for demonstrable harm.

The bitter irony of Judge Montali denying Iris Biotechnologies the chance to present its $100 million claim was not lost on those familiar with the case. This was a story not only of legal negligence but also of systemic failure— where those sworn to protect justice instead entrenched injustice under the guise of expediency.

It was this very ruling that left Heller without the necessary insurance to satisfy the $100 million in damages it owed to Iris Biotechnologies as a result of its egregious and well-documented malpractice.

Determined to seek justice, we appealed the bankruptcy court's patently unjust ruling to both the District Court and the Ninth Circuit Court of Appeals. In a final indignity, Iris Biotechnologies was denied even the basic right to an oral hearing—our pleas for fair, impartial review of the facts fell on deaf ears.

This legal stonewalling compounded the pain of losing our opportunity to recover $100 million in damages stemming from Heller Ehrman's clear malpractice. Even more egregiously, Heller had listed Iris Biotechnologies twice as creditors in its own bankruptcy filings—a tacit admission of financial liability.

Their malpractice devastated our company's ability to survive. On a personal level, I lost approximately $20 million across my brokerage and private investment accounts. It is easy for others to say, "learn, forgive, and move on," but when justice is denied despite overwhelming evidence, healing becomes far more difficult.

As I watch images of innocent families in Gaza mourning the loss of their children, parents, and loved ones—killed without cause—I feel their pain on a spiritual level. When lives are shattered and no accountability is offered, how can true healing begin?

Too often, the perpetrators of injustice remain untouched by the suffering they cause. They go on with their lives, never forced to confront the magnitude of the damage they've inflicted. In the case of Gaza, this cycle of unaccounted injustice has created generational trauma that will reverberate long after the last bomb has fallen.

Now, this war has begun spreading—to Yemen, Lebanon, Iraq, and Iran. The U.S. and some of its allies are complicit, having supplied the very weapons used to carry out the mass destruction. In response, Russia has pledged to supply advanced arms to Israel's opponents, threatening to tip the conflict into a broader war—one that could spiral into World War III.

Injustice has consequences. For my investors, and me the price was financial devastation. For the people of Gaza and conflict zones across the world, the cost is irreparable human loss—loved ones who can never return. Who will be held accountable for these killings? What recourse is available when the world's most powerful nations enable violence instead of stopping it?

The survivors of the Holocaust and their descendants have endured unimaginable suffering. But as I witness the scale of devastation in Gaza, I am forced to ask: are some among them now enacting atrocities of their own? Have the persecuted, in some cases, become persecutors?

This crisis is devastating not just for Palestinians—but for Jewish people as well. Their collective memory, rooted in millennia of exile and tragedy, now risks being tainted by state-led actions that contradict the moral legacy of "Never Again." How much longer will this bloodshed continue, and how much deeper will it scar both peoples?

And what about lawyers and judges who perpetuate injustice—who protects society when the legal system itself is broken? When those tasked with upholding truth instead twist it to shield the powerful, where does a person turn?

Chapter 72
Grappling with Major Injustices in Life

As I reflect on my life's journey, I am filled with gratitude for the blessings I've received—from humble beginnings in Burma to scientific innovation in Silicon Valley. Through every season—whether in academia, biotech entrepreneurship, or spiritual discovery—God's grace and the love of family, friends, and mentors have been my compass.

Grieving is not weakness—it is a holy, necessary process when the soul has endured great loss. It is my faith in God that sustains me.

I have many questions still. Some may never be answered. But I hope that by sharing my story—both the triumphs and the heartbreak—I can inspire others to persevere with passion, purpose, and compassion. Let us never lose sight of what truly matters: faith, family, science, and an unshakable desire to make this world more just and humane.

As I prepare for the next chapter in my journey, I do so with a heart open to what lies ahead. Even when the night feels endless, the dawn is always near.

When I reflect on the grave miscarriage of justice inflicted by Judge Dennis Montali in the legal battle between Iris Biotechnologies and Heller Ehrman, I am reminded of Pontius Pilate—who, despite knowing Jesus' innocence, condemned Him under pressure. In both cases, truth was sacrificed on the altar of convenience, and justice was denied to the innocent.

Chapter 73
Lessons from History and Culture

Life on earth is temporary. Egyptian pharaoh Thutmose III, often called the "Napoleon of Egypt," was one of the most militarily accomplished rulers in history. He led at least 17 military campaigns during his reign in the 15th century BCE and greatly expanded Egypt's empire, recording his victories on the walls of the Temple of Karnak. He accumulated immense spoils and tributes from conquered lands.

Ramses II, known as Ramses the Great, ruled for 67 years during the 19th Dynasty and is widely regarded as Egypt's most powerful and celebrated pharaoh. He commissioned grand temples, including Abu Simbel, and statues that projected his divine status. Despite their earthly wealth and power, these pharaohs believed the afterlife mattered more than the present, and their elaborate tombs in the Valley of the Kings reflect this worldview.

We're going to look at some of the richest people in history to see if you know who they are.

As of September 30, 2024, the richest person in history is not Elon Musk ($244 billion), Jeff Bezos ($197 billion), or Mark Zuckerberg ($181 billion). It is Mansa Musa, the 14th-century emperor of the Mali Empire. Mansa Musa was born around 1280 CE and ruled from 1312 to 1337. During his famous pilgrimage to Mecca in 1324, he traveled with a caravan reportedly numbering 60,000 people, giving away so much gold in Cairo that it caused inflation lasting over a decade. Mansa Musa's wealth is estimated to be over $400 billion in today's dollars, though his fortune is difficult to quantify precisely due to the vast gold reserves of the Mali Empire.

As of December 2024, Elon Musk is the wealthiest living person, with an estimated net worth of US $486 billion, according to the Bloomberg Billionaires Index, and $464 billion, according to Forbes. His fortune comes largely from Tesla, SpaceX, and X (formerly Twitter). Here are some of the wealthiest historical figures before Elon Musk:

- **Genghis Khan (1206–1227)** – founder of the Mongol Empire, controlling the largest contiguous empire in history. While his wealth is hard to quantify, he controlled immense resources.
- **Zhao Xu (1048–1085)** – known as Emperor Shenzong of Song, ruler of China during the Song Dynasty, which accounted for nearly 30% of global GDP at its peak.
- **Akbar the Great (1542–1605)** – ruler of the **Mughal Empire** at its height in India, known for religious tolerance and administrative reforms.
- **Amenhotep III (c. 1388–1351 BCE)** – father of **Akhenaten**, he presided over Egypt's diplomatic and architectural golden age.
- **Augustus Caesar (63 BCE–14 AD)** – Rome's first emperor, who transformed it into a powerful imperial state, reportedly controlling wealth equivalent to 20% of the Roman world's GDP.
- **King Solomon (970–931 BCE)** – the biblical king of Israel, famed for his wisdom and massive wealth, including gold imports equating to hundreds of tons annually, though much of this is symbolic and difficult to historicize.
- **Mansa Musa (1280–1337)** – as mentioned, considered the richest human in recorded history.

Most Americans today have likely never heard of Zhao Xu, Akbar the Great, or Mansa Musa, despite their immense global significance. This highlights how Western-centric education can obscure global history.

Have you ever heard of Emperor Shenzong of Song, who ruled during a time when China's economic and technological output was unmatched globally?

Now consider figures remembered not for wealth but for moral or spiritual influence:

- Jesus Christ, born in Bethlehem, lived humbly, died crucified by the Romans, and is revered for His resurrection.
- Siddhartha Gautama (the Buddha) gave up royal luxury to seek enlightenment and taught liberation from suffering through the Eightfold Path.
- Muhammad, born in Mecca in 570 CE, worked as a merchant and became wealthy through his marriage to Khadijah, but is best remembered for receiving the Qur'an, the holy text of Islam.
- Andrew Carnegie, who said, "The man who dies rich dies disgraced," gave away over 90% of his wealth, funding more than 3,500 libraries worldwide. His adjusted net worth peaked at $372 billion in today's dollars.
- Nikola Tesla, despite revolutionizing modern electricity and inventing the AC system, radio, and induction motor, died in a New York hotel room in poverty. His inventions underpin much of the technology we rely on today.
- Socrates, the ancient Greek philosopher, left no writings himself but taught Plato, influencing Western philosophy's foundations.
- Leonardo da Vinci, the Renaissance polymath, painted the Mona Lisa and conceptualized early prototypes of helicopters, tanks, and solar machines.

- Michelangelo, who sculpted David and painted the Sistine Chapel ceiling, produced works considered masterpieces of human creativity. I saw his sketches—surprisingly more intricate than the finished paintings—and was reminded of my own childhood drawings.

When I was in Chaco Canyon, New Mexico, I saw the remarkable architecture and celestial knowledge of the Ancestral Puebloans, dating from 850 to 1250 CE. Their buildings align with solar and lunar cycles and remain marvels of engineering.

The Great Gallery, located in Horseshoe Canyon, Utah, is one of the most magnificent pictograph panels in North America. It spans nearly 200 feet and features anthropomorphic figures with intricate designs. Some archaeologists believe the art dates back 1,500 to 4,000 years, made by archaic hunter-gatherer societies, predating the Ancestral Puebloans.

The Barrier Canyon Style figures, often appearing as larger-than-life spiritual beings, evoke themes of shamanism and cosmic observation. The Museum of Modern Art and Denver Museum of Nature & Science have featured reproductions of this art, emphasizing its aesthetic and anthropological value.

To see the original Great Gallery, I drove 47 miles on a rugged dirt road, then hiked five hours in temperatures exceeding 100°F (38°C), descending 780 feet (237 meters) at the start and climbing back at the end—all alone. But the effort was worth it. The silence and grandeur of that sacred site moved me deeply.

What is God calling you to do while on this earth? People tend to believe what aligns with their upbringing, experiences, and worldview. When two groups believe that the other's belief system is nonsense, meaningful communication becomes nearly impossible.

In 2023, the Pew Research Center estimated the global religious distribution as follows:

- **Christianity**: 2.382 billion (31.0%)
- **Islam**: 1.907 billion (24.9%)
- **Secular/Nonreligious/Agnostic/Atheist**: 1.193 billion (15.6%)
- **Hinduism**: 1.161 billion (15.2%)
- **Buddhism**: 506 million (6.6%)
- **Chinese traditional religions**: 394 million (5.6%)
- **Ethnic religions** (not otherwise categorized): 300 million (3.0%)
- **Other religions**: around 800 million (8.0%)

Those unaffiliated with any religion now comprise over 1 in 7 people worldwide. As of 2024, with the world population at approximately 8.1 billion, Christianity remains the largest religious group.

The non-religious population stands at around 1.13 billion (16%), including atheists, agnostics, and others who do not identify with any faith. Islam follows closely with 1.6 billion adherents (23%).

Religious conflict has plagued humanity for centuries. One of the most hotly contested religious sites is the Temple Mount in Jerusalem, sacred to both Jews and Muslims. It is the location of the First and Second Jewish Temples, the latter destroyed in 70 AD by Roman forces. Muslims revere the same site as Haram al-Sharif, home to the Dome of the

Rock and the Al-Aqsa Mosque, considered the third holiest site in Islam.

The tensions between Jews and Muslims over this sacred space are part of a conflict stretching back more than 1,300 years, of which the Israel-Gaza war is only the most recent eruption.

Islam was founded by the Prophet Muhammad in the 7th century, with the Hijra (migration to Medina) in 622 CE marking the beginning of the Islamic calendar. Sunnis believe the rightful successors to Muhammad were the first four caliphs, while Shias hold that Ali, Muhammad's cousin and son-in-law, and his descendants were the only legitimate successors.

In 1517, Martin Luther, a German monk and theologian, famously nailed his 95 Theses to the door of the church in Wittenberg, protesting the Catholic Church's sale of indulgences. This act ignited the Protestant Reformation, which emphasized direct engagement with scripture. Many Protestant denominations developed a more literal interpretation of the Bible, especially books like Revelation, shaping apocalyptic theology still prevalent today.

Due to light pollution, we've become disconnected from the cosmos. Before 1920, astronomers believed the Milky Way was the entire universe. That changed when Edwin Hubble confirmed that Andromeda was a separate galaxy in the 1920s. His findings expanded our understanding of the universe and our place in it.

I remember vividly when I first saw Andromeda through a 22-inch telescope. It was a coincidence. While driving from San Francisco to California City during my engineering internship at U.S. Borax in Boron, CA, I decided to take a seemingly shorter route through Los Padres National Forest via Highway 33.

The map suggested a shortcut, but the winding mountain roads slowed my speed to 15 mph in some stretches. After more than eight hours of driving in pitch darkness, I was exhausted. With no cell phones or GPS back then, I knew there'd be no rescue if my car failed.

At a remote 8,800-foot summit, I pulled into a parking lot to rest—only to be greeted by a crowd of angry people shouting for me to turn off my headlights. I had inadvertently stumbled upon a gathering of astronomers. After they calmed down, I learned that they were part of an amateur astronomy club observing the crystal-clear night sky.

Many of them had Celestron 8-inch or homemade telescopes. One kind observer invited me to look through his 22-inch scope focused on Andromeda (M31). I peered into the eyepiece and saw, with my own eyes, a spiral galaxy over 2.5 million light-years away. I forgot my fatigue, energized by the awe of witnessing another galaxy.

Most amateur astronomers use telescopes with 8-inch mirrors or smaller. Beginners often use models like the Celestron NexStar 5SE, a Schmidt-Cassegrain telescope with a 5-inch mirror. Coincidentally, I once met the CEO of Celestron on a small boat while traveling upriver in the Philippines—we were the only tourists on board.

Thanks to powerful tools like the Hubble Space Telescope and the James Webb Space Telescope (JWST), astronomers have discovered that the observable universe contains between 200 billion to 2 trillion galaxies. Most galaxies are 1,000 to 100,000 parsecs in diameter (3,000 to 300,000 light-years) and are separated by millions of parsecs (megaparsecs) in distance.

We live in a vast, almost unfathomable universe. In the face of such scale, we are humbled—and reminded of how small yet significant our lives are.

Most people today live under artificial lighting, which has significant consequences for human health, wildlife, and the natural environment. Major contributors to environmental pollution include vehicle emissions, industrial discharges, open burning by farmers and rangers, forest fires, and emissions from residential and industrial chimneys.

Light pollution, caused primarily by outdoor artificial light, disrupts ecosystems, affects human circadian rhythms, and obscures our view of the stars. It is one of the least discussed but most pervasive forms of pollution.

In 2016, scientists published the World Atlas of Night Sky Brightness, a computer-generated map created using thousands of satellite images. The Atlas reveals widespread light pollution across North America, Europe, the Middle East, and large parts of Asia. Only a few remote wilderness regions, such as Siberia, the Amazon Rainforest, parts of the Sahara Desert, and the deep interior of Australia, remain largely free from artificial nighttime illumination.

Singapore, Qatar, and Kuwait rank among the most light-polluted nations, where nearly 100% of the population lives under skies too bright to observe the Milky Way. In fact, over 80% of the world's population—and 99% of Americans and Europeans—can no longer see the natural night sky due to persistent skyglow.

Humans and animals have evolved to rely on the natural light–dark cycle, governed by the 24-hour circadian rhythm. Exposure to light at night disrupts this biological rhythm, interfering with sleep, hormone production, and immune system function. Melatonin, a hormone critical for regulating sleep, is produced in darkness and suppressed by light, especially blue light, which is common in LED lighting and electronic screens.

Chronic exposure to artificial light at night (ALAN) has been linked to sleep disorders, increased stress, anxiety, fatigue, and even elevated risks for obesity, depression, and certain types of cancer. The American Medical Association (AMA) has formally recognized these risks and supports policies to reduce light pollution and encourage further research into its long-term health effects.

Blue light, in particular, emitted from smartphones, tablets, TVs, and LED bulbs, has been shown to significantly lower melatonin levels. Reducing blue light exposure before bedtime has become a key recommendation by sleep specialists.

The night sky is not only a scientific frontier but also a source of spiritual and emotional awe. Living in the San Francisco Bay Area, I could only see a handful of stars at night. But in Yellowstone National Park and remote regions of New Mexico, the sky came alive with thousands of stars and a breathtaking view of the Milky Way. It's a profound difference—one that many city dwellers have never experienced.

To witness the stars more clearly, I often visited Lick Observatory, nestled in the Santa Cruz Mountains, just an hour's drive from my home in Saratoga, CA. The observatory, established in 1888, is a renowned research facility operated by the University of California and has played a pivotal role in astronomical discoveries. Lick's "Evenings with the Stars" programs invite the public to engage with space science firsthand. During my time at UC Berkeley, I chose to study astronomy—not because it was required, but because I was genuinely compelled to understand the universe.

As we explore beyond our atmosphere, another form of environmental degradation emerges: space pollution. The growing number of decommissioned satellites, defunct rocket stages, and orbital debris now clutter Earth's orbit. NASA and global space agencies track over 20,000 large pieces of space junk, while more than a million smaller fragments are believed to orbit the planet. Some interfere with telescopes, hampering our ability to study the cosmos.

The uncontrolled expansion of satellites—especially from large-scale projects like Starlink—poses both observational and navigational risks, and raises ethical questions about how we treat space as an extension of Earth's environment. Just as we've polluted our land, sea, and air, we now risk turning low-Earth orbit into a junkyard.

These concerns connect with broader themes—how the greatest empires in history rose and fell, how spiritual teachers have left legacies far more enduring than kings, and how nature's grandeur continues to offer insight and humility. Whether it's light pollution, space debris, or climate change, these problems point to a larger imbalance between modern human development and our spiritual, ecological, and ethical responsibilities.

Chapter 74
Personal Journeys and Reflections

Now, I want to share with you a two-week vacation in Alaska that was really refreshing for my soul. I saw many bears in Alaska, even a wild bear within 15 feet. It was pure joy to see a mother and calf Humpback whale breaching simultaneously several times from a ship sailing from Ketchikan, Alaska, through the Inside Passage to Anchorage. The ship stopped over at Juneau, where I visited the Mendenhall Glacier, a 13.6-mile-long glacier located in the Tongass National Forest. I took a helicopter ride to the top of Mendenhall Glacier, which is a mile wide. It was amazing to see virgin snow in the mountains. It was simply beautiful. I also walked to the edge of a crevice and looked down at the blue ice. I watched the water flow down several thousand feet.

It was pure joy watching Orcas (killer whales), the largest members of the dolphin family, which can reach speeds up to 34 mph, in the wild. They were simply majestic, unlike seeing their cousins in captivity performing for audiences. When I visited a fishery in Juneau, I saw many salmon swimming close to each other in the water. When I looked up, I thought, "Oh my God, this is the Pacific Ocean!"

In Ketchikan, I was walking around town and saw people fishing in a river from 30 feet above. What was unusual about this scene was that they were standing in the river and trying to catch salmon with fishing nets.

So, I asked some people what was going on. I was told that once a year, the authorities open the river for four hours for people to catch salmon with nets. This tradition is part of the community's subsistence fishing culture, and the temporary opening reflects state-regulated conservation efforts. It looked fun. So, I walked to a store that sells fishing gear and bought a net roughly 14 to 16 inches wide on a 6 or 7-foot pole, and started fishing. The river currents were strong, and the salmon moved fast. Catching a fish with a net was more challenging than I thought. The river was open for just four hours per year, and I felt lucky to be there.

After fishing for about fifteen minutes, a lady and her children came over and said, "You are having so much fun. You'll do better with a larger net. We are heading home now. You can use our net." I thanked them and took down their address. With the larger net, I fished for about half an hour. I caught a fifteen-pound salmon plus two smaller ones. I then walked to a gas station and asked how to get to the lady's house.

As it turned out, the map on the gas station wall did not include the newly developed street where she and her family lived. This reflects how some rapidly expanding communities in remote Alaska can outpace local infrastructure updates. The station attendant knew the general area and said, "You can take the map." I couldn't believe the lady and the station attendant's generosity. That would never have happened in California.

I decided to take the three salmon I caught to a fish market. They took out the roes and cleaned the fish. I then drove around to find the lady's address.

I returned her net and gave her the three salmon and roes. She was cleaning her fish when I got there. She then invited me to stay and enjoy salmon pasta with her family. I did, and we enjoyed our conversation. It was really refreshing to meet such kind people in Alaska.

Ketchikan averages about 153 inches of rain per year, making it one of the rainiest cities in the U.S. Once, I was talking to a young lady, and she said, "We are depressed because we haven't had any rain lately." I asked, "How long has it been?" She said, "3 days." I smiled. In California, droughts can last more than 20 years.

When I was in Haines, Alaska, where many eagles live, I took a picture of an eagle flying out of its nest. It looked just like the picture on a US postage stamp of an eagle with partially spread wings. The Bald Eagle population is particularly dense in Haines, where the Alaska Chilkat Bald Eagle Preserve is home to over 3,000 eagles each fall.

I also enjoyed Anchorage, Glacier Bay, Fairbanks and Denali Park. The 350-mile Denali Star Train, operated by Alaska Railroad, connects Anchorage to Fairbanks and offers stunning views from its glass-domed observation cars. I especially enjoyed Denali Park's Wonder Lake, from which you can see a very clear view of Mount McKinley, also known as Denali, the tallest mountain in the United States at 20,310 feet (6,190 meters). I had so much fun on my first Alaska vacation in 1994 that I forgot to eat for 24 hours.

On a different trip to Alaska with my parents, I took them to see a glacier that I saw on my previous trip from the visitor center. That was more than a decade earlier. Because of global warming, my parents and I could no longer see the glacier from the visitor center specifically built for viewing the glacier. We had to walk two miles to the glacier.

The path had signs showing where the glacier had receded annually due to global warming. This phenomenon is part of a well-documented pattern—Alaska has warmed more than twice as fast as the rest of the U.S. since the mid-20th century, according to NOAA.

While I was with my parents in Denali Park, we saw Bighorn sheep and many other animals. My parents and I also took a small plane to fly over Mount McKinley's base camp. We also flew very close to the summit, which was about 18 miles from the base camp, with a vertical gain of around 13,500 feet. Denali's base elevation is approximately 7,200 feet, and most summit expeditions start from the Kahiltna Glacier base camp. The views from the small plane were magnificent. My parents and I also saw many humpback whales and orcas in Alaska. I cherish the wonderful times I shared with my parents.

Just as the roots of a tree must strike deep beneath the surface to gather nutrients that feed the foliage, eventually reaching for the heavens, my own root system was anchored in academia and corporate facilities before ever branching out in service of more metaphysical practices to heal the soul. The material realms served as a fertile training ground for refining the discernment and skills that would later be redirected towards uplifting human consciousness itself through spiritual cultivation.

In this light, my work launching biotech with AI and medical companies dedicated to revolutionizing personalized medicine and analyzing genomic data can be seen as the transitional bridge—developing tools to map and optimize the body's biology was but a precursor for the realization that our minds and souls held the master codes still awaiting decryption. Having fulfilled my erstwhile pioneering roles in the physical domains, the final phase of reorienting fully towards the realm of spirit represented the ultimate culmination rather than a diversion from my path's overarching purpose.

For those whose spiritual senses have been awoken, we come to recognize our secular occupations and ambitions as merely support beams upholding the enlightened construction of our life's true magnum opus—an edifice of divine wisdom and transcendent service to be meticulously assembled unto glorification of the Creator's majesty. Only once our physical labors have erected a stable foundation can we then divert our full focus to the higher alchemy of spiritual work that truly elevates humanity towards its noble apex as bearers of God's light.

These personal journeys and reflections have shaped my understanding of the interplay between science, faith, and the human experience. They've taught me that our pursuits in the material world can serve as stepping stones to deeper spiritual insights and that every experience, whether in the laboratory or the wilderness, can be an opportunity for growth and connection with the divine.

Chapter 75
The Path Forward – Faith
and Perseverance

As we stand at the crossroads of history, technology, and spirituality, it's crucial to reflect on the lessons we've learned and chart a path forward that honors our spiritual heritage while addressing the challenges of our modern world. We are living in a time of unprecedented connectivity, scientific advancement, and social unrest, which makes the search for enduring meaning even more urgent. This chapter will explore how we can apply the insights gained from our journey to live more purposeful, faithful lives in the face of adversity.

The Importance of Faith in a Changing World

In a world that often seems to prioritize material success and technological advancement over spiritual growth, maintaining a strong faith can be challenging. Global surveys by institutions like Pew Research Center confirm a steady rise in secularism, especially in developed nations, yet spiritual longing persists across all cultures. Yet, as we've seen through the examples of historical and spiritual figures, it's often those who hold fast to their beliefs and values who leave the most lasting impact on the world.

In Matthew 16:26 (NKJV), Jesus said, *"For what profit is it to a man if he gains the whole world, and loses his own soul?"* This can be rephrased as: *"What will it profit a man if he gains the whole world and loses his own soul?"* This profound statement reminds us that our spiritual well-being should be our highest priority, even as we navigate the complexities of modern life.

In an age dominated by consumerism and instant gratification, this scripture serves as a timeless counterpoint to the fleeting rewards of worldly ambition. We must strive to balance our engagement with the world and our spiritual growth, always keeping in mind that our ultimate goal is not earthly success but alignment with God's purpose for our lives.

Perseverance in the Face of Injustice

As my personal experience with the legal system demonstrated, we often face injustices and setbacks that can shake our faith in human institutions. From wrongful convictions to systemic inequalities in bankruptcy and civil law, many people suffer outcomes that challenge their trust in man-made systems. However, it's in these moments of trial that our faith in a higher power becomes most crucial. Romans 8:28 says, *"And we know that for those who love God, all things work together for good according to His purpose."*

This doesn't mean that we should passively accept injustice. Rather, we should strive to be agents of positive change in the world, always guided by our faith and moral convictions. As Dr. Martin Luther King Jr. reminded us, *"The arc of the moral universe is long, but it bends toward justice."*

This quote, popularized during the Civil Rights Movement and originally paraphrased from a sermon by abolitionist Theodore Parker in the 19th century, continues to resonate in global struggles for fairness and reform. Our role is to actively participate in bending that arc, even when the path is difficult and the progress seems slow. Faith gives us the courage to endure injustice without becoming bitter and the strength to keep striving for righteousness when others lose hope.

Chapter 75
The Path Forward – Faith
and Perseverance

Embracing Our Calling

Each of us has a unique calling, a purpose that God has designed us to fulfill. Discovering and embracing this calling is a crucial part of our spiritual journey. As we saw in the story of Moses, God often calls us to tasks that may seem beyond our capabilities. Exodus 3–4 recounts Moses' reluctance, citing his lack of eloquence and authority, but God empowered him through signs and the support of Aaron. With faith and perseverance, and by relying on God's strength rather than our own, we can accomplish great things.

Jeremiah 29:11 reminds us, *"For I know the plans I have for you,"* declares the Lord, *"plans to prosper you and not to harm you, plans to give you hope and a future."* This promise can give us courage as we step out in faith to pursue our God-given purpose, even when the path ahead seems uncertain. Biblical history and modern testimony alike affirm that divine purpose often emerges most clearly in seasons of uncertainty and surrender.

Cultivating Compassion and Service

My experiences in Nazca, Peru, and elsewhere around the world have shown me the profound impact that compassion and service can have, both on those we serve and on our own spiritual growth. Jesus commands us to *"love your neighbor as yourself"* (Mark 12:31), and this love is most powerfully expressed through acts of service and compassion.

As we move forward, we must look for opportunities to serve others, especially those who are marginalized or in need. In Matthew 25:40, Jesus reminds us that whatever we do for the "least of these," we do for Him. This might involve volunteering, donating to worthy causes, or simply being more attentive to the needs of those around us in our daily lives. By doing so, we not only help others but also grow in our own faith and understanding of God's love. Compassion is not a passive emotion—it is love in action.

Stewardship of Creation

Our exploration of the night sky and the issues of light pollution remind us of our responsibility as stewards of God's creation. Genesis 1:28 gives humans dominion over the earth, but this is a call to responsible stewardship, not exploitation. The original Hebrew word for "dominion" (radah) in this context implies guardianship, not domination.

As we face environmental challenges like climate change and pollution, we must see our efforts to protect and preserve the natural world as an expression of our faith. Pope Francis' 2015 encyclical *Laudato* Si' echoed this, urging Christians to view care for the earth as a moral obligation. This might involve making more environmentally conscious choices in our daily lives, supporting conservation efforts, or advocating for policies that protect our planet. Protecting creation honors the Creator.

Nurturing Spiritual Growth in a Technological Age

As we've seen, technology has brought many benefits and challenges to our spiritual lives. The constant connectivity and information overload of our digital age can make it difficult to find quiet time for prayer, reflection, and spiritual growth. According to a 2023 Barna Group study, 56% of Christians reported that digital distraction frequently interferes with their spiritual practices.

Moving forward, we must be intentional about creating space in our lives for spiritual practices. This might involve setting aside specific times for prayer and Bible study, periodically "unplugging" from our devices to connect with God and nature, or using technology in ways that enhance rather than detract from our spiritual lives. Apps like YouVersion and Lectio 365 show how technology can be redeemed for spiritual formation.

Bridging Divides and Promoting Understanding

In a world often divided by religious, cultural, and ideological differences, we have a responsibility as people of faith to be bridge-builders. Jesus calls us to be peacemakers (Matthew 5:9), and this involves actively working to promote understanding and reconciliation across divides.

This might involve engaging in interfaith dialogue, working to address social injustices, or simply striving to show Christ's love to those who are different from us. Paul's teaching in Galatians 3:28—"There is neither Jew nor Greek... for you are all one in Christ Jesus"—points to the radical unity possible in the kingdom of God. By doing so, we can be a positive force for unity and peace in our communities and the world at large.

Preparing for the Future

As we look to the future, we must remain grounded in our faith while also being prepared for the challenges and opportunities that lie ahead. This includes staying informed about technological advancements and their potential impacts on society, engaging thoughtfully with ethics and policy issues, and continually seeking God's wisdom in navigating complex issues.

We should also be mindful of the potential for significant societal changes or challenges, as hinted at in biblical prophecies. Matthew 24, Revelation, and Daniel provide apocalyptic visions not to spark fear, but to encourage readiness, vigilance, and moral clarity. While we can't predict the future, we can prepare ourselves spiritually to face whatever may come with faith, courage, and hope.

As we conclude this journey through personal experiences, historical insights, and spiritual reflections, let us remember that our path forward is ultimately one of faith and perseverance. In a world that often seems chaotic and unpredictable, our anchor is our relationship with God and our commitment to living out His purposes for our lives.

The apostle Paul's words in Philippians 3:13–14 offer a fitting conclusion and challenge for us: *"Brothers and sisters, I do not consider myself yet to have taken hold of it. But one thing I do: Forgetting what is behind and straining toward what is ahead, I press on toward the goal to win the prize for which God has called me heavenward in Christ Jesus."*

Let us, too, press on, always seeking to grow in our faith, to serve others with compassion, to be responsible stewards of God's creation, and to be beacons of hope and love in a world that desperately needs it. May our lives reflect not just belief, but transformation—becoming living testimonies of the grace, truth, and power of God in a rapidly changing world.

Section 10: End Times Prophesies

Chapter 76
Signs of the End Times

The Bible provides a prophetic roadmap of events that will unfold before the return of Jesus Christ. These signs of the end times include global conflict, disease outbreaks, and a decline in genuine faith. Perhaps most troubling is the rise of an apostate church—one that strays from divine truth and embraces false doctrines.

While these prophecies may seem alarming, Scripture reminds us that God remains sovereign over all history. In light of this, our response should be to deepen our trust in Him and to continue demonstrating love—for God and for one another.

Perilous Times Foretold

Paul's second letter to Timothy offers a sobering description of the last days:

"This know also, that in the last days perilous times shall come. For men shall be lovers of their own selves, covetous, boasters, proud, blasphemers, disobedient to parents, unthankful, unholy, Without natural affection, trucebreakers, false accusers, incontinent, fierce, despisers of those that are good, Traitors, heady, high-minded, lovers of pleasures more than lovers of God; Ever learning, and never able to come to the knowledge of the truth."

(2 Timothy 3:1–7)

This portrait mirrors much of what we see today—moral confusion, pride, and rebellion against truth.

Christianity and Islam: Irreconcilable Doctrines

The Qur'an denies both the crucifixion and resurrection of Jesus—central tenets of the Christian faith. Islam teaches that Jesus was not crucified but that someone else was made to look like him. In contrast, Christianity affirms that Jesus died on the cross as the ultimate sacrifice for humanity's sins and rose again to conquer death, just as He foretold.

This fundamental contradiction raises serious questions: How can Pope Francis claim these faiths are equal or compatible? How many Catholic churches have held Ramadan celebrations, and why? Some believe Pope Francis played a role in the resignation of Pope Benedict XVI—an event marked, symbolically, by two lightning strikes hitting the Vatican on that very day.

The Irreplaceable Truth of Jesus Christ

As of 2023, there were 2.3 billion Christians, comprising 31.5% of the global population. Jesus Christ's ministry spanned just three years, yet He remains the most impactful figure in human history. Despite scholarly debate, historical evidence overwhelmingly supports the reality of His life, crucifixion, and the explosion of the early church based on His resurrection.

While Christians and Muslims can and do live in peace, their religious foundations are irreconcilable. There is only one truth. I have had meaningful and respectful conversations with Muslim neighbors in Saratoga and even shared a Ramadan meal with Muslims in Egypt. But theological honesty requires acknowledging that peaceful coexistence does not equal doctrinal agreement.

The Jewish Temple and the Sacred Rock

The Jewish people built their first temple during the reign of King Solomon, son of King David, completing it in 957 BCE. It was constructed around the rock where Abraham prepared to sacrifice his son Isaac, according to the Torah.

The Qur'an, written between 610–632 CE, claims it was Ishmael—not Isaac—who was to be sacrificed. But if that were true, it seems unlikely that the Jewish temple would have been built on that specific site. This discrepancy underscores the theological and historical tension over this sacred ground.

The Rock of Contention

The Prophet Muhammad is traditionally believed to have ascended into heaven from the same rock. Today, this relatively small outcrop in Jerusalem is the most contested piece of land on Earth, holding the potential to ignite a global conflict—possibly even World War III.

Buddhism and the Pursuit of Nirvana

Buddhism teaches that human effort can lead to Nirvana, an enlightened state free from suffering and rebirth. Unlike Christianity, Buddhism does not require belief in a personal God. Instead, followers are encouraged to rely on themselves—a stark contrast to the Christian call to rely wholly on God's grace through Jesus Christ.

Resurrection and the Call to Love

Jesus said, "Love one another as I have loved you." That command, given two thousand years ago, remains radically relevant today. He also urged His followers to pray for those who persecute them and to forgive freely.

As Revelation 22:10–11 proclaims:

"And he saith unto me, Seal not the sayings of the prophecy of this book: for the time is at hand. He that is unjust, let him be unjust still: and he which is filthy, let him be filthy still: and he that is righteous, let him be righteous still: and he that is holy, let him be holy still."

These words remind us that each person must choose whom they will serve, especially as the final days draw near.

End-Time Deception and Emerging Technologies

Scripture warns repeatedly of increasing deception in the last days. Today, that deception is amplified by biased media, manipulated narratives, and the rise of artificial intelligence. With AI and artificial general intelligence (AGI) advancing rapidly, it is becoming more difficult to distinguish between truth and fabrication.

This mirrors biblical warnings: "Even the elect could be deceived, if that were possible" (Matthew 24:24). The very tools meant to inform may, ironically, become instruments of mass deception, reinforcing the urgency of staying rooted in God's Word.

Natural Disasters and the Fragility of Creation

Natural disasters are also highlighted in biblical end-times prophecies. We can expect to see an uptick in devastating events such as wildfires, prolonged droughts, catastrophic floods, global pandemics, powerful earthquakes, and violent volcanic eruptions. These events serve as stark reminders of the earth's fragility and humanity's profound need for divine protection and intervention.

The Rise of the Beasts and Spiritual Deception

The Bible speaks of two metaphorical beasts that will arise in the last days, aiming to mislead humanity and draw people away from faith in God. These prophetic passages underscore the urgent need for spiritual discernment in these perilous times. Believers must exercise caution regarding whom they trust, testing every teaching against the authority of Scripture.

Even in the face of deception, we take comfort in knowing that Jesus Christ's power far surpasses that of any earthly or spiritual adversary.

The Mark of the Beast and the Coming Economic System

A pivotal end-time prophecy involves the implementation of a "mark of the beast", a sign required to participate in commercial transactions. This mark is described in Revelation 13, and while its form may become ubiquitous, the greater concern lies in its use as a tool for controlling personal freedom and economic access.

"And he causeth all, both small and great, rich and poor, free and bond, to receive a mark in their right hand, or in their foreheads: And that no man might buy or sell, save he that had the mark, or the name of the beast, or the number of his name. Here is wisdom. Let him that hath understanding count the number of the beast: for it is the number of a man; and his number is Six hundred threescore and six." (Revelation 13:16–18)

Revelation 13:11–12 further describes the second beast:

"And I beheld another beast coming up out of the earth; and he had two horns like a lamb, and he spake as a dragon. And he exerciseth all the power of the first beast before him, and causeth the earth and them which dwell therein to worship the first beast, whose deadly wound was healed."

These cryptic verses raise important questions about the identities of these entities and the nature of the "healed wound", suggesting a political or spiritual figure whose return to power will be globally recognized and worshipped.

As we witness the rise of digital currencies, real-time facial recognition, biometric scanning, location tracking, and centralized QR-based payment systems, the infrastructure for a system of global surveillance and control already exists. Authorities could potentially block individuals from buying or selling, track associations, and monitor dissent—all in real time.

Sober Watchfulness Anchored in Hope

While these developments are sobering, the Bible calls us to approach prophecy with both alertness and hope. These signs paint a picture of mounting global tension, but they also affirm God's ultimate control over history and His promise to redeem those who remain faithful.

Fulfillment of prophecy affirms the divine inspiration of Scripture and strengthens our faith. As global events align with ancient predictions, we are called to respond not with fear, but with spiritual preparation.

Spiritual Readiness in a Time of Testing

Believers are urged to maintain a constant state of **spiritual readiness**. This includes:

- **Cultivating a strong prayer life**
- **Diligent study of Scripture**
- **Active participation in a community of faith**
- **Standing firm in biblical convictions**, even under **societal pressure or persecution**

We are not called merely to survive, but to live with courage, compassion, and conviction—bearing witness to God's truth in a darkening world.

The Apostate Church and Compromised Faith

One of the more concerning signs of the end times is the rise of the Apostate Church—a religious institution that outwardly resembles Christianity but has inwardly departed from the truth. Such institutions may prioritize worldly acceptance over biblical fidelity, compromising core doctrines to align with cultural trends. Believers must remain discerning, evaluating all teachings and practices against the unchanging standard of God's Word.

False Prophets, Signs, and Deception

The proliferation of false prophets and teachers is another unmistakable sign. These individuals may perform signs and wonders, appearing convincing and trustworthy. Jesus warned that their deception would be so potent that, if it were possible, even the elect could be deceived (Matthew 24:24). This underscores the need for deep biblical knowledge and continual discernment through the Holy Spirit.

Globalism and the Coming World Order

The Bible also points to a future global government and economic system that will exert control over all buying and selling. Current movements toward globalization, centralized authority, and international economic regulation provide a possible pathway for this system to emerge.

Environmental Shifts and the Birth Pains

Scripture refers to "birth pains"—environmental and social upheavals that precede the end. These include:

- **Increased intensity and frequency of natural disasters**
- **Extreme weather events and climate shifts**
- **Widespread ecological degradation**

These changes reflect the groaning of creation and point toward the coming redemption described in Romans 8:22–23, "For we know that the whole creation groaneth and travaileth in pain together until now. And not only they, but ourselves also, which have the firstfruits of the Spirit, even we ourselves groan within ourselves, waiting for the adoption, to wit, the redemption of our body."

Technology, Surveillance, and Control

While technology offers many benefits, it also presents unique challenges in light of prophecy. The development of:

- **Advanced surveillance systems**
- **Implantable identification technologies**
- **Complete digitization of personal and financial data**

This raises serious concerns about how authoritarian regimes could wield these tools for global control, as predicted in Revelation.

Hope Anchored in Christ's Return

Despite all these signs and warnings, the Bible consistently encourages us to remain hopeful. The same prophecies that foretell tribulation also promise victory—the return of Christ, the defeat of evil, and the establishment of God's eternal kingdom.

We are not merely passive observers—we are called to be Christ's ambassadors. As we await His return, we must share His truth, embody His love, and live with steadfast hope, no matter how chaotic the world becomes.

Are we witnessing the beginning of the 3-1/2-year tribulation revealed in the Book of Revelation? The US is the Second Beast in Revelation. The Trump administration, whose term will end in about 3-1/2-years, has

1. Caused or is causing 300,000 deaths by reducing or eliminating USAIDS. According to reports from NPR, Democracy Now!, and The Times, modeling from Boston University estimates that reductions in funding for the U.S. Agency for International Development (USAID) may have led to nearly 300,000 deaths.

2. Attacked Iran, focusing on nuclear facilities, using some 125 military aircrafts, including 7 B-2 stealth bombers and fourth- and fifth-generation fighters, 14 30,000 pounds deep penetrating bombs, 75 precision guided weapons and more than two dozen Tomahawk missiles fired from a nuclear submarine primarily focused on surface infrastructures. In joining the Israel-Iran war, US made deliberate choices to violate the US Constitution and UN Charter, and acted lawlessly to risk overt and covert

retaliatory attacks that could lead to WWIII for generations to come.

3. Accelerated the impending global climate change disaster by withdrawing from the 2015 Paris Accords adopted by 195 Parties at the UN Climate Change Conference (COP21) in Paris, France, on December 12, 2015. It entered into force on November 4, 2016.

4. Withdrawn from World Health Organization (WHO) and jeopardized critical health initiatives, weakened international cooperation and undermined efforts to address pressing global challenges.

5. Increased chaos and anxiety globally by imposing unfair tariffs that would have to be paid mostly by Americans and billions of people globally could suffer due to disruption in trade and other non-economic activities.

6. Has taken authoritarian actions resulting in the "No Kings" movement that has produced thousands of protests nationwide and globally involving millions of people largely against the policies and actions of Donald Trump's second presidency.

7. Deployed the National Guard troops and Marines to confront protesters in Los Angeles, where protests against immigration raids by the Trump administration led to clashes in the streets.

8. Revoked $11 billion in funding for addiction and mental health care. More than 209,000 Americans die each year from alcohol, suicide and drug overdoses. These conditions account for an estimated $700 billion annually.

9. Conspired with the Netanyahu administration to enable the Israel IDF to bomb Iran and assassinate their leaders and scientists, resulting in severe retaliation by Iran. This could easily lead to WWIII.

10. Allowed the genocide and ethnic cleansing in Gaza to continue and more than 55,000 Palestinians,

mostly women and children, have died and 1.8 million displaced people don't have enough food, water, and medicine due to Israel blockades.

11. Tried to end the Ukraine-Russia war, but failed so far, and the intensity of the damages done by both sides is increasing. The EU, NATO, and G-7 don't want to end this war, which has caused more than a million deaths or injured and turned more than 6.9 million Ukrainians into refugees in other countries, primarily in Europe. As of mid-2025, approximately 3.7 million people are internally displaced within Ukraine.

12. Demanded NATO partners to increase defense spending to 5% on the backs of the European citizens, who will see continued degradation in their quality of life.

13. Taken actions to legitimize cryptocurrency, which is backed by nothing. In time, this Ponzi scheme will become self-evident. Like the creation of the Federal Reserve that facilitated perpetual wars resulting in over 100 million deaths, cryptocurrency can do a lot of damage to billions of people around the world.

14. Fired many scientists who have contributed to improving health for millions of people and pushing to cut assistance to those who can least afford healthcare and food.

15. Ignored homelessness, which is fast becoming a crisis nationally. In 2024, there were approximately 771,400 homeless individuals in the country, an increase of 118,300. Many famous tourist destinations such as the Fishermen's Wharf in San Francisco have turned into abandoned neighborhoods.

16. Made the US economy worse, people are suffering and some countries refuse to accept "US Dollar" as payment for their goods. Annualized GDP growth

for the first quarter of 2025 was -0.3%, a significant drop from the previous quarter's 2.4%.

17. In 2025, the U.S. dollar is down almost 10% year-to-date against the euro, down more than 8% against the Mexican peso, and down 8% against the Japanese yen. The US government is projected to spend over $1 trillion annually on interest payments for its national debt of more than $37 trillion.

18. Taken actions to limit "Free Speech," especially on university campuses and cut billions in funding to universities such as Harvard, which is older than the US.

19. Governed through executive orders, issuing 142 orders in the first 100 days. These orders are about 5 times more than most presidents since Franklin D. Roosevelt issued 99 to rescue the nation from the Great Depression.

20. Caused significant decline in truth. According to The Washington Post, Trump made more than 30,000 false and misleading claims in his first term and it hasn't stopped.

21. Selected his cabinet based upon loyalty to Trump and not merit. What would be the consequence of this approach?

22. Devalued and politicized the justice system significantly. President Trump is the first convicted felon ever to become president and he has regularly attacked the courts, and wielded investigative powers to target his opponents.

23. Used the powers of government against major institutions Trump dislikes, like law firms, universities and the media, in ways that are often transparently political.

24. Instilled a culture of fear that leads to violence.

25. Caused a decline in trust by allies and not only adversaries. This could accelerate the decline of the US dollar as the global currency.

Politics, Prophecy, and the Call for Spiritual Vigilance

I did not vote in the 2024 presidential election because I believed I would regret casting a ballot for either Donald Trump or Kamala Harris. Instead, I pray that God would protect President Trump and guide him to make decisions that promote justice, humility, and global peace.

Whether one believes in the divine inspiration of the Bible or not, there is timeless practical wisdom in its pages. America's political obsession with containing the rise of China, for example, is not just geopolitical—it reflects the age-old sin of envy, one of the seven deadly sins.

The Bible tells us that envy triggered the first recorded murder, when Cain killed his brother Abel. It was also envy that drove Satan to rebel against God, ultimately leading to his fall from heaven to Earth. This spiritual condition continues to shape global conflict and policy today.

Historical Interventions and the Cost of Global Leadership

A brief survey of history illustrates how power and envy have long driven international conflicts, often under the banner of humanitarianism. NATO, formed in 1949, did not conduct its first military intervention until 1995—46 years after its inception. That action, heavily influenced by President Bill Clinton, marked the beginning of a series of U.S.-led interventions without UN authorization, raising serious questions of international legality and moral consistency.

In March 1999, the U.S.-led NATO bombing of Yugoslavia was justified by the Kosovo crisis. The 78-day campaign targeted military and civilian infrastructure alike—oil refineries, factories, and even television stations—without the approval of the UN Security Council or General Assembly. It resulted in the deaths of 2,500 people and displaced over a million civilians.

Meanwhile, during this period, President Clinton admitted to an affair with Monica Lewinsky—an act that violated the sanctity of the Oval Office. Just three days after this public admission, on August 20, 1998, Clinton authorized Tomahawk missile strikes in Afghanistan and Sudan, one of which reportedly killed Osama bin Laden's daughter. The political timing of that strike raised questions that history has not forgotten.

The War on Terror and Its Fallout

In 2001, Osama bin Laden orchestrated the 9/11 attacks, which brought down the World Trade Center towers, damaged the Pentagon, and resulted in 2,977 deaths—the deadliest terrorist attack in history. The U.S. response—the global War on Terror—spanned decades, cost millions of lives, and fundamentally reshaped international law, surveillance, and military engagement.

The Proxy War in Ukraine

Today's Ukraine-Russia conflict is widely seen as a proxy war between NATO and Russia. Its origins trace back to President Clinton's decision to expand NATO eastward in 1999, a move that broke earlier promises made to Russian leaders in exchange for agreeing to the reunification of Germany. This act, though viewed as strategic by some, is seen by others as the seed of provocation that led to the unfolding catastrophe—a war that has killed or injured over a million people.

Gaza, Genocide, and Global Hypocrisy

The current war in Gaza has been labeled by numerous international watchdogs—including Amnesty International and UN Special Rapporteurs—as a genocidal campaign. These organizations cite large-scale civilian deaths, the destruction of critical infrastructure, and the targeting of healthcare facilities. In many documented cases, there is no evidence that these strikes had legitimate military objectives.

In contrast to its intervention in Yugoslavia, NATO has taken no action in Gaza, despite the scale and intensity of the violence far surpassing that in Kosovo. This disparity in response raises difficult moral questions about selective justice, geopolitical bias, and who gets to define humanitarian crisis.

June 2025: The Israel-Iran War

In June 2025, Israel bombed Iran, killing at least 865 people and wounding more than 3,300. These attacks were carried out without UN authorization, violating international law. The United States subsequently joined the conflict, launching its own strikes against Iran—again bypassing both the UN Security Council and the U.S. Constitution, which requires Congressional approval for acts of war.

This act mirrors earlier unilateral actions and reflects a growing pattern of executive overreach, contributing to the destabilization of already fragile global relationships.

A Spiritual Lens on Global Events

Taken together, these modern events align eerily with biblical prophecy. The escalation of global warfare, selective justice, technological control, and institutional deception are all signs Scripture warns will precede the return of Christ.

While these signs may seem daunting, they serve as a call to spiritual vigilance, deeper faith, and active moral engagement. By understanding these prophecies and their real-world implications, believers can face the days ahead with wisdom, courage, and unshakable trust in God's sovereign plan.

Final Reflection: Living with Hope Amid Turmoil

As we confront political instability, war, and the increasing collapse of moral clarity, we must not lose heart. The believer's hope is not in worldly leaders or political alliances, but in the eternal kingship of Christ. These signs of the end times are not just warnings—they are also reminders of God's control, His justice, and His coming restoration.

"Therefore keep watch, because you do not know on what day your Lord will come." (Matthew 24:42)

Let this moment in history awaken our souls—not just to the fragility of human empires—but also to the enduring strength of divine truth.

Chapter 77
The Rise of Digital Currency

The world is rapidly transitioning away from physical money toward digital forms of currency. This shift marks a paradigm change in global finance, revolutionizing how individuals conduct transactions, store value, and engage with banking systems. While digital currency may seem novel to some, many Americans have already embraced digital banking services, with over 65% of U.S. adults using mobile banking apps as of 2024.

Financial institutions continue to introduce advanced digital tools, including AI-powered budgeting assistants, biometric authentication, and contactless payment systems. These innovations reflect a broader move toward a cashless society, driven by convenience, security, and efficiency.

Digital currencies exist primarily in electronic form and are exchanged through secure computer networks, especially those connected to the Internet. Unlike credit card payments or online transfers tied to fiat currencies, true digital currencies—such as cryptocurrencies and Central Bank Digital Currencies (CBDCs)—never exist in physical form. This distinction makes them uniquely compatible with our increasingly interconnected, cash-optional world.

Central banks in major economies—including Brazil, China, the Eurozone, India, Nigeria, and the United Kingdom—are leading the way in exploring or implementing CBDCs. As of 2024, over 130 countries, representing more than 98% of global GDP, are researching or developing CBDCs, according to the Atlantic Council's GeoEconomics Center.

China's digital yuan (e-CNY) has already been launched in pilot programs across dozens of cities, with over 260 million wallets opened and billions in transaction volume. The Bahamas became the first nation to roll out a fully operational CBDC, the Sand Dollar, in 2020, setting a precedent for small economies leveraging digital finance to enhance accessibility.

In the U.S., a pilot initiative led by the Federal Reserve Bank of New York is testing digital dollar tokens for interbank settlements. Participants include major institutions such as BNY Mellon, Citi, HSBC, Mastercard, PNC, TD Bank, Truist, U.S. Bank, and Wells Fargo. This signals strong institutional momentum toward mainstream adoption of digital currency infrastructure.

JPMorgan Chase, meanwhile, has already implemented JPM Coin—its own internal digital token used for wholesale transactions. The system enables real-time, blockchain-based transfers of dollar and euro balances, processing over $1 billion per day as of 2023. While this represents just a small share of JPMorgan's daily $10 trillion flow, it highlights the efficiency gains digital tokens offer in high-volume environments.

However, CBDCs raise major privacy concerns. Unlike cash, which offers anonymity, digital currencies can allow governments to track every transaction. Critics argue this could pave the way for surveillance and limit personal freedom. In authoritarian regimes, programmable money could even be used to restrict purchases or freeze dissenters' accounts. These risks warrant close attention as the U.S. considers its digital dollar roadmap.

The QR code (Quick Response code)—invented by Japan's Denso Wave in 1994 to track auto parts—has now become a staple in digital commerce. It facilitates contactless payments and can store up to 4,000 alphanumeric characters. In countries like China, QR-based payment apps like WeChat Pay and Alipay dominate daily life, processing trillions of dollars in transactions annually.

In a fully digitized economy, it's conceivable that QR codes or similar technologies—possibly embedded in digital IDs or mobile wallets—will become required for purchases. This has sparked debate among civil liberties advocates, who warn that tying personal identification to every transaction could threaten financial autonomy and increase vulnerability to cybercrime.

The global digital currency shift also carries macroeconomic and geopolitical consequences. By digitizing national currencies, central banks may gain unprecedented control over monetary policy, enabling real-time interventions such as direct stimulus payments or programmable interest rates. However, this centralization could reduce the role of commercial banks, forcing a structural evolution in traditional banking.

Moreover, CBDCs have the potential to challenge the U.S. dollar's role as the world's reserve currency. For example, China's efforts to internationalize the digital yuan in cross-border trade—particularly through its Belt and Road Initiative—signal a strategic bid to reduce dependence on dollar-based systems.

At the same time, digital finance could deepen global inequalities. Roughly 2.6 billion people remain offline, and many lack access to smartphones, secure IDs, or digital literacy. If digital currency becomes the primary medium of exchange, those without access risk being excluded from essential services. This raises urgent questions about digital inclusion and infrastructure development.

The environmental impact is another critical issue. While CBDCs are typically energy-efficient, many cryptocurrencies like Bitcoin rely on proof-of-work mining, which consumes immense amounts of electricity. Bitcoin alone is estimated to use more energy annually than the entire country of Argentina. Transitioning to greener blockchain protocols (e.g., proof-of-stake) and regulating crypto mining could help reduce digital finance's carbon footprint.

Cybersecurity is also paramount. As digital assets become more valuable and widely used, they become bigger targets for hackers. High-profile breaches—such as the $600 million Poly Network hack—underscore the risks. Governments and private firms must invest in secure infrastructure and establish clear consumer protections.

Emerging technologies like AI and the Internet of Things (IoT) are accelerating the shift. For instance, IoT-enabled smart devices could trigger automated payments for services like energy or tolls, while AI could manage investment portfolios or detect fraud in real time. These developments offer convenience—but also increase system complexity and the stakes of failure.

On an individual level, navigating the transition to digital finance will require new skills and habits. Financial literacy education—especially around cybersecurity, personal data protection, and digital wallets—will become essential. Tools like biometric authentication and multi-factor verification can enhance security, but they also raise concerns about data sovereignty and surveillance.

In conclusion, the rise of digital currencies represents a transformative moment in human economic history. If implemented thoughtfully, they can improve efficiency, expand financial access, and enable smarter policy responses. But if rushed or poorly regulated, they could increase inequality, infringe on freedoms, and destabilize traditional institutions. Balancing innovation with oversight will be key.

As we enter this new financial era, individuals, businesses, and policymakers must approach digital currencies with open eyes, informed by both the promise and the perils of this powerful technology.

Chapter 78
Global Adoption of QR Codes
and Digital Payments

China and India are leading the world in the adoption of QR code payments, with China having the largest number of mobile payment users globally. As of 2024, over 90% of urban Chinese consumers regularly use QR codes for everyday transactions via platforms like WeChat Pay and Alipay, which together process trillions of dollars annually. India's Unified Payments Interface (UPI) has also facilitated a massive surge in QR code payments, particularly through apps like PhonePe, Paytm, and Google Pay India, which are now widely accepted even by street vendors and small businesses.

Several Asia-Pacific nations, including Indonesia, Singapore, Malaysia, Thailand, and the Philippines, are also embracing QR code technology. For example, Singapore's PayNow and Thailand's PromptPay have national QR code standards to ensure interoperability between banks and merchants. It's worth noting that these nations are not predominantly Christian, although Christian communities do exist within them. Many people in these regions may be unfamiliar with or skeptical of Biblical prophecies—particularly those concerning the end times—due to differing cultural and religious perspectives.

The slower adoption of QR codes in the United States was primarily due to early smartphones lacking native QR scanning capabilities, which required users to download third-party applications. However, this barrier began to fade after 2017, when Apple integrated QR code scanning directly into its iPhone camera app (iOS 11), followed by similar changes in Android devices. Since then, QR code usage in the U.S. has increased significantly, particularly in response to COVID-19, which normalized touch-free payments in restaurants, retail, and healthcare.

QR codes have the ability to store more information than traditional barcodes—up to 4,296 alphanumeric characters or 7,089 numeric characters—and their ease of use has driven adoption across various industries, especially in digital payments. The convenience is undeniable: customers scan a code, are directed to a secure payment page, and complete the transaction using options like Apple Pay, Google Pay, Samsung Pay, or linked bank accounts. This process is fast, secure, and eliminates the need for cash or physical credit cards.

However, the rise of QR code payments is not without risks. Phishing attacks and malicious QR codes—referred to as "quishing"—are on the rise. Hackers can embed dangerous URLs in QR codes that, once scanned, can prompt the download of malware or redirect users to counterfeit websites to steal personal and banking information. In 2022, the FBI issued an alert about this growing threat, emphasizing the need for vigilance and encryption in QR-based systems.

QR code payments also depend on a smartphone with a camera and reliable Internet access, which may not be consistently available in all locations. This dependence can present challenges in rural areas or for populations that are technologically underserved. Additionally, older generations or individuals uncomfortable with smartphones may find the shift to QR payments confusing or alienating.

To address such challenges, the U.S. government offers subsidized mobile devices and Internet access to low-income individuals through programs like Lifeline and the

Affordable Connectivity Program. While these initiatives are intended to bridge the digital divide, they also unintentionally accelerate the mass adoption of digital technologies, including QR-based payment platforms, and may hasten the transition away from traditional cash-based transactions.

This global shift toward QR code payments raises profound questions about financial inclusion, accessibility, and ethical oversight. While QR codes simplify and speed up transactions, they may also exclude or marginalize those without smartphones, banking access, or digital literacy. Ensuring that vulnerable populations aren't left behind will be crucial as societies increasingly depend on these systems.

Given the ease and ubiquity of QR code scanning, it is not inconceivable that one day individuals might be asked—or required—to carry a personalized, scannable identifier. This could take the form of a wearable chip, digital ID, or even an invisible ink QR code on the hand or forehead— technologies that are already being explored for secure identification and health tracking purposes. Though speculative, such developments echo concerns raised by theologians and privacy advocates alike.

From a Biblical perspective, these technological advancements bring renewed attention to prophecies in the Book of Revelation concerning the "mark of the beast." The Bible warns that receiving this mark—whether in the hand or forehead—will have eternal spiritual consequences.

Revelation 14:9–11 (KJV) warns:

"If any man worship the beast and his image, and receive his mark in his forehead, or in his hand, The same shall drink of the wine of the wrath of God... and he shall be tormented with fire and brimstone... And the smoke of their torment ascendeth up for ever and ever..."

Revelation 16:1–2 describes divine judgment upon those who accept this mark:

"...there fell a noisome and grievous sore upon the men which had the mark of the beast, and upon them which worshipped his image."

These verses highlight that the decision to accept such a mark—no matter how practical or socially necessary it may seem—has deeper spiritual implications. What may appear as a harmless digital ID or payment method could, in the eyes of believers, represent a submission to systems that oppose God's authority.

As digital payment technologies become more ubiquitous and potentially mandatory for participation in economic life, believers may soon be forced to discern carefully how to engage with these systems. This will involve not just technical or financial decisions, but spiritual ones, grounded in scripture and guided by prayer, community, and conviction.

It's important to note that not all digital payment systems or technological advancements inherently represent the mark of the beast as described in the Book of Revelation. Many of these innovations—such as contactless payments, mobile wallets, and blockchain-based transfers—serve practical purposes and improve convenience and efficiency. However, the accelerating trend toward cashless societies and the integration of biometric and digital identity systems into everyday commerce certainly creates an environment in which such a system could, one day, be implemented.

As we observe these technological and economic transformations, it is crucial for believers to remain informed, spiritually grounded, and discerning. While we cannot know precisely how end-time prophecies will unfold, Scripture urges vigilance and preparedness. Matthew 24:42 reminds us, "Therefore keep watch, because you do not know on what day your Lord will come." This means staying alert, not fearful—prioritizing our allegiance to God over convenience, comfort, or conformity to emerging economic systems.

The adoption of QR codes and digital payments also raises urgent concerns about data privacy, surveillance, and control. Every transaction completed through these systems can be digitally logged, timestamped, and associated with personal identifiers. Companies and governments are increasingly capable of analyzing this data to derive behavioral insights and implement predictive analytics. While this can help prevent fraud, optimize services, and combat illegal activity, it also opens the door to highly invasive forms of monitoring, censorship, and social control—especially in authoritarian regimes or under unchecked corporate influence.

Moreover, the shift to digital payment infrastructures has major implications for monetary policy and global financial stability. Central banks may acquire more precise tools to monitor consumer activity or implement targeted stimulus, such as programmable CBDCs (Central Bank Digital Currencies). However, managing a hybrid system of physical and digital currencies poses technical and regulatory challenges, including risks of cyberattacks, liquidity imbalances, and increased dependency on centralized databases.

The cross-border nature of digital currencies and mobile payment platforms also introduces new complexities for international finance. Interoperability between national digital currencies (such as China's digital yuan, India's e-rupee, or the proposed digital euro) will likely require coordinated international frameworks, potentially under institutions like the IMF, BIS, or G20. These developments could reshape geopolitical power dynamics and the role of traditional reserve currencies like the U.S. dollar.

For businesses, especially small and medium-sized enterprises, the adoption of QR code and digital payment systems presents both opportunities and growing pains. On the one hand, such systems can reduce overhead costs, improve transaction speed, and attract younger, tech-savvy customers. On the other hand, they often require upfront investment in software, hardware, employee training, and cybersecurity protocols—a burden that not all small vendors can easily afford.

The rise of digital payments also has implications for charitable giving, community fundraising, and informal economies. As physical cash becomes less common, cash-based donations or off-the-grid financial support may decline. Churches, non-profits, and grassroots causes will need to adapt by offering secure, user-friendly digital giving platforms, or risk alienating supporters who are moving away from physical currency.

These rapid changes raise a host of philosophical and spiritual questions. How will they impact our financial literacy and awareness of how money functions? Will they alter our psychological relationship to value, ownership, and generosity? Digital payments may render money "invisible" and abstract, potentially disconnecting people from mindful spending or budgeting practices.

In conclusion, the global adoption of QR codes and digital payments represents a watershed moment in the history of human commerce. While these technologies offer significant benefits—efficiency, accessibility, and innovation—they also bring profound challenges in terms of privacy, equity, cybersecurity, and spiritual discernment. As we move into this new financial frontier, it will be imperative for individuals, families, communities, and policymakers to proceed with wisdom, humility, and a commitment to ensuring that technological progress serves the best interests of all people—not just the powerful.

Let us remember the words of Proverbs 4:7:

"Wisdom is the principal thing; therefore get wisdom: and with all thy getting get understanding." In doing so, we can navigate these shifts faithfully—eyes open, hearts anchored, and souls aligned with eternal truths.

Chapter 79
The Teachings of Jesus Christ

Jesus' teachings form the cornerstone of the Christian faith and provide timeless guidance for navigating life's challenges—especially the turbulent times prophesied in Scripture concerning the end days. His words offer not only wisdom and moral clarity, but also comfort and direction for believers facing an uncertain future.

Jesus imparted crucial wisdom about life's priorities and eternity. He asked, in Matthew 16:26, KJV, "For what is a man profited, if he shall gain the whole world, and lose his own soul?" He also said, "For where your treasure is, there will your heart be also" (Matthew 6:21). These statements encourage deep reflection on what truly matters in life, urging followers to prioritize spiritual integrity and eternal values over temporary material gain.

Regarding His divine authority and the mission entrusted to His followers, Jesus declared in what is known as the Great Commission (Matthew 28:18–20): "All power is given unto me in heaven and in earth. Go ye therefore, and teach all nations, baptizing them in the name of the Father, and of the Son, and of the Holy Ghost: Teaching them to observe all things whatsoever I have commanded you: and, lo, I am with you always, even unto the end of the world." This commission emphasizes both the global scope of the Christian message and the comforting promise of Christ's abiding presence.

Jesus also warned of widespread deception (Matthew 24:4–5): "Take heed that no man deceive you. For many shall come in my name, saying, I am Christ; and shall deceive many". This warning is particularly relevant in an age of mass communication, digital misinformation, and spiritual confusion. Discernment, grounded in Biblical truth, is essential for navigating such times.

He foretold signs of the end, including wars, natural disasters, and betrayal among people (Matthew 24:6–8): "And ye shall hear of wars and rumours of wars: see that ye be not troubled: for all these things must come to pass, but the end is not yet... and there shall be famines, and pestilences, and earthquakes, in divers places. All these are the beginning of sorrows". He continued, "Then shall many be offended, and shall betray one another, and shall hate one another" (Matthew 24:10).

Jesus also predicted the rise of false prophets and a moral collapse:
"And many false prophets shall rise, and shall deceive many. And because iniquity shall abound, the love of many shall wax cold" (Matthew 24:11–12). Yet, despite these grim developments, He offered hope (Matthew 24:13): "But he that shall endure unto the end, the same shall be saved." This endurance is not passive but active—rooted in faith, love, and obedience to God's Word.

Speaking of the Great Tribulation, Jesus said (Matthew 24:21–22): "For then shall be great tribulation, such as was not since the beginning of the world to this time, no, nor ever shall be. And except those days should be shortened, there should no flesh be saved: but for the elect's sake those days shall be shortened." This passage underscores both the severity of the end times and God's mercy in limiting the duration for the sake of the faithful.

Vigilance is a central theme in Jesus' teaching (Matthew 24:42–44): "Watch therefore: for ye know not what hour your Lord doth come... Therefore be ye also ready: for in such an hour as ye think not the Son of man cometh". Just as a watchman stays alert to guard against a thief, so believers must remain spiritually prepared for Christ's return.

These teachings provide a comprehensive framework for end-time readiness. They emphasize spiritual preparedness, discernment, and unwavering faith. Jesus' words remind believers that while difficult times may arise, ultimate victory and eternal security belong to those who remain faithful to Him.

Moreover, Jesus' command to love remains central—even in tribulation (John 13:34). He said, "A new commandment I give unto you, That ye love one another; as I have loved you." This radical love includes forgiveness, grace, and even loving one's enemies (Matthew 5:44). In a world increasingly marked by division and hostility, such Christlike love becomes a powerful witness.

Jesus also clarified the nature of His kingdom (John 18:36): "My kingdom is not of this world". This truth reminds Christians that their primary allegiance is to God's eternal reign—not to earthly governments, ideologies, or systems of control. As global power structures become more centralized and potentially coercive, this teaching takes on renewed relevance.

In the Sermon on the Mount, Jesus laid out principles for righteous living: integrity, humility, mercy, peacemaking, and trusting God's provision (see Matthew 5–7). These teachings form a moral compass for navigating a corrupt world, offering peace and spiritual stability even amid global unrest.

Many of Jesus' parables carried eschatological significance. In the Parable of the Ten Virgins (Matthew 25:1–13), He emphasized the importance of being spiritually prepared. In the Parable of the Talents (Matthew 25:14–30), He urged faithful stewardship of time, resources, and spiritual gifts—especially in times of uncertainty and testing.

Throughout His ministry, Jesus spoke of judgment and accountability. He warned of a day when the Son of Man would return to separate the righteous from the unrighteous, as illustrated in the Parable of the Sheep and the Goats (Matthew 25:31–46). These teachings underscore the eternal consequences of earthly choices and the need for continual repentance and obedience.

Finally, Jesus gave believers a model for how to pray:

"Thy kingdom come, Thy will be done in earth, as it is in heaven" (Matthew 6:10, KJV). The Lord's Prayer aligns the believer's desires with God's redemptive plan for humanity, acknowledging both His sovereignty and our dependence on Him.

In all of these teachings, Jesus provides enduring hope. While He doesn't shy away from describing the hardships and tribulations that will precede His return, He consistently reassures His followers of God's abiding love, sovereign power, and ultimate victory over evil. His promise in John 16:33 — "In this world ye shall have tribulation: but be of good cheer; I have overcome the world" —offers profound comfort and encouragement to believers as they navigate the uncertainties of the end times.

Jesus' teachings also emphasize the centrality of faith. He often praised individuals who demonstrated unwavering trust in God—such as the centurion who believed Jesus could heal from a distance (Matthew 8:10) or the woman with the issue of blood (Luke 8:48). In the context of end-time prophecies, this call to bold, persevering faith becomes even more vital. Believers are urged to trust God's plan and stand firm on His promises, even when chaos, persecution, or global upheaval tempt them toward fear or compromise.

The concept of forgiveness is another foundational theme in Jesus' ministry. He instructed His followers to forgive "seventy times seven" times (Matthew 18:22), signifying boundless grace. This radical forgiveness is especially relevant during times of tribulation, when betrayal, conflict, and moral collapse will tempt people toward bitterness and revenge. Practicing Christ-like forgiveness will be a powerful witness in an age of offense.

Jesus also promised the coming of the Holy Spirit, who would guide, comfort, and empower believers after His ascension. In John 14:26, He assured them: "But the Comforter, which is the Holy Ghost... shall teach you all things, and bring all things to your remembrance, whatsoever I have said unto you." This divine presence equips believers with wisdom and spiritual strength, especially during the trials of the end times.

In His teachings about the end of the age, Jesus stressed the suddenness and unpredictability of His return. In Matthew 24:37–39, He compared it to the days of Noah, when people were "eating and drinking, marrying and giving in marriage" until the flood came and took them all away. This comparison underscores the need for constant spiritual vigilance and moral readiness.

Jesus' teachings on stewardship and eternal priorities also carry weight for those preparing for the end times. In Matthew 6:19–21, He warned, "Lay not up for yourselves treasures upon earth... But lay up for yourselves treasures in heaven... For where your treasure is, there will your heart be also." Wise use of time, resources, and influence becomes even more critical as the world moves toward instability and spiritual deception.

Throughout His ministry, Jesus displayed profound compassion for the marginalized and oppressed—whether healing lepers (Luke 17:12–19), defending women (John 8:1–11), or dining with sinners (Luke 5:29–32). His actions challenged societal and religious norms, offering a model of inclusive love and justice. In times of fear and division, this radical compassion must remain a guiding principle for believers, who are called to reflect His love even in a hostile world.

Jesus also taught candidly about the cost of discipleship. In Luke 9:23–24, He said, "If any man will come after me, let him deny himself, and take up his cross daily, and follow me." He warned that following Him would not always lead to worldly comfort or acceptance, but would often bring reproach, sacrifice, and even persecution. This sobering truth prepares the faithful to endure with courage and conviction when opposition arises.

In conclusion, Jesus' teachings offer a complete and timeless blueprint for living faithfully—not just in ordinary times, but also especially during the prophesied end times. They offer wisdom for navigating deception, encouragement for enduring trials, and hope for final redemption. As the world grows darker and more uncertain, His words remain an unshakable foundation for those seeking clarity, strength, and eternal purpose.

Chapter 80
The Four Horsemen and the
Beasts of Revelation

The seven seals in the Book of Revelation represent a prophetic sequence of events unfolding on earth during the end times. Revelation 6:1–8 introduces the first four seals, describing the Four Horsemen of the Apocalypse, each symbolizing a distinct form of tribulation. These figures have long captivated the imagination of scholars and believers, offering both symbolic and potentially literal interpretations of the trials preceding Christ's return.

The First Horseman – White Horse (Conquest)

Revelation 6:2 says: "And I saw, and behold a white horse: and he that sat on him had a bow; and a crown was given unto him: and he went forth conquering, and to conquer." This rider has often been debated. While some view this figure as representing the Antichrist or deceptive conquest, others, as noted in this interpretation, see it as symbolic of imperial expansion, especially that of the British Empire and other colonial powers.

Historical records estimate that British colonial policies contributed to over 100 million deaths in India between 1881 and 1920, primarily through enforced famines, economic exploitation, and institutional neglect. The crown and bow could symbolize a form of conquest masked as civilization or order—dominance without immediate destruction.

The Second Horseman – Red Horse (War)

Revelation 6:4 states: "And there went out another horse that was red: and power was given to him that sat thereon to take peace from the earth, and that they should kill one another: and there was given unto him a great sword." This horseman is widely recognized as symbolizing global warfare.

The 20th century alone saw two world wars that killed over 100 million people, along with prolonged conflicts in Vietnam, Korea, Iraq, and Afghanistan. Many of these wars were facilitated or sustained through global financial systems, including those influenced by central banks such as the U.S. Federal Reserve, founded in 1913. The "great sword" represents both the physical violence and political powers that instigate widespread bloodshed.

The Third Horseman – Black Horse (Famine and Inequality)

Revelation 6:5–6 describes: "And I beheld, and lo a black horse; and he that sat on him had a pair of balances in his hand... A measure of wheat for a penny, and three measures of barley for a penny." The balances symbolize economic disparity and scarcity.

This rider is linked to major famines in the 20th century, such as the Great Leap Forward in China (1958–1962), which led to an estimated 30–45 million deaths, the Holodomor in Ukraine, and Bengal famine during WWII, each a product of economic mismanagement, ideological extremism, or colonial neglect.

The 2007–2009 Global Financial Crisis, caused largely by reckless banking practices in the U.S., affected billions worldwide, exacerbating poverty and inequality. The symbolism of this rider can extend beyond famine to include economic oppression and the global struggle for equity and justice.

The Fourth Horseman – Pale (or Green) Horse (Death)

Revelation 6:8 says: "And I looked, and behold a pale horse: and his name that sat on him was Death, and Hell followed with him..." The Greek word for "pale" is chloros, which can also be translated as green, symbolizing decay or disease. This rider brings mass death through plague, pandemic, and systemic collapse.

Pandemics like the Spanish Flu (1918, ~50 million deaths), HIV/AIDS (over 36 million deaths), and COVID-19 (which caused over 7 million confirmed deaths globally as of 2024) illustrate this prophecy's modern resonance. If geopolitical tensions in places like Ukraine and the Middle East escalate into global warfare, the potential for World War III raises the terrifying prospect of billions of casualties through nuclear or biological conflict.

The Beast from the Sea – Revelation 13:1–2

Beyond the seals, Revelation 13:1–2 introduces a powerful and ominous symbol: a beast with seven heads and ten horns, rising from the sea. It reads: "And I stood upon the sand of the sea, and saw a beast rise up out of the sea, having seven heads and ten horns, and upon his horns ten crowns, and upon his heads the name of blasphemy. And the beast which I saw was like unto a leopard, and his feet were as the feet of a bear, and his mouth as the mouth of a lion: and the dragon gave him his power, and his seat, and great authority."

Some scholars associate this beast (7 heads, 10 horns, and 10 crowns), 27 nations with a revived global political system, as possibly the European Union (EU).

The EU currently consists of 27 member nations. Notably, 10 countries. Czechia, Estonia, Cyprus, Latvia, Lithuania, Hungary, Malta, Poland, Slovenia, and Slovakia, joined simultaneously in 2004, aligning with the "ten kings" in Revelation 17:12, who "receive power... one hour with the beast."

The Vatican and the Healing of the Wound

Revelation 13:3 says: "And I saw one of his heads as it were wounded to death; and his deadly wound was healed." Some interpretations connect this with the Papacy, particularly the events of 1798, when Napoleon's general captured Pope Pius VI, temporarily stripping the Church of political power.

This "wound" was arguably "healed" with the 1929 Lateran Treaty, which reestablished Vatican City as a sovereign state under Mussolini's regime. Further prophetic resonance is seen in 1984, when the U.S. and the Holy See formally established diplomatic ties under President Ronald Reagan and Pope John Paul II, restoring Vatican influence on a global scale.

Today, the Holy See exercises global influence as both a religious and sovereign entity, participating in international diplomacy and intergovernmental organizations. Its unique dual status—governing both a religion and a state—sets it apart and aligns with Revelation's depiction of the beast with blasphemous power and broad authority.

Economic Control and the Power to Buy and Sell

Revelation 13:16–17 warns that the second beast, the United States, will eventually enforce a system in which "no man might buy or sell, save he that had the mark."

Modern developments in digital currency, Central Bank Digital Currencies (CBDCs), and cashless payment technologies are being scrutinized through this prophetic lens.

The global rise of QR code payments, biometric authentication, and digital ID systems create a potential infrastructure for economic exclusion—fulfilling the prophecy that economic participation could one day be conditional.

Diverse Interpretations and Ongoing Relevance

It's important to acknowledge that interpretations of Revelation vary widely. While some see direct historical and institutional correlations, others view the prophecies as symbolic, spiritual, or yet to be fulfilled in a future global system not yet fully realized.

Whether read as historical allegory, spiritual metaphor, or future roadmap, the messages in Revelation compel believers to remain watchful, discerning, and rooted in Christ.

The imagery of the beasts in Revelation draws rich parallels with the visions recorded in the Book of Daniel, forming a composite tapestry of prophetic symbolism that spans both the Old and New Testaments. This interconnectedness underscores the unity of biblical prophecy and reinforces the consistent themes of God's sovereignty, the rise and fall of earthly empires, and His ultimate victory over evil.

As we consider these prophecies, it's important to remember their central purpose: not to incite fear, but to offer hope, strength, and encouragement to believers facing persecution or uncertainty. Revelation was written to a church under duress, and its message remains timeless: God is in control, Christ has triumphed, and justice will prevail. While interpretations of how these prophecies will unfold may differ, the unshakable foundation of Christian faith is that God will overcome the forces of darkness and restore righteousness.

The beasts of Revelation—fearsome in appearance and commanding in influence—serve as stark warnings about the corrupting nature of unchecked power and the dangers of systems that rebel against God's authority. These beasts personify institutions, ideologies, and coalitions that seek to dominate politically, economically, and spiritually, often presenting themselves as saviors while working in opposition to divine truth. They remind believers to remain spiritually discerning and not place ultimate trust in human systems, political alliances, or charismatic leaders, but in the sovereignty of God alone.

At the same time, these prophecies demand wisdom and interpretive humility. Throughout church history, well-intentioned believers have sometimes misread or misapplied apocalyptic texts, leading to unnecessary panic, divisiveness, or even harmful actions. It is vital to approach these prophecies with a heart that is anchored in Scripture, guided by the Holy Spirit, and centered on the hope of redemption rather than the fear of destruction.

The symbolism of the Four Horsemen and the beasts also invites deeper reflection on the spiritual dimensions of evil and its many manifestations throughout human history. While these figures may correlate with specific empires or historical events, they also represent enduring archetypes: conquest, war, famine, death, and systemic rebellion. These are not merely historical realities, but ongoing spiritual forces—consequences of sin that continue to afflict humanity.

In this light, the Four Horsemen can be understood as symbolic of the global suffering that arises from human rebellion and brokenness. Their emergence signals the intensification of spiritual warfare and the culmination of history's long battle between the kingdom of God and the kingdom of darkness. This perspective calls believers to active engagement in spiritual warfare—through prayer, discipleship, sacrificial love, and resistance to evil in every form.

The beast rising from the sea, with its seven heads and ten horns (Revelation 13), represents a culmination of oppressive worldly systems—political, economic, and religious powers that align in opposition to God's kingdom. This beast is not merely a figure of brute force; it is a sophisticated, seductive power structure that demands allegiance and wages war against the saints (Revelation 13:7).

Its authority, derived from the dragon (commonly interpreted as Satan), emphasizes the spiritual undercurrent behind earthly events, showing that the ultimate battle is not merely political but cosmic in scale.

The prophecy about the beast's "deadly wound" being healed—and the world's amazement and worship that follow—serves as a solemn warning about the allure of evil disguised as revival or peace. Evil may appear defeated, only to rise again in a more powerful or deceptive form. This scenario reminds believers that external appearances can be misleading and that discernment is required to recognize spiritual deception. Even when evil seems to prevail, the faithful are called to endure, resist, and remain rooted in truth.

The mark of the beast, also described in Revelation 13, is closely linked to these prophetic symbols. While its exact form is still debated—whether literal, symbolic, digital, or biometric—it represents a forced alignment with a system that denies God's authority. The mark is ultimately about allegiance. To accept the mark is to willfully participate in a system that exalts itself above God and suppresses His people. The gravity of this decision is underscored by Revelation's severe warnings about its eternal consequences.

In contrast to the beast's system of coercion and fear, God's people are marked with His seal—a sign of ownership, protection, and divine purpose (Revelation 7:3; 14:1). This spiritual seal reflects the believer's identity in Christ and serves as a counter-testimony to the world's systems of control.

Conclusion

The Four Horsemen and the beasts of Revelation form a dramatic and sobering depiction of the final stages of history—a time when spiritual deception, suffering, and opposition to God will reach their peak. Yet they are not merely about catastrophe—they are also about the clarity that comes in contrast. Evil will be fully revealed, but so too will the faithfulness of God's people and the certainty of Christ's return.

Though their specific identities and timing are debated, the message is clear: Remain faithful. Remain alert. Do not be deceived. These prophecies are a divine call to resilient discipleship, reminding the Church of its role as a witness to truth in a world increasingly bent on falsehood.

In the end, Revelation is not a book of despair—it is a book of victory, worship, and hope. It closes not with destruction but with renewal: a new heaven and a new earth, where God dwells with His people and wipes away every tear. That is the promise believers hold fast to—even as the storm clouds gather.

Let's be clear that the first beast of Revelation is EU and the Vatican. The second beast is the United States. NATO is mostly an alliance between the United States and EU.

Chapter 81
The Reign of Seven Kings and the
Future of Faith

Revelation 17 describes the reign of seven kings, which can be interpreted as a succession of empires that have shaped world history. This prophetic timeline provides a framework for understanding the progression of global powers leading up to the end times. Let's explore this succession of empires and their potential significance:

1. Egypt (Ramses 1279-1213 BCE): The ancient Egyptian empire, known for its advanced civilization and monumental architecture, represents the beginning of this prophetic timeline.
2. Assyria (Ashurbanipal 669-631 BCE): With its military might and cultural influence, the Assyrian Empire succeeded Egypt as a dominant power in the ancient Near East.
3. Babylon (Nebuchadnezzar the Great 605-562 BCE): The Babylonian Empire, famous for its hanging gardens and the exile of the Jewish people, followed Assyria.
4. Mede-Persia (Cyrus the Great? -530 BCE): This vast empire, which allowed the Jews to return from exile and rebuild Jerusalem, came next in succession.
5. Greece (Alexander the Great 336 – 323 BCE): The Greek Empire, spreading Hellenistic culture across a vast territory, followed the Mede-Persian Empire.
6. Rome (Augustus 31 BCE – 14 AD): The Roman Empire, which was in power during the time of Christ and the writing of the New Testament, succeeded Greece.
7. Ottoman Empire (36 Sultans between 1299-1922): Some interpretations suggest that the Ottoman

Empire, which controlled much of the Middle East and southeastern Europe for centuries, represents the seventh king.

This progression of empires provides a historical context for understanding biblical prophecy and the development of world powers. While interpretations of the "seven kings" or empires described in the Book of Revelation may differ, many scholars agree that successive world empires have played defining roles in shaping prophetic history, especially those that oppressed God's people or opposed divine authority.

Turkiye, the modern successor to the Ottoman Empire, holds a unique and often overlooked position in both ancient and prophetic history. As the custodian of land that hosted some of the earliest known human settlements, Turkiye is home to Göbekli Tepe, a Neolithic site dated around 8000 BCE. Its massive stone structures and enigmatic carvings challenge conventional timelines and raise profound questions about early human spirituality and civilization.

Turkiye is also a land where Christianity and Islam intersect, a nation with deep Christian roots that now identifies primarily as an Islamic republic. This spiritual tension adds to its prophetic significance.

The region's connection to early Christianity is especially notable. The Council of Nicaea, convened in 325 AD under Emperor Constantine, was held in what is now northwestern Turkiye. This council was foundational in shaping orthodox Christian doctrine. Additionally, the island of Patmos, where the Apostle John received the visions recorded in Revelation, lies just off Turkiye's Aegean coast. Most significantly, the seven churches addressed in Revelation (Ephesus, Smyrna, Pergamum, Thyatira, Sardis,

Philadelphia, and Laodicea) were all located in ancient Asia Minor—modern-day Turkiye. This geographic detail grounds key portions of Revelation in a very real and enduring region, suggesting Turkiye may yet play a pivotal role in end-time developments.

Turkiye's historical role as the seat of two influential empires—the Eastern Roman (Byzantine) Empire and the Ottoman Empire—continues to shape its identity. The Byzantine Empire preserved Christianity in the East for over a millennium, while the Ottoman Empire, following its conquest of Constantinople in 1453, expanded Islamic rule across three continents, enduring until 1923. This imperial legacy contributes to modern Turkiye's sense of destiny and geopolitical assertiveness.

In today's world, Turkiye stands at a strategic crossroads—literally and figuratively between East and West. It has been a member of NATO since February 18, 1952, traditionally aligning with Western powers. However, recent actions under its current leadership suggest a recalibration. Turkiye has expressed interest in joining BRICS, a growing economic coalition led by Brazil, Russia, India, China, and South Africa. BRICS extended invitations in 2024 to include Egypt, Iran, Saudi Arabia, Ethiopia, the UAE, and Argentina. This possible realignment could drastically reshape regional alliances and global power dynamics.

At the same time, Turkiye has spent over two decades attempting to join the European Union, with progress continually stalled. As the country with the second-largest military in NATO, Turkiye's shifting loyalties and potential dual alignments—with both BRICS and Western institutions—could signal the emergence of a new global coalition or transitional phase.

This brings us back to the mysterious "woman in purple and scarlet" who rides the beast in Revelation 17—a symbol widely interpreted as a powerful religious-political system that dominates and influences kings and nations. Revelation 17:11 states:

"And the beast that was, and is not, even he is the eighth, and is of the seven, and goeth into perdition."

This enigmatic passage suggests the rise of an eighth kingdom, emerging from one of the previous seven, yet distinct in its nature and ultimate fate. While this eighth system is not fully manifest, we may be witnessing its formation in transitional alliances, centralized global authority, or unprecedented technological control.

Revelation continues:

"These have one mind, and shall give their power and strength unto the beast. These shall make war with the Lamb, and the Lamb shall overcome them: for He is Lord of lords, and King of kings: and they that are with Him are called, and chosen, and faithful." (Revelation 17:13–14)

This prophecy reveals a final coalition of powers that will unite in opposition to Christ and His followers. Though they appear invincible, they are destined for defeat. The Lamb— Jesus Christ—will overcome them, and with Him will be the faithful remnants, who did not yield their allegiance to the beast.

Chapter 81
The Reign of Seven Kings and the
Future of Faith

Conclusion

Turkiye's spiritual and historical legacy, its imperial past, and its modern geopolitical evolution all contribute to its potential significance in end-time prophecy. Its proximity to ancient Christian sites, its pivotal role in early church history, and its central place in current global affairs may signal that it is more than a bystander in the unfolding drama of Revelation.

In 2025, Turkiye is host to peace negotiations between Russia and Ukraine to end the war that already killed or injured more than a million people.

Whether Turkiye emerges as a central player in the eighth kingdom or remains a bridge between power blocs, its story reminds us that biblical prophecy is not abstract—it is rooted in real lands, real leaders, and real decisions. And above all, the battle lines drawn in Revelation are not merely geopolitical—they are spiritual, calling every believer to stand firm, remain faithful, and keep their eyes fixed on the Lamb who will ultimately reign over all.

Chapter 82
The Catholic Church and Pedophilia

Revelation 17:3–4 declares:

"So he carried me away in the spirit into the wilderness: and I saw a woman sit upon a scarlet coloured beast, full of names of blasphemy, having seven heads and ten horns. And the woman was arrayed in purple and scarlet color, and decked with gold and precious stones and pearls, having a golden cup in her hand full of abominations and filthiness of her fornication."

These verses offer some of the most vivid and controversial imagery in the Book of Revelation, often interpreted as a symbol of a powerful religious system intertwined with worldly political powers. The purple and scarlet attire is especially significant—colors traditionally associated with authority, wealth, and religious ceremony.

In the Roman Catholic Church, bishops wear purple, while cardinals wear scarlet red. This combination is unique among religious institutions and aligns strikingly with the description in Revelation. Additionally, the adornment with gold, precious stones, and pearls mirrors the opulence and pageantry often observed in ecclesiastical traditions, particularly in the Vatican.

The City on Seven Hills

Revelation 17:9 adds another clue:

"And here is the mind which hath wisdom. The seven heads are seven mountains, on which the woman sitteth."

Many scholars and commentators throughout church history have connected this verse to Rome, which historically was built upon seven hills: Palatine, Aventine, Caelian, Esquiline, Viminal, Quirinal, and Capitoline. This geographical detail is often cited as evidence that the prophetic "woman" represents a religious-political power headquartered in Rome.

Revelation 17:18 strengthens this association:

"And the woman which thou sawest is that great city, which reigneth over the kings of the earth."

The only city in the world that functions as both a religious headquarters and an independent sovereign state is Vatican City, the administrative and spiritual center of the Roman Catholic Church. It maintains diplomatic relations with over 180 countries and holds observer status in the United Nations.

Systemic Abuse and the Apostate Church

Despite its spiritual claims, the Catholic Church has been plagued by decades of sexual abuse scandals, involving thousands of priests and devastating hundreds of thousands of victims.

- On February 28, 2024, a search for "How many pedophile priests have gone to jail since Pope Francis took over the Vatican?" revealed virtually none.
- On October 5, 2021, Al Jazeera reported that an investigation in France revealed more than 216,000 victims of sexual abuse by around 3,000 clergy and church workers between 1950 and 2020. According to Jean-Marc Sauvé, the abuse was "systemic"—and 80% of the victims were boys.

- On October 27, 2023, the BBC published a damning report: more than 200,000 children were sexually abused by Spain's Catholic clergy. Spain's ombudsman condemned the Church's culture of denial and cover-up, calling it a "devastating" breach of trust.
- In the United States, Catholic dioceses have paid over $3 billion in settlements. The infamous Boston Globe investigation (2002), later dramatized in the film Spotlight, uncovered widespread abuse and systematic concealment by Church leadership.

In 2001, Pope John Paul II apologized for abuse within the Church, calling it a "profound contradiction" of Christ's teachings. But his critics argue he did not act swiftly or decisively to protect victims. His offer of indulgences in 2000, just before the flood of lawsuits and settlements began, raises troubling questions about the Church's awareness of the scandal's scope.

Pope Benedict XVI expressed "shame" over the crisis and met with some victims, but under his leadership, the Church's accountability mechanisms remained opaque.

When Pope Francis was elected in 2013, hopes ran high that he would finally implement reform. But when asked shortly after his election whether the new Pope would truly "clean up the mess," a Jesuit priest at Santa Clara University responded simply: "Nothing will happen."

Indeed, the problem persists. In 2024, survivors and advocates continue to protest the lack of criminal accountability and the Church's slow progress on transparency.

Apostasy and False Teaching

Pope Francis has also drawn criticism for interfaith statements that appear to contradict foundational Christian doctrine. In a September 2024 visit to Singapore, a first papal visit in nearly 40 years, he told a group of young people:

"All religions are paths to God… There is only one God, and religions are like languages, paths to reach God."

This statement directly contradicts Jesus' words in John 14:6:

"Jesus saith unto him, I am the way, the truth, and the life: no man cometh unto the Father, but by me."

Such ecumenical pluralism aligns with the "apostate church" described in Scripture—a church that claims religious authority but denies essential truths. In 2 Thessalonians 2:3, Paul warned that a great falling away—apostasia—would precede the rise of the antichrist. Apostasia is a Greek word that translates to apostasy, meaning the abandonment of a religious or political belief, or a formal renunciation of a faith.

The Catholic Church is not alone in this drift. Many Protestant denominations have also embraced doctrines and ideologies contrary to Scripture. The apostate church, therefore, is not defined by denomination but by its rejection of Biblical truth, even while maintaining a religious appearance.

Revival in the Global South

Amid this darkness, the light of the Gospel continues to spread. Christianity is growing rapidly in regions often overlooked by Western institutions:

- In **Africa**, the number of Christians has risen from about **10 million in 1900 to 734 million in 2024**.
- In **Latin America**, millions remain faithful to Christian doctrine despite cultural and institutional challenges.
- Even in **China**, where Christianity is heavily persecuted, **underground house churches** are thriving.

According to the PEW Research Center, "As of 2010, about a quarter of the global Christian population was in Europe (26%), a quarter in Latin America and the Caribbean (25%) and a quarter in sub-Saharan Africa (24%). Significant numbers of Christians also live in Asia and the Pacific (13%) and North America (12%). Less than 1% live in the Middle East-North Africa region, where Christianity began. Sub-Saharan Africa is predicted to have the largest share of the world's Christians, rising from 24% in 2010 to 38% in 2050, Europe's share is projected to drop to about 16% from 26%. Latin America and North America are also expected to see modest declines in their respective shares of the global Christian population.

Conclusion

The prophetic image of the woman clothed in scarlet and purple, seated on a beast with seven heads and ten horns, is one of seduction, corruption, and spiritual deception. Many elements—color symbolism, geographic location, global influence, religious opulence, and systemic abuse—appear to converge in the modern Roman Catholic Church, though this interpretation is not without controversy.

Yet, the warning is not limited to any one institution. The real danger lies in any religious system that rejects the authority of Christ, prioritizes power over truth, and covers sin rather than repenting of it.

Still, the Gospel continues to advance. And as Revelation promises, the Lamb—Jesus Christ—will overcome every false power. His followers, the "called, and chosen, and faithful," will stand victorious with Him.

"Come out of her, my people, that ye be not partakers of her sins." (Revelation 18:4)

Chapter 83
Evidence of End-Time Prophecies

Isaac Newton, one of the most outstanding scientists, needs no introduction. Here, we'll focus not on all the remarkable achievements he had in physics and mathematics but on his research rooted in a personal pursuit of God. In fact, Newton wrote more than a million words on theology—far more than he ever published on science. He devoted more time to biblical exegesis and alchemical studies than to physics and mathematics.

In particular, he spent most of his time studying the Book of Daniel and the Book of Revelation in the Bible. These two books are about end-time prophecies, and I'll share with you some of the prophecies that have been fulfilled. I'll also discuss what is being or about to be fulfilled. Newton saw these prophetic texts not merely as spiritual metaphors, but as cryptic records of real historical processes and future world events. He approached them with the same logical rigor he applied to science.

Newton has been the most influential scientist of all time, formulating the laws of mechanics, the law of universal gravity, and other laws to describe everyday objects in motion to rockets taking off for space exploration. Without Isaac Newton's insights, we could not have the Hubble Telescope or the James Webb Space Telescope (JWST). His invention of the reflecting telescope in 1668 eliminated chromatic aberration found in refracting telescopes and became the foundation for modern astronomical instruments.

I would not have seen the Andromeda Galaxy without looking through a 22" Reflector Telescope if not for his

695

reflector telescope invention. The Andromeda Galaxy, our nearest galactic neighbor, is located about 2.5 million light-years away and can be observed using large reflecting telescopes—technology made possible by Newton's design innovations.

He also co-invented calculus, without which there would be no physics or engineering. Although the invention of calculus was contested by Gottfried Wilhelm Leibniz, both Newton and Leibniz independently developed the framework that underpins modern mathematics, physics, and engineering.

Isaac Newton started his most prolific scientific work during the Great Plague of London, lasting from 1665 to 1666. The plague killed an estimated 100,000 people, almost a quarter of London's population, in 18 months. He spent two years at his mother's farm in Woolsthorpe, hiding from an outbreak of the plague that took hold of Cambridge, where he was a student. This period became known as Newton's "Annus Mirabilis" (year of wonders), during which he developed early ideas about gravity, optics, and the foundations of calculus.

To Isaac Newton, science was a portal to God's mind, a bridge between humans and the Divine. Newton, a name that represents the quintessential rationalist, was, in fact, a rational mystic. He believed that science was like a religious practice, a meeting with God's mind. In Newton's view, natural laws were not autonomous—they were decrees issued and upheld by God.

In studying the Book of Daniel and the Book of Revelation during the Great Plague, Newton surely wondered whether he was living in the end times. He calculated the year when the end would come. The most definitive date he set for the apocalypse was 2060. This calculation was based on his interpretation of Daniel 12:7 and Revelation 11:2–3, and his belief that the end times would follow 1,260 years after the establishment of the papacy—placing the culmination around the year 2060. Newton was careful to state that this was not a prediction of the end of the world, but the end of ecclesiastical corruption and the beginning of a renewed era.

On September 23, 2017, the Bible's Book of Revelation 12:1-2 was fulfilled in a spectacular event that could be seen by millions of people. On that date, many Christian prophecy watchers noted a rare astronomical alignment involving the constellation Virgo (the Virgin), the sun, the moon, and the planet Jupiter. They interpreted this as a fulfillment of Revelation 12:1–2, which describes a 'woman clothed with the sun, with the moon under her feet, and a crown of twelve stars on her head.'

The previous time such an event happened was seven thousand years ago. Astronomers clarified that while elements of this alignment have occurred before, the 2017 arrangement was especially unique, prompting much public attention. However, mainstream scientists and theologians remain divided on whether this event constituted an actual prophetic fulfillment.

The prophecy said, "And there appeared a great wonder in heaven; a woman (Constellation Virgo) clothed with the sun, and the moon under her feet, and upon her head a crown of twelve stars (nine stars of Constellation Leo with Mercury, Venus, and Mars). And she is being with child (Jupiter inside Virgo) cried, travailing in birth, and pained to be delivered." This exact cosmological alignment was a sign of the end time prophesied in the Bible about two thousand years ago.

A total solar eclipse is a rare event that happens once every 1–2 years somewhere on Earth. However, the path of totality—a narrow corridor where the total eclipse is visible—only covers a small portion of the planet each time, making any specific location's experience of a total eclipse extremely rare. Two total solar eclipses (seven years apart) that mark an X in the middle of a specific country that will play a key role in the end time is extremely rare.

The total solar eclipse of August 21, 2017, moving from northwest (Oregon) to southeast (South Carolina), and the total solar eclipse of April 8, 2024, spanning from southwest (Texas) to northeast (Vermont), mark an X in the middle of the United States. This intersection occurs near the town of Makanda in southern Illinois, close to the New Madrid Seismic Zone—an area known for catastrophic earthquakes in the early 1800s. In 2017, I drove from California to Yellowstone Park in Wyoming to watch the total solar eclipse, which blew my mind. When I saw the total eclipse, I experienced something like a flash (not exactly) in my mind. The most accurate way to describe that experience is that it literally blew my mind.

On April 8, 2024, millions of people watched the total eclipse in person in Mexico, the US, and Canada. Mazatlán, Mexico, was the first place it arrived at, and many people enjoyed it. The Indianapolis Motor Speedway, where the world-famous Indy 500 annual automobile race is held, and Niagara Falls were some of the most popular spots where people watched the total eclipse. NASA reported that the eclipse path covered over 15 U.S. states and was observed by more than 30 million people in person, with tens of millions more watching online. I liked the beaches in Mazatlán, the Indy 500 race, and Niagara Falls when I was there years ago.

One of the most important end-time prophecies soon to be fulfilled is written in the Book of Revelation 13:14–18: "And deceiveth them that dwell on the earth by means of those miracles which he had the power to do in the sight of the beast; saying to them that dwell on the earth, that they should make an image to the beast, which had the wound by a sword, and did live."

The United States is the Second Beast of Revelation. It is the beast coming up out of the earth, and he has two horns like a lamb, and he spake like a dragon. Coming up out of the earth means a place lightly populated when Revelation was written almost two thousand years ago. Jesus Christ was the Lamb of God. Christians founded the United States of America. The Continental Congress adopted the Declaration of Independence on July 4, 1776.

On June 21, 1788, the Constitution was ratified. In 1791, the First Amendment to the Constitution affirmed the separation of church and state. Unlike the first beast with ten crowns, the lamb has no crown. The US has no king. It is a republic. The two horns may symbolize its dual founding principles— civil liberty and religious freedom. In 1955, 92% of the US population was Christian. He spoke as a dragon, which is in line with being the only superpower in the world since the early 1990s. A multi-polar world, however, is emerging. Since the collapse of the Soviet Union in 1991, the United States has maintained global military and economic dominance, though the rise of China, Russia, and BRICS nations now signals a shift toward multipolarity.

The United States created the Federal Reserve, which facilitated WWI, WWII, and many other wars, killing over 100 million people. The Federal Reserve was created in 1913 as the central banking system of the United States. Though not directly causing wars, its monetary policies have played roles in funding U.S. military engagements and shaping global economic trends. The US caused the "Great Depression" and "Great Recession."

The Great Depression began with the U.S. stock market crash in 1929, spreading globally. The 2008 financial crisis, triggered by risky mortgage-backed securities and lax regulation in the U.S. housing market, nearly collapsed the global financial system. The Federal Reserve and major U.S. banks were central actors in both crises.

The global financial system was only days away from a total meltdown in 2008 due to America's corruption. Former U.S. Treasury Secretary Henry Paulson and Federal Reserve Chair Ben Bernanke later admitted the global financial system was within hours of collapse during the Lehman Brothers failure. It is guilty of the seven deadly sins.

Greed, pride, envy, wrath, lust, gluttony, and sloth—often considered symbolic rather than literal indictments—are increasingly cited by moral commentators as descriptors of modern Western excess, particularly in critiques of American materialism, military expansion, and cultural exports.

Greed – The top 1% of Americans have 10 times more wealth than the bottom 50%.

Anger – highest homicide rate in the world

Lust – Adultery, fornication, and pornography are rampant

Gluttony – Eating and drinking excessively, with two-thirds of the people obese or overweight

Envy – People having a mostly negative feeling of desire for something that someone else has, and you do not

Sloth – Apathy, indifference, and depression that can lead to unfulfilled duties and obligations for many people

Pride – The strongest military superpower in the world and the country with the highest GDP, neglecting its $37 trillion debt.

The first Beast of Revelation is more complicated. There are 27 nations in the beast with seven heads, ten horns, and ten crowns. The only entity with 27 nations is the European Union. In Revelation 13:1–2, the beast is described as rising from the sea—often interpreted symbolically as peoples, nations, and languages. The seven heads and ten horns echo imagery in the Book of Daniel, linking ancient empires with modern systems of global governance. Vatican City, an independent country and the smallest nation in the world, is surrounded by Italy. Though not a member of the European Union, Vatican City maintains diplomatic relations with the EU and shares geographical, cultural, and economic ties with Europe.

The European Union Constitution doesn't even mention God, as if God doesn't exist. This omission was a deliberate decision made during the drafting of the Treaty establishing a Constitution for Europe in 2004, following contentious debates among member states about the role of Christianity in Europe's history. Instead, it refers only to the "cultural, religious and humanist inheritance of Europe."

If there is no God, there are no Ten Commandments. Does it make it easier for the EU to steal Russian government money from its banks? In 2022, following Russia's invasion of Ukraine, the EU froze over €300 billion of Russian Central

Bank assets. As of 2024, the EU is considering using interest generated from these frozen assets to support Ukraine, which has sparked international legal debates.

Nowadays, worshippers of God rarely attend most of the great churches in Europe. Church attendance has declined across Western Europe, with Pew Research (2018) showing that only around 10–20% of Christians in countries like Germany, France, and the Netherlands attend church weekly.

Will the EU be better off under Darwinism or under God? While secularism and evolutionary science dominate public discourse in much of Europe, discussions about moral authority, tradition, and spiritual meaning continue in various academic and religious circles. In the US, within the last 25 years, the number of respondents who say that religion is "very important" to them has declined from 62 percent to 39 percent. (Source: Gallup, 2023.) It is hard to imagine that the EU has become a common thief.

The European Union has had a unique partnership with 27 European countries since 2013, known as Member States or EU countries. Together, they cover much of the European continent. The EU is home to around 447 million people, which is around 6% of the world's population. The UK left the EU in 2016, and Sweden joined the NATO military alliance on March 7, 2024, not the EU, as it has been a member of the European Union since 1995.

The EU countries are: Austria, Belgium, Bulgaria, Croatia, the Republic of Cyprus, Czech Republic, Denmark, Estonia, Finland, France, Germany, Greece, Hungary, Ireland, Italy, Latvia, Lithuania, Luxembourg, Malta, Netherlands, Poland, Portugal, Romania, Slovakia, Slovenia, Spain, and Sweden.

Are the European Union and the Papacy parts of the first beast of Revelation? Many Bible prophecy scholars debate this. Some view the EU and Papacy as symbolic of revived Roman imperial and religious power. Revelation 17:9 mentions "seven mountains," which some interpret as the seven hills of Rome—possibly referencing the Vatican's influence. However, interpretations vary across denominations and theological traditions.

Since Donald J. Trump became US President for the second time on January 20, 2025, the world has been turned upside down. The United States withdrew from the World Health Organization and the Paris Climate Accords. He instigated a trade war with Canada and Mexico by proposing tariffs and installing new tariffs on China. He wants to annex Canada, Greenland, and the Panama Canal. He proposed forcibly relocating the Palestinian population from the Gaza Strip to other Arab states and rebuilding Gaza into a tourist destination. These policy developments, though hypothetical as of mid-2025, reflect an imagined second Trump term based on his previous foreign policy postures.

No official record exists of annexation plans for Canada or Greenland, though Trump expressed interest in purchasing Greenland during his first term.

The Trump administration suspended all military aid to Ukraine, offered concessions to Russia, and requested half of Ukraine's oil and minerals as payment for U.S. support. On February 28, 2025, he publicly had a blowup with Ukraine President Zelenskyy in the Oval Office in front of the press.

Trump accused Zelenskyy of potentially causing WWIII. He also established the Department of Government Efficiency, or DOGE, led by Elon Musk, to cut spending, limit federal bureaucracy, and oversee mass layoffs across federal agencies.

DOGE is a fictional agency; however, Elon Musk has expressed interest in government reform and efficiency. His administration has also publicly rebuked NATO and the EU. These are just some of the glaring changes in just six weeks. It is an understatement to say turbulent times are ahead for the world.

Europe's three most powerful nations, Germany, France, and the United Kingdom, are afraid that, without the US's full commitment to NATO, Russia will invade them in the future. Are they afraid because the armies of Hitler and Napoleon invaded Russia in the past, and Winston Churchill asked for American participation in invading Russia in 1945? There is no confirmed historical evidence that Churchill formally proposed invading the Soviet Union in 1945, but Operation Unthinkable was a British military contingency plan to confront the USSR, declassified in 1998.

In Germany's 2025 election, Friedrich Merz's conservatives won, but Alternative for Germany (AfD) doubled its support in just four years to 20.8% and became the second biggest political force in parliament. Outgoing Chancellor Olaf Scholz's SPD had its worst performance in decades, with only 16.4%. Under Merz's leadership, Germany's relationship with the US may not be smooth. Merz, as a CDU leader, advocates a strong transatlantic alliance, but internal EU divisions and populist pressures, such as the rise of AfD, complicate foreign policy consistency.

The EU is trying to create an army of its own, despite opposition from some of its members. The concept of a European Union defense force has been debated for decades. The EU's Permanent Structured Cooperation (PESCO), launched in 2017, is a step in that direction. However, NATO remains the dominant defense alliance, and several EU members—like Poland and the Baltic states—prefer continued reliance on NATO over a unified EU army.

We are living in challenging times. We must strive and pray for peace.

When you need God's protection, provision, and guidance, remember Psalm 23 (A Psalm of David).

"The LORD is my shepherd; I shall not want. He maketh me to lie down in green pastures: he leadeth me beside the still waters. He restoreth my soul: he leadeth me in the paths of righteousness for his name's sake. Yea, though I walk through the valley of the shadow of death, I will fear no evil: for thou art with me; thy rod and thy staff they comfort me. Thou preparest a table before me in the presence of mine enemies: thou anointest my head with oil; my cup runneth over. Surely goodness and mercy shall follow me all the days of my life: and I will dwell in the house of the LORD forever." (Psalm 23, KJV)

Your journey on this earth began as a vulnerable baby, and your life itself is a miracle. Be grateful to all who cared for you—parents, guardians, mentors, and even strangers who played a part in your survival and growth.

Globally, an estimated 73 million abortions occur each year, according to the World Health Organization (WHO). This number eclipses the death tolls from wars and pandemics combined. Congratulations to you—you are among those who were given the gift of life. You get to live.

You were fortunate to be born. You are alive, breathing, and hopefully healthy. That, in itself, is cause for gratitude and celebration. Every choice you make transforms both yourself and your surroundings. How you live defines and reveals the identity of your soul.

Everyone is entitled to believe what he or she wants. Respecting others' beliefs is part of living in harmony. But that does not mean we abandon judgment altogether. Sound, righteous judgment—rooted in wisdom—is vital. The Bible itself teaches that we are to judge rightly (John 7:24), though ultimate judgment belongs to God.

This book is, in essence, about discovering the truth—not just as an abstract concept, but also as a personal journey of identity and purpose. It is also about rediscovering why we were born—why we are here on this earth.

Some people don't believe they were born with a purpose. But those who know who they are often have a clear sense of why they are here. Identity and purpose are intertwined. If your identity has been defined by others—parents, peers, society, or institutions—then your purpose often feels uncertain, even irrelevant.

Many people live with confusion about their true nature. Lost in the mist of illusion and the seduction of the material world, they wander without spiritual direction. These are the souls who grieve not just their spirit, but also the Spirit that longs to guide them. They are lost in the forest of worldly desires on the island of death, unaware that just beyond the shadows lies the sea of eternity—an endless ocean of divine meaning, purpose, and peace.

For those deeply immersed in fantasy worlds—whether through video games, media, or addictive distractions—truth can seem distant or irrelevant. If this describes you, my prayer is that you find clarity, even if only in a single moment of silence. That spark can ignite a lifetime of change.

We can choose what to believe, but our choices do not alter the truth itself. Each of us sees reality through a particular lens. Some call it God. Others call it Nature. Some invoke the Universe, Consciousness, or even Chaos. But all of us, in our own way, are trying to describe the same mysterious Source behind existence.

I once visited a Mayan temple complex with a long stone path. As I walked, I noticed I couldn't see the path clearly ahead. The sun was behind me, and my own shadow fell directly in front of me—obscuring my view. It struck me then: I was the one blocking my path. How often do we do the same in life—letting our own fears, ego, and doubts obscure the way forward? Physically, emotionally, spiritually—what parts of you are casting shadows on your destiny?

Jesus said, "I am the way, the truth, and the life." What is blocking you from seeing Him for who He claimed to be? We often hear that there are many ways to do something— and yes, in tasks, that may be true. But in life's most essential matters—truth, salvation, peace—the path may not be as wide as we assume.

Climbers ascending Mount Everest from the Nepal side follow a narrow, treacherous route through the Hillary Step. On the Tibet side, the route is different, equally dangerous, and also leads to the summit. The lesson? There may be multiple paths to the mountain, but each requires commitment, humility, and perseverance. Your spiritual path is no different. Choose wisely.

Chapter 83
Evidence of End-Time Prophecies

Each day, I practice spiritual surrender. I let go of everything—beliefs, relationships, material possessions, even my identity—if only for a moment. I release my anxieties, shame, pride, and every form of emotional weight. I release the seven deadly sins: pride, greed, wrath, lust, gluttony, envy, and sloth. In doing so, I clear the space in my soul to start again—with God as my foundation.

I reflect on the moments when I have truly felt God's presence—quiet, powerful, unmistakable—and also the moments when God seemed silent or absent. I carry both. They shape my faith. In prayer, I open my day. Not with demands, but with awareness. Not for blessings alone, but for the strength to walk in God's presence—step by step, moment by moment.

Chapter 84
Reflections on Major Transitions
and the Future

Would Catholicism continue perpetually, or will "The Last Pope" exist in our lifetime? According to the 12th-century "Prophecy of the Popes" attributed to Saint Malachy, some believe Pope Francis was the final pontiff before the end times—though the Vatican has never endorsed this prophecy. Among the one billion Catholics, how many can name even twelve cardinals in the top hierarchy of the Vatican? There are currently over 200 cardinals globally, but only about 120 are eligible to vote in a papal conclave. The average lay Catholic remains largely unfamiliar with the College of Cardinals, reflecting the distance between hierarchy and laity.

Pope Francis, who changed the church forever, died on April 21, 2025, and many books will be written on what he did to change the world. He was known for championing environmental stewardship (as seen in Laudato Si'), interfaith dialogue, and progressive stances on social issues like LGBTQ+ inclusion and economic inequality. His efforts to decentralize church authority and push for synodality redefined papal leadership in the modern age.

When the enlightened Dalai Lama dies, who can adequately replace him as the spokesman for Buddhism? The current 14th Dalai Lama, Tenzin Gyatso, has suggested the possibility that his reincarnation may not continue or could be chosen outside Tibet—potentially even as a woman—challenging centuries-old tradition. If the previous Dalai Lamas keep reincarnating, how would their souls ever achieve Nirvana? In Tibetan Buddhism, the bodhisattva ideal explains this: enlightened beings voluntarily delay their own final liberation in order to serve others across lifetimes. This self-sacrificial cycle is not seen as failure to attain Nirvana, but as the highest spiritual service. The Dalai Lama is already in his nineties and many books will also be written about him. Born in 1935, his impact on global peace, Tibetan identity, and interreligious harmony will endure long after his passing.

When the Dalai Lama came to Santa Clara University in 2014 to speak on the topic of "Compassion, Business, and Ethics," I received an invitation to attend as an advisory board member of the distinguished Ignatian Center for Jesuit Education and I learned from the wisdom he shared.

The Nature of Life and Remembrance

Empires rise and fall. Many have lived, but few are remembered. We are cast members in an eternal play, seeing scenes and characters change, but the play remains. From Rome to the British Empire, history shows that no human institution is eternal. Yet ideas and legacies—spiritual, cultural, and moral—can outlast stone and steel.

The Challenges of the 21st Century

We live in an age when global disasters increase. The 21st century has been a turbulent one to date, and it could get worse. As people cope with, all at the same time, global heat waves, floods, drought, and wildfires, as well as financial pressures and health issues, things can get out of control very quickly in an age of instant news and reactions to that news.

The United Nations has declared climate change the defining issue of our time, citing that over 3.6 billion people live in areas highly vulnerable to environmental disasters. The World Health Organization has also noted that climate-related health threats are on the rise, including malnutrition, heatstroke, and infectious disease.

At a spiritual level, many Christians see our time as signs of the Second Coming of Christ as they witness many prophesies getting fulfilled according to the books such as Matthew, John, Daniel, and Revelation in the Bible. Events like wars, famines, and earthquakes—mentioned in Matthew 24—are often cited by eschatologists as signs, though such interpretations have existed across centuries.

Political institutions are also facing scrutiny and change. With Queen Elizabeth II's passing, will the monarchy survive in the United Kingdom? King Charles III ascended the throne in 2022, but support for the monarchy has declined among younger generations, particularly in Commonwealth countries like Jamaica, Australia, and Canada, where republican movements are gaining traction.

Will countries worldwide demand reparations from the Monarchy for profits gained from slavery and natural resources in Africa, the plundering of $450 trillion from India, and the opium wars that humiliated China for a century?

While the $450 trillion figure regarding India is debated, economic historians like Utsa Patnaik estimate Britain extracted roughly $45 trillion in wealth during colonial rule. Recent years have seen growing calls for reparations and formal apologies, especially from Caribbean nations and former African colonies.

These issues reflect ongoing debates about historical injustices and the role of traditional institutions in the modern world. These questions reflect a quest for truth in an era where omissions and lies often obscure reality.

Chapter 84
Reflections on Major Transitions
and the Future

The 21st century has indeed been marked by turbulence and rapid change. We've witnessed increased global disasters, from devastating natural events to worldwide health crises. The interconnectedness of our world means that local events can quickly have global repercussions, leading to financial pressures and societal upheavals that affect people across the planet. The COVID-19 pandemic starkly demonstrated this interconnectedness, triggering a global economic slowdown and transforming education, labor markets, and healthcare systems.

It's crucial to approach the challenges of the 21st century with wisdom and discernment. Throughout history, believers have seen signs of the End Times in their own era, and it's important to balance prophetic understanding with practical engagement in the world around us. Theologians like N.T. Wright and John Piper emphasize preparedness over speculation, urging Christians to live faithfully regardless of eschatological timelines.

As we consider the future of faith and the potential fulfillment of biblical prophecies, it's essential to remember that our primary calling as believers is to love God and love our neighbors. Regardless of how end-time events unfold, we are called to be lights in the darkness, sharing God's love and truth with those around us.

The rapid pace of technological advancement also raises questions about the nature of humanity and our relationship with God. As artificial intelligence and biotechnology progress, we may face new ethical dilemmas and challenges to traditional understandings of human nature and consciousness.

Bioethicists and theologians alike are now debating issues such as brain-computer interfaces, CRISPR gene editing, and AI-driven worship content. The Vatican's Pontifical Academy for Life has even issued guidelines on AI ethics. How will faith traditions adapt to these changes while maintaining their core principles?

Climate change and environmental degradation present another set of challenges that intersect with religious and ethical considerations. Many faith traditions emphasize stewardship of the Earth, but how will they respond to the urgent need for global action on climate issues? Pope Francis's encyclical Laudato Si' and initiatives like GreenFaith and Islamic Declaration on Climate Change signal a growing interfaith environmental movement. Will we see a rise in eco-theology or new religious movements centered on environmental concerns?

The increasing secularization of many societies, particularly in the West, raises questions about the future role of organized religion in public life. Will we see a continued decline in religious affiliation, or might there be a resurgence of faith in response to global challenges and uncertainties?

Surveys by Pew Research suggest that "nones" (religiously unaffiliated) are growing, especially in the U.S. and Europe, but global religiosity remains high, particularly in Africa, the Middle East, and Southeast Asia. Crises often prompt a spiritual revival, even in secular cultures.

At the same time, we're witnessing the rapid growth of Christianity in the Global South, particularly in Africa and parts of Asia. This shift in the center of gravity of the Christian world may lead to new theological emphases and cultural expressions of faith. By 2050, Africa is expected to be home to over 40% of the world's Christians, and Pentecostal movements are rapidly shaping doctrines, worship, and political engagement across the continent. How will this impact global Christianity and its relationship with other world religions?

Interfaith dialogue and cooperation may become increasingly important in a world facing global challenges that require collective action. Will we see more collaboration between different faith traditions, or will religious differences continue to be a source of conflict? Organizations like the Parliament of the World's Religions and the United Nations Alliance of Civilizations are fostering such cooperation, especially on issues like climate justice, refugee crises, and human rights.

The digital age has also transformed how people engage with faith and spirituality. Online communities, virtual church services, and religious apps have become increasingly common. How will these technological changes shape religious practice and community in the coming decades?

Digital platforms have enabled hybrid forms of worship, real-time Bible study, and global outreach, but they also pose risks of disconnection, disinformation, and consumeristic spirituality. Future generations may redefine "sacred space" entirely.

The war in Ukraine has severely impacted Germany's industrial base, which had long relied on cheap Russian gas for energy-intensive sectors such as manufacturing and chemicals. The rapid pivot away from Russian energy imports following the invasion led to skyrocketing energy costs, supply chain disruptions, and factory closures across Europe. Many Europeans, especially during the winter months, experienced energy rationing and rising living costs.

The lessons from this conflict will shape international relations for decades to come. It's up to leaders and citizens alike to ensure these lessons lead to a more stable and peaceful world, rather than a continuation of the cycle of mistrust and conflict.

On June 24, 2024, leaders of the Group of Seven (G-7) wealthy democracies agreed to engineer a $50 billion loan package for Ukraine, funded by interest accrued on approximately $300 billion in frozen Russian central bank assets held in Western jurisdictions. This represents an unprecedented financial maneuver, raising significant legal and geopolitical concerns. President Vladimir Putin called the G7 deal "theft" and warned of long-term consequences. Critics argue that the move risks undermining global confidence in Western financial institutions, particularly among countries outside the G-7 bloc.

If the G-7 leaders have set a precedent of repurposing sovereign assets, could it create a chilling effect for other countries—especially emerging economies—holding reserves in G-7 banks? Observers warn of potential capital flight toward non-Western financial hubs or alternative currencies such as the Chinese yuan or BRICS-backed instruments.

NATO's strategic pivot toward the Indo-Pacific and recent rhetoric targeting China's role in supporting Russia has heightened tensions. On July 10, 2024, NATO formally labeled China a "decisive enabler" of Russia's war in Ukraine, a significant diplomatic escalation. This suggests NATO's ambition to expand its focus into Asia, particularly via security cooperation with countries like Japan, South Korea, and Australia. Many analysts view this as a risky entanglement that could spark broader conflict in the Pacific.

China remains the world's longest continuous civilization, with over 4,000 years of recorded history. The trauma of the

"Century of Humiliation" (1842–1949), which included the Opium Wars, unequal treaties, and foreign occupation, has shaped national identity and policy. Modern Chinese leaders, especially under Xi Jinping, have vowed to prevent a repeat of this period, emphasizing national sovereignty, military modernization, and economic self-reliance.

For centuries, China was the leading power in science, technology, and trade, until dynastic stagnation and foreign intervention eroded its dominance. Today, China's population, technological progress, and military strength make it a formidable global actor. In the event of a direct military confrontation, particularly one involving nuclear powers, mutual destruction would be nearly inevitable— making de-escalation a critical global priority.

Shifting Political Landscapes in 2024–2025

The second half of 2024 saw sweeping political change within the G-7 countries. Germany's ruling Social Democratic Party (SPD), led by Chancellor Olaf Scholz, received only 14% in the European Parliament elections on June 9, signaling waning public confidence. In Italy, Prime Minister Giorgia Meloni's Brothers of Italy party dominated with 28%, consolidating her far-right coalition's power.

In the UK, Keir Starmer became the seventh prime minister in 17 years after leading the Labour Party to victory on July 5. France's President Emmanuel Macron suffered a major defeat in the July 7 legislative elections, resulting in the resignation of his government on July 16. Though Macron remains president, his political influence has significantly weakened.

On July 21, U.S. President Joe Biden withdrew from the 2024 presidential race, citing health concerns and political pressure. Vice President Kamala Harris was named the Democratic nominee, assuming Biden's campaign funds despite not winning a single primary.

In Japan, Shigeru Ishiba replaced Fumio Kishida as Prime Minister on October 1, making him the ninth PM since 2008, reflecting ongoing political instability in the country. In Canada, following repeated by-election losses and growing dissent within his own party, Prime Minister Justin Trudeau resigned on January 7, 2025.

Mark Joseph Carney, a former central banker and climate finance advocate, became the new leader of Canada's Liberal Party on March 9 and assumed office as prime minister on March 14, 2025. Carney previously served as Governor of the Bank of Canada (2008–2013) and Governor of the Bank of England (2013–2020), making him the only person to have led both institutions.

Economic Upheaval: Trump's Return and Global Fallout

Donald Trump won the U.S. presidential election on November 5, 2024, in a landslide, campaigning on economic nationalism and trade protectionism. On April 2, 2025, shortly after markets closed, President Trump announced a sweeping global tariff plan. The Dow Jones Industrial Average (DJIA) closed at 42,215.24 and the NASDAQ at 17,596.71. Within three trading days, the DJIA plummeted to 37,965.60 (-10.07%) and the NASDAQ fell to 15,603.26 (-11.33%), wiping out over $10 trillion in global stock market capitalization.

Tech stocks were among the hardest hit: Dell Technologies dropped 19%, HP fell 15%, and Western Digital plunged 18%. Analysts described the tariff plan as an economic act of aggression that triggered global panic. In response, Canadian Prime Minister Mark Carney declared an end to Canada's post-WWII economic reliance on the U.S., stating:

"The system of global trade anchored on the United States... is over. Our old relationship of steadily deepening integration with the United States is over."

EU Commission President Ursula von der Leyen warned that Trump's tariffs would cause consumer prices to rise, disrupt supply chains, and spike inflation, especially in critical sectors like pharmaceuticals and agriculture.

China, facing a proposed 54% U.S. tariff on its exports, retaliated with a 34% levy on American goods. Trump threatened to nearly double U.S. tariffs in response. According to the BBC:

"Beijing's Commerce Ministry vowed to 'fight till the end,' calling Washington's move blackmail."

U.S. firms importing from China could soon face up to 104% total import taxes, making business unsustainable for many companies reliant on Chinese manufacturing. Economists warn this tit-for-tat cycle risks igniting a full-scale trade war between the world's two largest economies, with long-lasting consequences for global inflation, recession risks, and supply chain stability.

The most recent BRICS summit was held in Kazan, Russia, on October 24, 2024. According to Reuters, here's a brief summary of the outcome of the BRICS summit:

- *XI AND MODI:* Chinese President Xi Jinping and Indian Prime Minister Narendra Modi met just two days after New Delhi announced that it had reached

718

a deal with Beijing to resolve a four-year military stand-off on their Himalayan frontier.

- **LOTS OF LEADERS ATTENDED:** Putin, who wanted to show that the West's attempt to isolate Russia over the Ukraine war has failed, was able to attract major leaders such as Xi and Modi and nearly 20 others, including Turkey's Tayyip Erdogan. U.N. Secretary-General Antonio Guterres also attended.
- **BRICS WAITING LIST:** Putin said more than 30 countries had expressed a desire to join the BRICS, though there was little immediate clarity on how the expansion would work.
- **UKRAINE WAR:** BRICS leaders did raise the Ukraine war with Putin in different formats, but there was no sign that anything specific would be done to end the conflict.
- **BRICS MONEY:** BRICS predicted its influence would grow and outlined common projects ranging from a grain exchange to a cross-border payments system.
- **MIDDLE EAST:** Putin told BRICS leaders that the Middle East was on the brink of a full-scale war after a sharp rise in tension between Israel and Iran.

In December 2024, Assad's regime in Syria fell to terrorists. This was a surprise to most countries. Syria is now ruled under Sharia law. As we navigate complex issues and political power shifts, including de-dollarization, I believe faith will continue to play a vital role in shaping human societies and individual lives. Whether through traditional religious institutions or new expressions of spirituality, people will continue to seek meaning, purpose, and connection to the divine.

In conclusion, while the future of faith may be uncertain in many ways, the human quest for spiritual truth and meaning remains constant. As we face the challenges of our time, may we approach the future with hope, wisdom, and a commitment to living out our deepest values and beliefs! Regardless of one's specific faith tradition, the call to love, compassion, and justice can guide us as we work towards a better future for all of humanity.

Chapter 85
My Perspective on the Origin of
Life and Health Scare

The origin of life, according to science, began with the formation of simple inorganic molecules into organic compounds—amino acids, sugars, and lipids—eventually leading to the complex macromolecules that make up DNA and cells. While scientists have recreated basic building blocks in laboratory conditions, such as through the Miller-Urey experiment, the precise mechanisms that formed the first living cells remain unknown. Even the simplest single-cell organisms like bacteria contain a genetic program far too complex to be the product of random molecular collisions. To date, no laboratory or theory has successfully explained how functional DNA coding sequences arose spontaneously.

All computer programs are designed by programmers using binary code—combinations of "0" and "1." Similarly, the genetic code of all living beings is written using four chemical bases: Adenine (A), Cytosine (C), Guanine (G), and Thymine (T). These genetic "letters" form a language far more intricate than binary. DNA provides the instructions for each cell, telling it where it fits within the body and what functions to perform—whether to become a liver cell, a neuron, or part of a limb. The probability of this happening by pure chance is infinitesimally small.

Humans have souls, which are separate from their bodies. I know this for sure because I saw my mother's soul leave her body about five minutes after she stopped breathing. It was a moment of deep spiritual clarity, beyond scientific explanation. Evolution did not create souls. I know that God is real because I experienced God's spirit when I least expected it. No hallucination, dream, or emotion compares to the transformative certainty that comes with experiencing God's presence.

The authors of the Bible did not have a full scientific understanding when they wrote in the Gospel of John (Chapter 1:1–4):

"In the beginning was the Word, and the Word was with God, and the Word was God. The same was in the beginning with God. All things were made by Him, and without Him was not anything made that was made. In Him was life, and the life was the light of men."

We now understand that the "Word" can be metaphorically linked to the genetic code (DNA)—the foundational language of life. All organisms are built from this intricate code. Such complexity and precision reflect design, not random accident. I have examined all possible scientific origins of life, and the only consistent truth is that God is the Creator. No matter how many billions of years you allow, the DNA code will not randomly generate itself—because information always comes from intelligence.

God is real, and people need to know this truth. They need the assurance that God loves them deeply and is available to guide, comfort, and help those who seek Him in prayer and supplication.

The Federal Reserve and Global Control

The Federal Reserve has enabled wars that killed over 100 million people and injured many more. Its influence on monetary policy and military-industrial funding has shaped world conflicts from WWI to Iraq. Beyond war, it has fostered a system of global debt, trapping entire nations and populations under compounding interest systems. The Fed is the most lethal financial weapon ever created—and it must be abolished. A decentralized, transparent financial system could foster true economic justice. When people refuse to be slaves to fiat currencies, we will begin to move toward a more enlightened world.

An Age of Converging Crises

The world is overwhelmed by simultaneous challenges: financial instability, wars, pandemics, climate change, artificial intelligence, misinformation, social media overload, and civil unrest. Take time to refocus your mind. Many advanced civilizations—like the Maya, Khmer, and Akkadian—collapsed due to climate shifts that led to water and food scarcity, triggering wars and migration crises. We must act before it's too late.

AI can be incredibly helpful to humanity. It can also become an existential threat. Unchecked AI development could outpace ethical and spiritual guidance. We must learn to discern—for ourselves and for future generations.

Cryptocurrencies, while innovative in design, are increasingly showing signs of speculative instability and unsustainable energy use. Many will likely be revealed as Ponzi schemes in time. They also consume immense amounts of electricity and water, harming the environment. Bitcoin alone consumes more energy annually than some countries.

When no one can buy or sell without digital currency and QR codes, we risk sacrificing our freedoms, privacy, and even lives. Digital control can become totalitarian if misused. Lifelong learning will be essential to thrive in the 21st century and beyond. Entrepreneurship, creativity, and global collaboration will shape the new economy.

Chapter 85
My Perspective on the Origin of
Life and Health Scare

When your body dies, your soul will live on eternally. Don't be short-sighted. Plan your horizon beyond this world. Prophecies from various traditions—Christian, Jewish, Islamic, Indigenous, and others—foretell a dramatic transformation of the world. The end of this age may be nearer than most expect. Share scripture and love one another as God loves us. Worship only God.

Reflections on Health, Food, and Gratitude

Today in the U.S., people have to pay far more for organic food. Some organic produce costs 3 to 4 times more than non-organic food. Most Americans cannot afford to eat organic all the time. Ironically, in the poorest country— Burma (now Myanmar)—where I was born, all the food I ate growing up was organic by default. We didn't have pesticides or GMOs, just clean, natural produce and fresh meals.

When I came to the United States at age 14, I was about 5'4" and weighed 110 pounds. I was fit and healthy. But within a year of arriving in San Francisco, I was almost obese. I was eating processed and fried foods like Kentucky Fried Chicken (before it was called KFC), curry chicken wings, donuts, fried rice, and drinking Coca-Cola. I gained at least 50 pounds in one year. In college, I worked to bring my weight down to 129 pounds. This taught me that lifestyle and food choices make a profound difference. Choose wisely.

During the pandemic, I checked my mother's blood pressure and blood sugar almost every day, administering medications three times daily. I got only 3 to 4 hours of interrupted sleep most nights, as she often needed help at night. I neglected my own health during that time. Before the pandemic, I had no health issues. But one evening in 2022, after washing the dinner dishes, my left fingers felt tingly. I was surprised—and I immediately knew something wasn't right.

I called my sister Grace to inform her of the situation. Then I lay down and felt numbness spreading across the left side of my body. Within 30 minutes, I was in the ICU at Sutter Hospital. The next day, I underwent open-heart surgery. Without Grace and her eldest son, Joshua, I may not have survived. Their quick action saved my life.

I thank God and the medical professionals at Sutter Medical for preserving my life. I am still here today, writing this book, and I remain grateful every single day.

Chapter 86
Blessings, Well Wishes, and
One Last Thing

Appreciate what you have, cherish your loved ones, and embrace the journey of lifelong learning. May the God who formed you in love—the One who knows you more deeply than you know yourself—bless you and keep you always. May He make His face shine upon you and fill you with His peace, joy, and strength. May He grant you the courage to be brave, the wisdom to make good choices, and the heart to love others as He loves you.

Moreover, may you always remember that you are a masterpiece—created by the God who hung the stars in the sky and painted the colors of the rainbow. You were fearfully and wonderfully made (Psalm 139:14), and there is no one else exactly like you. May you never doubt how special and precious you are to Him, or how much He delights in every part of who you are.

So go out into the world with a smile on your face, knowing that you are loved beyond measure and that your life is in the hands of the God who created you for a divine purpose. Keep running the race of faith with perseverance and passion, as encouraged in Hebrews 12:1–2, knowing that Jesus is right beside you, cheering you on every step of the way.

Furthermore, may you be a light in the darkness, a voice for the voiceless, and a friend to the friendless. May you use your unique gifts and talents to make the world a brighter, better, and more beautiful place. Your life matters. Your story matters. Your kindness and faith have the power to impact generations. May you always know that your life has meaning, value, and the power to change lives.

May the grace of our Lord Jesus Christ, the love of God, and the fellowship of the Holy Spirit be with you always (2 Corinthians 13:14).

Jesus said that nothing should distract us from preaching the gospel to all nations. In Matthew 28:19, He gave the Great Commission: "Go therefore and make disciples of all nations." Today, with the Internet and smartphones, billions of people have access to the Bible online—many in their native languages. With this book, it is my hope and prayer that more people will come to know Jesus Christ as their Lord and Savior.

I know that this book may be banned in some countries precisely because of this final paragraph. But I also believe that God can open doors that no one can shut (Revelation 3:8), and that He can soften hearts, even in the most unlikely places. I trust in God's timing, His wisdom, and His power to reach people wherever they are.

In life there are many mountains to climb. Whether you are climbing for fun, justice or other reasons, choose wisely. To God be all the glory.

Glossary

AI	Artificial Intelligence is the simulation of human intelligence processes by machines.
AIPAC	American Israel Public Affairs Committee, a pro-Israel lobbying group.
Amtrak	The National Railroad Passenger Corporation is a passenger railroad service that operates in the United States.
Angel Investor	An individual who provides capital for a business start-up, usually in exchange for convertible debt or ownership equity.
Anthropogenic	Caused or produced by humans
Anunnaki	Deities in ancient Mesopotamian cultures.
Apostate Church	A religious institution that outwardly appears Christian but has departed from true faith.
Artificial General Intelligence (AGI)	AI that matches or exceeds human intelligence across all domains.
Artificial Intelligence (AI)	The simulation of human intelligence processes by machines, especially computer systems.
Artificial Superintelligence (ASI)	AI that vastly surpasses human cognitive abilities in virtually all domains.
Big Bang	The theoretical starting point of the universe involves a massive expansion of space and time from an initial singularity.
Biodiversity	The variety of life in a particular habitat or ecosystem

Glossary

Biotech

Short for biotechnology, the use of biological processes, organisms, or systems to manufacture products intended to improve the quality of human life.

Biotechnology

The use of biological processes, organisms, or systems to manufacture products intended to improve the quality of human life.

Blockchain

A decentralized, distributed ledger technology that records transactions across many computers.

BRICS

An economic group initially consisting of Brazil, Russia, India, China, and South Africa.

Carbon pricing

A method of charging for carbon emissions to reduce their production

Central Bank Digital Currency (CBDC)

A digital form of a country's fiat currency issued and regulated by the central bank.

Chronic Disease

A condition that lasts one year or more and requires ongoing medical attention or limits activities of daily living or both.

Circadian rhythm

The natural, internal process that regulates the sleep-wake cycle and repeats roughly every 24 hours.

Climate change

Long-term alterations in temperature and weather patterns

Computer Vision

The field of AI that trains computers to interpret and understand visual information from the world.

Cosmic microwave background radiation

The leftover heat from the early stages of the universe which provides evidence for the Big Bang theory.

CRISPR-Cas9	A unique technology that enables geneticists and medical researchers to edit parts of the genome by removing, adding, or altering sections of the DNA sequence. It is currently the simplest, most versatile, and precise method of genetic manipulation to modify gene function.
CRISPR	A gene-editing technology that allows for precise modifications of DNA sequences.
Cuneiform	An ancient writing system using wedge-shaped marks
Deep Learning	A type of machine learning based on artificial neural networks, capable of learning from large amounts of unstructured data.
Deepfake	Synthetic media is where a person in an existing image or video is replaced with someone else's likeness using AI techniques.
Digital Currency	Money that exists only in electronic form and is managed through computer systems.
Digital Literacy	The ability to use digital technologies effectively and appropriately.
Dim sum	It is a large range of small Chinese dishes that are traditionally enjoyed in restaurants for brunch.
DNA (deoxyribonucleic acid)	The molecule that carries the genetic instructions for the development, functioning, growth, and reproduction of all known organisms.
Ecosystem	A community of interacting organisms and their environment

Glossary

Edwards Air Force Base A United States Air Force installation located in southern California, known for its history of aeronautical research and development.

Enuma Elish An ancient Mesopotamian creation myth.

Epigenetics The study of changes in organisms caused by modification of gene expression rather than alteration of the genetic code itself.

ESL English as a Second Language, a program for non-native English speakers.

Ethics Moral principles that govern a person's behavior or the conducting of an activity.

Evolution The process by which species change over time through the inheritance of genetic variations across generations.

Exoplanet A planet that orbits a star other than the Sun.

Federal Reserve The central banking system of the United States

Fiat currency Money that is not backed by a physical commodity like gold

Four Horsemen of the Apocalypse Symbolic figures in the Book of Revelation represent conquest, war, famine, and death.

Generative AI AI systems that are capable of creating new content, such as text, images, or music.

Genomics The study of all of a person's genes (the genome), including interactions of those genes with each other and with the person's environment.

Glass-Steagall Act	A 1933 law that separated commercial and investment banking activities
Göbekli Tepe	An ancient archaeological site in Turkiye dated around 8000 BCE.
Global Financial Crisis (GFC) of 2007-2009	The most severe worldwide economic crisis since the Great Depression. Also known as the Great Recession
Great Depression (1929–1939)	The most severe global economic downturn that affected many countries across the world.
Great Recession	See Global Financial Crisis (GFC) of 2007-2009
Growth Mindset	The belief is that abilities and intelligence can be developed through effort, learning, and persistence.
Gut-Brain Axis	The communication system between the gastrointestinal tract and the central nervous system.
Hamas	A Palestinian Sunni Islamist political and military organization
Hieroglyphs	A system of writing using pictorial symbols
High-Frequency Trading (HFT)	A type of algorithmic trading that uses powerful computers to transact a large number of orders in fractions of a second.
Homo sapiens	The scientific name for modern humans
Human Genome Project	An international scientific research project aimed at determining the sequence of the human genome and identifying and mapping all human genes.

Glossary

Ignatian Center

A center at Santa Clara University that promotes and enhances the distinctively Jesuit Catholic tradition of education.

Indulgences

In Catholic tradition, a way to reduce the amount of punishment one has to undergo for sins.

Initial Public Offering (IPO)

The process of offering shares of a private corporation to the public in a new stock issuance.

Intellectual Property (IP)

Creations of the mind, such as inventions, literary and artistic works, designs, symbols, names, and images, are used in commerce.

Internet of Things (IoT)

The network of physical objects is embedded with sensors, software, and other technologies for the purpose of connecting and exchanging data with other devices and systems over the Internet.

ISO Class 1 cleanroom

A controlled environment with a low level of pollutants, used in manufacturing or scientific research that requires a high level of air purity.

Lifelong Learning

The ongoing, voluntary, and self-motivated pursuit of knowledge for personal or professional reasons.

Light pollution

Excessive or inappropriate artificial light that affects the ability to see the night sky.

Machine Learning

A subset of AI that enables systems to learn and improve from experience without explicit programming.

Mansa Musa

The 14th-century emperor of the Mali Empire was considered the wealthiest person in history.

Mark of the Beast

A prophesied mark is required for buying and selling in the end times.

Glossary

Melatonin

A hormone that regulates the sleep-wake cycle, with its production increasing in the evening and decreasing in the morning.

Metabolites

Small molecules that are intermediates and products of metabolism.

Metacognition

Awareness and understanding of one's own thought processes.

Microbiome

The collection of all microbes, such as bacteria, fungi, and viruses, that live on and in our bodies.

Middle Passage

The stage of the Atlantic slave trade in which millions of enslaved Africans were transported to the Americas.

Military-industrial complex

The relationship between a country's military, defense industry, and political leadership

NATO

North Atlantic Treaty Organization, a military alliance between North American and European countries

Natural Language Processing (NLP)

The ability of computers to understand, interpret, and generate human language.

Natural selection

The process by which organisms with advantageous traits are more likely to survive and reproduce, leading to changes in species over time.

Neural Networks

Computing systems inspired by biological neural networks are capable of machine learning and pattern recognition.

Paleogenomics

The study of ancient DNA

Glossary

Patent

A government authority or license conferring a right or title for a set period, especially the sole right to exclude others from making, using, or selling an invention.

Pharmacogenomics

The study of how genes affect a person's response to drugs.

Photomask

A plate with holes or transparencies that allow light to shine through in a defined pattern is used in photolithography. A photomask set of 70 to 80 distinct layers is required to create an advanced semiconductor with features in the nanometers.

Precision Medicine

A medical model that proposes the customization of healthcare, with medical decisions, treatments, practices, or products being tailored to the individual patient's genes, environment, and lifestyle.

Proactive Healthcare

An approach that focuses on prevention and early intervention rather than treating symptoms after they appear.

Protestant Reformation

A 16th-century movement for the reform of the Roman Catholic Church that led to the establishment of Protestant churches.

QR Code

A type of matrix barcode that can store various types of data and is readable by smartphones.

Quantitative easing

A monetary policy where a central bank purchases securities to increase the money supply

Quantum mechanics

The branch of physics that describes the behavior of matter and energy at the atomic and subatomic levels.

Rangoon (Yangon)

It is the former capital city of Burma (Myanmar).

Rapture	The belief is that believers will be taken from Earth to meet Christ in the air before or during the tribulation.
Reinforcement Learning	A type of machine learning where an agent learns to make decisions by taking actions in an environment to maximize a reward.
Resilience	The capacity to recover quickly from difficulties; toughness.
Rotary International	A global non-profit service organization that brings together business and professional leaders to provide humanitarian service and advance goodwill and peace around the world.
Scientific method	The systematic process of observing, hypothesizing, experimenting, and analyzing that is used to investigate natural phenomena and acquire knowledge.
Semiconductor	A material with electrical conductivity between that of a conductor and an insulator.
Singularity	A point in space-time at which gravitational forces cause matter to have infinite density and infinitesimal volume and space and time to become infinitely distorted.
Soul	The spiritual part of a human being is regarded as immortal.
Spiritual walkabout	A journey of spiritual discovery and growth.
STEM	Science, Technology, Engineering, and Mathematics.
Sustainable	Able to be maintained at a certain rate or level without depleting resources

Glossary

Telemedicine	The remote diagnosis and treatment of patients by means of telecommunications technology.
Theory of relativity	Albert Einstein's theory that describes gravity as a warping of space-time caused by the presence of mass and energy.
Thingyan	The Burmese New Year festival is celebrated in mid-April.
Trail of Tears	A series of forced relocations of Native American tribes in the United States in the 1830s.
Tribulation	A period of great suffering and distress is predicted in biblical end-time prophecies.
UC Berkeley	University of California, Berkeley, a prestigious public research university
UNESCO	United Nations Educational, Scientific and Cultural Organization.
Universal basic income	A government program providing all citizens with a regular, unconditional sum of money.
US Borax	A mining company that supplies nearly half the world's refined borates, minerals essential to life and modern living.
USDA	United States Department of Agriculture, a federal agency responsible for developing and executing federal laws related to farming, forestry, rural economic development, and food.
Value-Based Care	A healthcare delivery model in which providers are paid based on patient health outcomes rather than on the amount of healthcare services they deliver.

Vatican City	An independent city-state and the headquarters of the Roman Catholic Church.
Venture Capital (VC)	A form of private equity financing that is provided by venture capital firms or funds to startups, early-stage, and emerging companies that have been deemed to have high growth potential.
YMCA	The Young Men's Christian Association is a worldwide organization with a mission to put Christian principles into practice through programs that build a healthy spirit, mind, and body for all.

www.ingramcontent.com/pod-product-compliance
Lightning Source LLC
Chambersburg PA
CBHW060123130626
46556CB00006B/2207